Current Topics in Membranes and Transport

Volume 7

Current Topics in Membranes and Transport

VOLUME 7

Edited by

Felix Bronner

Department of Oral Biology
University of Connecticut Health Center
Farmington, Connecticut

and

Arnost Kleinzeller

Department of Physiology
University of Pennsylvania School of Medicine
Philadelphia, Pennsylvania

1975

Academic Press   **New York   San Francisco   London**

A Subsidiary of Harcourt Brace Jovanovich, Publishers

QH
601
.C84
v.7

ACADEMIC PRESS, INC.
111 Fifth Avenue, New York, New York 10003

United Kingdom Edition published by
ACADEMIC PRESS, INC. (LONDON) LTD.
24/28 Oval Road, London NW1

LIBRARY OF CONGRESS CATALOG CARD NUMBER: 70-117091

ISBN 0-12-153307-7

PRINTED IN THE UNITED STATES OF AMERICA

Contents

Ion Transport and Short-Circuit Technique

WARREN S. REHM

List of Contributors

Numbers in parentheses indicate the pages on which the authors' contributions begin.

Richard A. Dilley, Department of Biological Sciences, Purdue University, West Lafayette, Indiana (49)

Robert T. Giaquinta, Department of Biological Sciences, Purdue University, West Lafayette, Indiana (49)

Paul G. LeFevre, Department of Physiology and Biophysics, Health Sciences Center, State University of New York at Stony Brook, Stony Brook, New York (109)

E. A. C. MacRobbie, Botany School, University of Cambridge, Cambridge, England (1)

Warren S. Rehm, Department of Physiology and Biophysics, University of Alabama in Birmingham, The Medical Center, University Station, Birmingham, Alabama (217)

Contents of Previous Volumes

Volume 6

Ion Transport in Plant Cells

E. A. C. MacROBBIE

Botany School
University of Cambridge
Cambridge, England

I. INTRODUCTION

A. General

This review is concerned with the processes of ion movement, both active and passive, across plant cell membranes. In particular, it aims to deal with the existence and identification of processes of active transport of ions in plant cells and tissues, the study of the mechanisms involved and of the consequences and functions of such transport. There are a number of recent reviews of ion transport in plant cells and tissues (Mac-Robbie, 1970a; Steward and Mott, 1970; Anderson, 1972; Higinbotham, 1973a,b); the present review is aimed at aspects not covered in detail in these, or at more recent findings. In particular it lays more emphasis on

1

the importance of vacuolar transport processes in the overall ionic relations of the cell.

The study of the ionic relations of a cell should start with the measurement of the dynamic state of the system (ion concentrations, potentials, and two-way fluxes associated with each intracellular compartment), the aim being to identify the nature of the ion transfers across all membranes and of their appropriate driving forces. In practice we have enough information about conditions at the external membrane, the plasmalemma, to define the main transport processes there, and to progress to the stage of characterization of the mechanisms. But our knowledge of the activities of intracellular membranes is much more imperfect; we have some information on ion movements across the thylakoid membranes of isolated chloroplasts, or the membranes of isolated mitochondria, but very little knowledge of such activities *in vivo*, and almost no information on ion transfers at most other intracellular membrane. The exception is the vacuolar membrane, the tonoplast; we have indirect indication of ion movements to the vacuole in a number of plant cells, but because of some uncertainties in the interpretation of intracellular kinetics, such information must be regarded as qualitative rather than quantitative in nature.

The most interesting transport processes are those described as active, but in fact we arrive at this description by elimination of simpler mechanisms involving a lesser dependence of the transport on influences outside the membrane. The simplest ion flux is by ordinary passive diffusion, with the electrochemical potential gradient ($\Delta\bar{\mu}$) of the ion concerned as the only driving force, and with independent movement of individual ions within the membrane. But it seems that such fluxes play a part in the movement of ions across the outer membranes of plant cells in the resting state only when the external concentrations are high or when the surface:volume ratio is not too large, or both. Rothstein (1964) has argued that such fluxes cannot be tolerated by small cells growing in dilute solution, where passive leakage must be minimal and all transport subject to control.

Other types of transport, involving chemical interaction of the transported ion with groups in the membrane, with so-called "translocators," deviate from the simple process to greater or lesser degree. In artificial membranes, various types of translocators are recognized—for example, mobile lipophilic carrier molecules, which may or may not contribute to the membrane conductance, according to whether the ion–carrier complex is charged or uncharged, and the induction in lipid bilayers of channels with finite lifetimes and associated discrete conductance steps, in the presence of various substances. Translocators of this type may be gated, if in some way their operation is sensitive to some feature of the membrane environment, for example to the potential across the membrane or to the chemical state of a membrane constituent. It may well be that similar processes occur naturally in cell membranes, but if so, they have not as yet been identified. In most of these instances, although the kinetics may differ radically from those of passive diffusion, the overall driving force on the ion remains its own electrochemical potential gradient, and after transport the ion and its translocator remain chemically unchanged. A clear distinction must be made between this and another form of deviation, in which not only the kinetics but also the direction of transport is aberrant, and we find either the net transfer of an ion against its own electrochemical potential gradient (uphill), or the maintenance of a nonequilibrium distribution of an ion across a membrane in the face of appreciable fluxes. This we define as active transport. It is then clear that we must look elsewhere for the main driving force for such ion transfer, and that some other constituent of the system must be undergoing concurrent translocation or chemical change. If we cannot find a concurrent flow of another substance across the membrane to which the apparently uphill ion flux could be coupled, then we envisage that vectorial chemical reactions within the membrane provide asymmetry to the chemical interactions between the ion and membrane constituents, and that a chemical constituent produced by some energy-yielding metabolic sequence is consumed in the overall process. Having identified the existence of processes of active transport within the cell, we move on to the problem of identifying the energy-yielding sequence to which the flux is linked, and the means of coupling this to the membrane transport, i.e., the agent by which energy is transferred to the pump. The last stage is that of describing the detailed molecular events in the membrane, but for most ion transport processes this is not yet within sight.

As was said earlier, our experimental information on ion movements in the intact cell allows us to discuss the nature of transport processes only at the plasmalemma and perhaps at the tonoplast, and our measurements refer to a limited number of experimental systems. A good deal of work has been done on various giant algal cells, both on freshwater species living in fairly dilute salt solutions and on marine species; such cells have the advantage of allowing separate analysis of cytoplasm and vacuole, thereby giving some direct information on ion movements at the tonoplast. In higher plant tissues the experimental possibilities are more limited. Vacuolar concentrations, cytoplasmic and vacuolar potentials, and net fluxes at the plasmalemma can be measured with fair certainty, but processes at the tonoplast are only indirectly accessible to experimental observation, with some consequent uncertainty in interpretation. However, the main transport activities of the plasmalemma seem to be accessible with present experimental methods, and we may discuss these before considering the possibilities of measuring tonoplast fluxes.

B. Transport Processes

1. ACTIVE TRANSPORT AT THE PLASMALEMMA

It is clear from the measured values of ion concentrations, electric potentials and net ion fluxes, that in a very wide range of plant cells three major transport processes are responsible for the ionic regulation observed. In giant algal cells it is clearly established that these are associated with the plasmalemma, although further regulation at the tonoplast is also indicated. In higher plant cells the less direct evidence also suggests that the three transport systems should be allocated to the plasmalemma. It appears that the plant cell is in general involved in: (a) the maintenance of a high K:Na ratio, particularly in the cytoplasm; (b) the active extrusion of H^+, presumably for the maintenance of a suitable cytoplasmic pH; (c) the net accumulation of salt to a high internal concentration, associated with the existence of a large central sap cavity, occupying some 90% or more of the cell volume, in which most of the salt is sequestered; since the cell potential is negative with respect to the outside, this must involve active transport of at least the anions.

The importance of the first and last of these processes has been generally recognized for many years, that of H^+ extrusion only in the last few years. The processes of cation transport seem to be of general occurrence in both eukaryotic and prokaryotic cells, but net salt accumulation is a process specific to vacuolate plant cells, and is their most striking characteristic. Such salt accumulation in the cell vacuole is responsible for a considerable fraction of the high osmotic pressure by which cell turgor is maintained. It is seen in two forms, as the accumulation of both anion and cation from the outside, most frequently as KCl or NaCl, or alternatively as the accumulation of an inorganic cation from the solution, with an internally generated organic acid anion of a type specific for the tissue, but frequently malate. In giant algal cells the osmotic pressure is almost entirely contributed by KCl and NaCl, but in higher plant cells the contents of the vacuole are a complex mixture of inorganic and organic solutes, and the relative contributions to the osmotic pressure of inorganic salts, organic acid anions, and sugars, depend on the age of the cell and the conditions of growth. The overall process involves entry of ions at the plasmalemma and transfer of salt from cytoplasm to vacuole, but we shall see that these may be interdependent processes, and experimentally there are considerable difficulties in distinguishing the sequential events leading to the vacuolar accumulation.

Before discussing the characteristics of the three transport processes we must consider an important technical problem in measuring fluxes in higher plant tissues. Meaningful influx measurements in such cells can be made only under conditions defined by the kinetics of ion movements within intracellular compartments, normally measured by the analysis of efflux kinetics. Thus the kinetic analysis of tracer movement into and out of the tissue as a whole is necessary to identify the conditions for the proper measurement of fluxes at the plasmalemma, and can also be used to estimate cytoplasmic concentrations and tonoplast fluxes. For the second purpose, the analysis involves making certain assumptions about intracellular compartmentation.

2. FLUX ANALYSIS

The kinetics of movement into and out of the tissue as a whole reflects the kinetics of specific activity changes in the cytoplasmic phase (or phases) which communicate with the external solution; this in turn depends both on plasmalemma fluxes and on fluxes to or from other internal phases of the cell, although there may be processes of intracellular exchange which have negligible effects on the specific activity of the flux crossing the plasmalemma during the experimental time periods, and are therefore masked in such measurements. In general the kinetics of loss of tracer to unlabeled solution is measured, a method first used by Pitman (1963, 1964) in beet tissue, but since used by others on a range of tissues; some of the results have been reviewed previously (MacRobbie, 1971a). In such washing-out curves, two exponential terms appear in addition to the free space exchange, and it is taken that these represent exchange of the cytoplasm and vacuole, generally assumed to be in series with one another. On this assumption, and on the further assumption that the ion content of the cytoplasmic phase Q_c is much less than that of the vacuole Q_v, fluxes at the plasmalemma and the tonoplast, and the cytoplasmic content, may be calculated. The fit of the model was tested to some degree by Pitman (1963) and Cram (1968a), but the possibility remains that the reduction of the cell or tissue to two phases in series may be a misleading simplification under some conditions, and some uncertainty remains over the calculated intracellular figures.

The importance of making influx measurements over time intervals chosen in the light of the efflux kinetics is shown by Cram and Laties (1971), who compared ^{36}Cl influx in barley roots under conditions of short uptake/short wash and of long uptake/long wash. The first measures

M_{oc}, the flux from outside to cytoplasm, and is the measure required for discussion of uptake mechanism, whereas the second is complicated by loss of tracer from the cytoplasm during uptake and wash. If either the uptake or the wash is "long," then the experiment measures the steady influx to the vacuole in a quasi-steady state of cytoplasmic specific activity, and is sometimes called M_N (or M_{ov}); it is equal to the quantity $M_{oc} \cdot M_{cv} / (M_{cv} + M_{co})$, where M_{cv} is the flux from cytoplasm to vacuole, and M_{co} is the flux from cytoplasm to outside. Since, when conditions change, any or all of these three fluxes may change, the effect on M_N is difficult to interpret. If we wish to draw inferences about the transport mechanism from the results, we must be sure that we are in fact measuring an individual flux at either plasmalemma or tonoplast. This means using short experimental times; for chloride in carrot or barley roots, for example, uptake times of 5–10 minutes were used.

II. POTASSIUM–SODIUM REGULATION

In a very wide range of cells the cytoplasmic sodium is maintained low by active extrusion of sodium at the plasmalemma; this may be accompanied by the active uptake of potassium (particularly at low external concentrations), but in other cells potassium is close to equilibrium with the outside solution. The results on giant algal cells are summarized in MacRobbie (1970a) and need not be further discussed.

The existence of active sodium extrusion, with or without concomitant active potassium uptake, inferred from concentration and potential measurements in a number of higher plant species (review by Higinbotham, 1973a), has been confirmed by flux measurements in some instances. Observations by Pitman and Saddler (1967), in which cytoplasmic levels of sodium and potassium were measured by means of efflux analysis, suggested that in excised barley roots at low external potassium concentrations there was both active uptake of potassium and extrusion of sodium. The probability of tracer loss from the cytoplasm to the solution by way of transport to the xylem and leakage from the cut end, rather than by direct efflux from the root cortical cells, led to criticism of this conclusion, but Jeschke (1970a, 1973) has confirmed the existence of potassium-dependent, active extrusion of sodium in barley roots, by separate measurement of the two components of ion efflux. There appeared to be little effect of ouabain on the sodium extrusion in this tissue (Pitman and Saddler, 1967; Jeschke, 1973). By contrast, the active sodium extrusion in carrot tissue was found to be sensitive to ouabain, although it did not depend on external potassium (Cram, 1968b), and Jeschke (1970b) reported some effect of strophantin on sodium efflux from *Elodea* in the dark (but not in the light). This is reminiscent of the variability in different species of characean cells, where sodium extrusion may or may not be coupled to potassium uptake, and may or may not be sensitive to ouabain.

The importance of chemically mediated potassium uptake is suggested by the work of Poole (1966, 1969, 1971, 1973) on red beet tissue, from which he argues that K^+ movement at the plasmalemma is not strongly influenced by the membrane potential as such, but that K^+ uptake is largely by K^+/H^+ exchange. He has discussed two possibilities for the effect of Cl^- on the uptake of K^+: the existence of a neutral mechanism for the uptake of KCl, or alternatively the stimulation of a K^+/H^+ exchange by a decrease in cytoplasmic pH induced by HCl transport across the plasmalemma.

Young and Sims (1972, 1973) showed that efflux of K^+ is negligible in growing fronds of *Lemna minor*, and suggested that K levels are maintained by an active K^+ influx, involving a membrane-bound ATPase. The transport activity is regulated in response to changes of external concentration. They have isolated a vesicular membrane fraction with rubidium-binding characteristics comparable with those of the uptake process, and suggest that this represents the first isolation of a transport system from a plant tissue.

In *Chlorella pyrenoidosa* also Barber (1968a, b) argues that passive diffusion plays little or no part in ion movements, for which exchange fluxes of one kind or another are responsible. His evidence suggests the presence of active K^+/H^+ exchange as well as the active K^+/Na^+ exchange involved in the maintenance of the high K/Na of the cell, but a K^+/K^+ exchange is also postulated. On the addition of potassium to sodium-rich, potassium-depleted cells (Shieh and Barber, 1971; Barber and Shieh, 1972), very rapid rates of K^+/Na^+ exchange are observed; the restoration of normal levels of cell potassium (160–170 mM) is achieved by fluxes many times higher than the steady-state rates involved in the maintenance of such normal levels. This is very similar to the picture suggested by studies of potassium transport in *Neurospora crassa* (Slayman and Tatum, 1964, 1965; C. L. Slayman and Slayman, 1968; C. W. Slayman and Slayman, 1970; C. W. Slayman, 1970), where again it is argued that the fluxes observed are carrier-mediated rather than passive diffusion.

Even in giant algal cells the importance of passive diffusive fluxes under normal conditions (in the presence of calcium in the solution) is not clearly established. In favor of the view that there are significant passive fluxes, are observations of the strong dependence on potential of potassium fluxes and sodium influx in voltage-clamped cells of *Nitella* and *Chara*

(Walker and Hope, 1969). Against this view, however, are the observations that major fractions of the cation influxes are linked to chloride entry, with light and inhibitor sensitivities parallel to those of the chloride pump (Raven, 1968; Smith, 1967, 1968; Findlay *et al.*, 1969), and that the insensitivity of the potential to the presence of chloride in the solution suggests chemical rather than electrogenic coupling of cations and chloride.

Further, it is now clear that there are conditions in which the cell potential is much more negative than the potassium equilibrium potential E_K, and that if the membrane is freely permeable to potassium the levels in the cell could be maintained only by the active extrusion of K^+. This was shown in various higher plant cells by Etherton (1963) and Higinbotham *et al.* (1967, 1970), in *Neurospora crassa* by Slayman (1965), in the giant marine alga *Acetabularia mediterranea* (Saddler, 1970a), in *Nitella translucens* in some conditions by Spanswick (1972). However, in *Acetabularia*, where fluxes have been measured in this state, the metabolic dependence of the K^+ influx and efflux is the reverse of that predicted from the electrochemical potential measurements (Saddler, 1970b), perhaps suggesting that passive potassium fluxes are very small.

The view that passive diffusion plays a negligible role in ion movements at the plasmalemma of higher plant cells is strongly held by Epstein. A very large body of work in support of his ideas of carrier-mediated uptake mechanisms is summarized in his recent book (Epstein, 1972). It should be clear from the previous discussion that the relative importance attributed to passive diffusive fluxes in the ionic relations of plant cells is diminishing, but there are two difficulties in interpreting many of the uptake studies in higher plant tissues. The first is that potassium uptake may be associated with all three of the major transport processes suggested earlier; the high K/Na may be achieved by K^+/Na^+ exchange, cell pH may be regulated by K^+/H^+ exchange, and net salt accumulation involves the uptake of chloride (or nitrate) plus cations, including K^+. The situation is further complicated by the possibility of links between these processes, which will be discussed later. It is nevertheless important to try to distinguish the contributions of these separate processes to any measured potassium influx. The second difficulty is the technical one already discussed, of ensuring that influx measurements do in fact measure the plasmalemma influx M_{oc}.

There are indications in a number of cells of further regulation of K/Na balance at the tonoplast. Thus in the four characean cells for which figures are available, the ratio of K/Na is even higher in the cytoplasm than in the vacuole (figures in Table 1 of MacRobbie, 1970a). It appears that the cytoplasm may extrude sodium actively both to the outside solution and to the vacuole. The same was argued for barley roots by Pitman and

Saddler (1967). Estimates of cytoplasmic concentrations and fluxes in the halophyte *Triglochin maritima*, growing in a wide range of different salinities, were given by Jefferies (1973); sodium was passively distributed at the tonoplast at all salinities, but at high external potassium concentrations there was evidence of active potassium movement from vacuole to cytoplasm.

We may argue therefore that the need to maintain high cytoplasmic levels of K/Na may be fulfilled by regulation of these ions at the tonoplast as well as at the plasmalemma. High K/Na is general in the cytoplasm of all cells. In plant cells Steward and Mott (1970) have stressed its importance for growth, particularly for growth by cell division; in a comparison of the ionic composition of different clones of carrot cells in culture they found a marked correlation between a high K/Na and a high capacity for rapid growth, particularly in the stages of growth by division.

III. ACTIVE H+ EXTRUSION

The evidence for the active extrusion of hydrogen ions from plant cells is of two kinds, the direct demonstration of acidification in the external solution, and indirect indications from anomalies in the electrical behavior which could be explained by electrogenic extrusion of cations.

That both roots and certain storage tissues are capable, in suitable conditions, of acidification of the surrounding medium has been recognized for many years, and the conditions in which the process is stimulated, namely at high pH outside, or in the presence of an anion absorbed with difficulty or metabolized, are also well established (Ulrich, 1941; Hurd, 1958; Jackson and Adams, 1963; Poole, 1966; Hiatt, 1967; Hiatt and Hendricks, 1967; Kirkby, 1969; Pitman, 1970). Under these conditions the excess cation uptake is balanced by release of H+ in the outside solution, and is associated with the accumulation of the cation together with an internally synthesized organic acid anion, such as malate, in the cell vacuole. In sulfate solutions it seems that K^+/H^+ exchange at the plasmalemma and secretion of K^+ and A^- to the vacuole are the processes involved. In bicarbonate solutions H+ may appear in the external solution by disturbance of the CO_2/HCO_3^- equilibrium as bicarbonate is taken up by the cells, as well as by outward translocation of H+ across the membrane. Since the cell is 80–200 mV negative with respect to the outside solution, it seems impossible that the pH of the cytoplasm should be low enough for the extrusion of H+ as a downhill process, and active transport of H+ out of the cell is indicated.

Direct evidence for H+ extrusion is also available for certain green cells, although in this instance the operation of an active H+ efflux pump may be masked by processes leading to alkalinization of the medium; thus Jeschke (1970b) found light-dependent H^+ extrusion by leaves of the water plant *Elodea densa* gassed with air or CO_2-free nitrogen, but in bicarbonate-containing solutions a light-dependent uptake of H^+ was observed (Brinckmann and Lüttge, 1972; Hope *et al.*, 1972). Two light-dependent processes capable of producing OH^- in the medium have been suggested: the disturbance to the HCO_3^-/CO_2 equilibrium by the removal of CO_2 in photosynthesis, and the inward movement of H^+ at the plasmalemma in response to pH changes in the cytoplasm arising from light-driven movement of H^+ into the chloroplast thylakoids. The extent to which the second process occurs in intact cells is not clear.

The existence of processes leading to both acidification and alkalinization of the medium is also recognized in characean cells, where for some reason localization of regions of acid-secretion and base-secretion leads to banded deposits of $CaCO_3$ on and in the cell wall, of the order of millimeters in length (Arens, 1939; Spear *et al.*, 1969). Lucas and Smith (1973) showed that alkalinization is associated with HCO_3^- uptake and consequent CO_2 fixation, and appears to be localized at specific points (although these points can migrate), whereas acidification is more uniform in its distribution. Localization on such a gross scale, rather than on a molecular scale, is difficult to explain; it is presumably associated with differences in the underlying fine structure, perhaps, for example, with the presence or the absence of aggregates of endoplasmic reticulum between the cell membrane and the chloroplasts, or with the presence of lomasomes, but it is surprising that the regions of localized base secretion should be so widely separated.

The indirect evidence for active H^+ extrusion comes from electrical indications of processes transferring positive charge out of the cell. These include observations in many cells of membrane potentials more negative than can be accounted for as a diffusion potential controlled by any or all of the major ions present, and which show large and rapid responses to interference with metabolism. The best example is perhaps *Neurospora crassa* studied by C. L. Slayman (1965, 1970), where the potential of -200 mV can be reduced to the level of a diffusion potential by a range of metabolic inhibitors (azide, dinitrophenol, carbon monoxide, cyanide, anoxia). During inhibition the decrease in membrane potential paralleled the decline in the level of ATP in the cell (Slayman *et al.*, 1970). Slayman suggested active H^+ extrusion by an electrogenic pump. In a number of higher plant cells also, in both green and nongreen tissues, there is evidence for a significant contribution of an electrogenic pump to the membrane potential, with H^+ extrusion as the most likely candidate (oat coleoptiles, pea epicotyls: Higinbotham *et al.*, 1970; beet tissue: Poole, 1966, 1973;

barley roots: Pitman *et al.*, 1970; *Elodea*: Jeschke, 1970b; Spanswick, 1973; *Mnium* leaves: Lüttge *et al.*, 1972; *Vallisneria* leaves: Bentrup *et al.*, 1973).

Electrogenic H^+ extrusion is also suggested in giant algal cells. Kitasato (1968) proposed that the discrepancy between the measured conductance of characean cell membranes and that calculated from the fluxes of K^+, Na^+, and Cl^- could be accounted for if the membrane were significantly permeable to H^+, but that an active extrusion of H^+ would then be required to maintain a reasonable cell pH in the face of considerable inward driving forces on H^+; the sensitivity of the potential in the presence of Ca^{2+} to pH, but not to K^+, would also be explained in these terms. Spanswick (1972, 1973) has modified this hypothesis, and rendered it compatible with more of the experimental observations, by postulating a voltage-dependent, light-stimulated, active H^+ extrusion, in which the charge-carrying pump contributes the major conductance. The effects of pH on potential are then considered to reflect control of the pump rate rather than to demand significant passive permeability to H^+. The observations of Bentrup *et al.* (1973) on *Vallisneria*, showing much greater sensitivity of the potential to pH in the light than in the dark, can also be explained in these terms. On the other hand, in *Elodea* the potential, although more negative than a diffusion potential, is insensitive to pH. This suggests that it is not determined by a voltage-dependent H^+ extrusion. Spanswick (1972) suggests that in *Nitella translucens* the active H^+ pump achieves a rate of about 22 pmoles cm^{-2} sec^{-1}, this being the current necessary to depolarize the cell to a level close to E_K. This is several times higher than any of the other transport systems in the cell, and is similar to two other estimates of the pump flux; from observations of the acidification of the medium Spear *et al.* (1969) suggested that in *Nitella clavata* the rate of H^+ extrusion was 5–20 pmoles cm^{-2} sec^{-1}, and Strunk's measurement (1971) of short-circuit current in perfused cells of the same species gave values of about 17 pmoles cm^{-2} sec^{-1}.

However it should be noted that although Spanswick's measurements represent the capability of the pump, the rate of H^+ extrusion under conditions of zero net current flow will be considerably less. Spanswick argues that the pump will extrude H^+ until the electrochemical gradient for H^+ is too large to be overcome by the energy available from the driving reaction. In this condition the charge transfer by the pump is balanced by the net passive flux of ions, and the working rate is not in fact higher than the passive fluxes of the major ions. In the steady state the charge transferred by the H^+ pump is balanced by passive influx of K^+, or another equivalent passive flux, and the H^+ efflux is balanced chemically by the production of H^+ internally, or by passive H^+ entry, or by OH^- release, perhaps by Cl^-/OH^- exchange. The chemical balance need not be exact, if the pH of the cytoplasm is changing, but the charge balance must be so.

In Spanswick's work the electrogenic pump is identified by the depolarization and increase in membrane resistance produced by turning the pump off, in the dark, or at low temperature, or in the presence of inhibitors. Spanswick has further shown (1974) that the pump can work in far-red light (710 nm), is relatively insensitive to dichlorophenyldimethylurea (DCMU) but is sensitive to carbonyl cyanide m-chlorophenyl hydrazone (CCCP) and to N,N'-dicyclohexylcarbodiimide (DCCD); he suggests therefore the operation of a membrane-bound ATPase leading to H^+ extrusion. He finds inhibition in the presence of CO_2 and suggests competition between the H^+ transport and CO_2 fixation for ATP; inhibition by CO_2 even in far-red light (when CO_2 fixation is abolished) is explained as an inhibition of cyclic phosphorylation by overoxidation of a component of the cyclic electron transport chain. It still remains to be explained why the sensitivity to CO_2 of an ATP-dependent H^+ extrusion should be greater than that of the ATP-dependent Na/K transport system.

Vredenberg (1973) and Vredenberg and Tonk (1973) interpreted some rather different results in terms of electrogenic pumping, probably of H^+. They measured current/voltage curves in *Nitella translucens* in the dark after varying periods of light pretreatment, and distinguished two effects of light on the electrical characteristics of the cell—a short-term effect (0.5–2 minutes), in which the membrane resistance is decreased, the potential generated by the electrogenic pump is thereby decreased, and the membrane depolarizes; and a longer-term effect (2–20 minutes) in which pretreatment in light puts the cell into a "higher energy state" in which the rate of electrogenic pumping is increased and the membrane hyperpolarizes. Their experimental conditions differ from those of Spanswick in that CO_2 is present in the solutions, and that they are concerned with the effect of light pretreatment on the subsequent activity of the electrogenic pump in the dark, but it is difficult to relate the two studies without further experimental work.

By contrast, although electrogenic ion transport seems to be responsible for the major part of the cell potential of -170 mV in *Acetabularia* (Gradmann and Bentrup, 1970; Saddler, 1970c), the evidence points to electrogenic chloride uptake rather than H^+ extrusion (Saddler, 1970c). It seems likely that there is active H^+ extrusion (since the cytoplasmic pH is likely to be above the value of 4.8 for passive distribution at the plasmalemma), but it seems to make a negligible contribution to the potential, compared with that made by the very large chloride influx (200–790 pmoles cm^{-2} sec^{-1}; Saddler, 1970b). If there is any link between H^+ extrusion and the mechanism of chloride uptake, as has been suggested by Spear *et al.* (1969)

and Smith (1970), then the difference in pattern may reflect the different problems faced by cells growing in the high salt, high pH environment of the sea and the low salt, low pH environment faced by most plant cells. The active extrusion of H^+ from the cell to the environment may be a basic function of cell membranes in a wide range of cells, from *Streptococcus faecalis* upward (Harold and Papineau, 1972). Raven and Smith (1973) have suggested that H^+ (and OH^-) pumps have a fundamental role in the regulation of intracellular pH, that active H^+ extrusion arose very early in cellular evolution, and that coupling of H^+ transport to other solute transport processes, or to energy generation by organelle membranes, were secondary events. They pointed out that growth on CO_2 or hexose as carbon source, and NH_4^+ as nitrogen source, must be associated with net production of H^+ in the cell; the reverse is true of the use of HCO_3^- as carbon source and NO_3^- as nitrogen source, which will result in increasing alkalinity in the cytoplasm. They therefore argue that mechanisms for the control of intracellular pH are essential for balanced growth in a range of conditions, and that this is achieved by active transport systems at the plasmalemma.

We might note, however, that transfers to and from the vacuole may also be involved in the regulation of cytoplasmic pH. It seems unlikely that H^+ is in equilibrium at the tonoplast any more than at the plasmalemma. For example, in *Nitella* the vacuole has a pH of 5.5 (Hirakawa and Yoshimura, 1964), and a potential of about $+17$ mV with respect to the cytoplasm (Spanswick and Williams, 1964; Kishimoto and Tazawa, 1965). Unless the cytoplasmic pH is as low as 5.2, active movement of H^+ out of the cytoplasm at both plasmalemma and tonoplast may be indicated. In higher plant cells, conditions that are associated with H^+ movements at the plasmalemma also involve movements at the tonoplast. For example, we have seen that excess cation uptake (from whatever cause) is associated both with H^+ extrusion at the plasmalemma, and with synthesis and vacuolar accumulation of organic acid anions to balance the absorbed cation. In such cases, however, the pH of the sap rises as that of the medium falls (Hiatt, 1967; Hiatt and Hendricks, 1967); nevertheless it is again doubtful whether H^+ can be in equilibrium at the tonoplast, with a sap pH of 5–5.5 and only a small potential at this membrane.

IV. SALT ACCUMULATION IN THE VACUOLE

A. General

The simplest experimental system is provided by giant algal cells, where the process involves entry and accumulation of (chloride + cations), and where cytoplasm and vacuole may be separated for tracer analysis, opening the way to study of the interrelations of ion fluxes at the plasmalemma and ion transfer from cytoplasm to vacuole. In higher plant cells the greater diversity of solutes accumulated in the vacuole, and the mutual adjustment of the various solutes accumulated in response to changing conditions, combine with the lack of any direct experimental access to cytoplasmic conditions to make study of the process much more difficult. Although detailed understanding of the greater complexities of processes in higher plant cells is obviously essential in the long run, the hope for short-term progress may well lie with the easier experimental systems. Before considering the details of the experimental observations, we should, however, consider some general characteristics of the process of solute accumulation to which Steward and Mott (1970) have recently drawn attention.

Steward's thesis is that salt accumulation is only one facet of a much more general process, namely the ability of plant cells to secrete solutes internally in their vacuoles, and that the creation of vacuoles *de novo* during the growth and development of the cell is of primary importance. In support of his case for the consideration of a more general process of solute accumulation is a very considerable body of experimental information on the solute relations of carrot tissue in culture, in a range of conditions chosen to stress one or other facet of the complex of processes of vacuolar accumulation (Mott and Steward, 1972a,b,c; Craven *et al.*, 1972). The conclusions and inferences from this work were discussed in two review papers (Steward and Mott, 1970; Mott and Steward, 1972d). Such comparisons between carrot tissues growing in culture in different media highlight the differences in the solute relations of rapidly growing (dividing) cells, of rapidly expanding cells, and of nongrowing mature cells, particularly in respect to the balances between potassium and sodium, between inorganic anions absorbed from outside and internally synthesized organic anions, and between salts and nonelectrolytes as osmotic solutes.

For example, solute contents were determined in cells grown for 18 days in culture, in media ranging from simple inorganic salt solutions to a full organic and inorganic nutrient medium supplemented with a full range of growth factors; comparisons were made of fresh weight increase, cell size, cell number, total solutes and of the relative contributions of inorganic salts, organic salts, and nonelectrolytes. Two important general principles are established from this work—that reversible replacements of one type of solute for another may result from changing conditions, and that the pattern changes considerably during growth. The reversible replacements are most clearly seen in nongrowing conditions; in $CaCl_2$ cells maintain their levels of solutes, but given either sugar or salt ($KCl + NaCl$) they accumulate the solute supplied; on transfer from sugar to salt or salt to sugar they lose sugar and gain salt or vice versa. Thus the importance of maintaining a high solute concentration in the vacuole emerges, but the form of solute accumulated appears to be relatively unimportant, and much the same total solute level may be reached in different ways in different conditions.

The changing pattern of solute relations during the phases of growth was seen by following the changes during the growth initiated by adding to the minimal nutrient medium graded supplements of salts, nutrients, and growth factors. In suitable conditions the induction period is followed by a phase of rapid growth by cell division, producing small cells with incipient vacuoles, in which potassium organate accounts for most of the osmotic pressure, and sugars, sodium, and chloride contribute very little; the potassium level in the cell is particularly critical during this phase of growth. A phase of growth by cell enlargement follows, involving a considerable increase in total solute content and enlargement of vacuoles, with a large contribution of organic nonelectrolytes (sugars, amino acids, amides), and increasing values of sodium and chloride. Steward therefore stresses that absorption of inorganic ions is only one aspect of the overall problem of the solute relations of plant cells. It happens to be the aspect of importance in mature cells in the presence of salt outside, but potassium organates and organic nonelectrolytes are more characteristic of the solute relations of growing cells, and of many mature cells in the intact plant.

The complexities of the solute relations are further seen in the effects of the amount and form of nitrogen supplied to the cells. Nitrate-grown cells are characterized by high levels of potassium organates in their vacuoles, achieved by the uptake of K^+ and NO_3^- to the cytoplasm, the reduction of nitrate to organic form (with the generation of OH^-), the stimulation of organic acid synthesis, and the transfer of organic acid anions with potassium to the vacuole; in growing cells in nitrate media the protein content is particularly high. Cells supplied with nitrogen in the form of ammonium salts have a quite different pattern, with higher levels of soluble nitrogen but lower protein N; the most striking effect, however, is a marked reduction in the osmotic value, largely the result of reduced levels of organic acids, which more than compensate for the increased levels of amides and amino acids. Steward therefore argues that our concepts of the process of salt accumulation must be extended, to include the regulation of nonelectrolytes as well as electrolytes, and their interrelations, and to cover the changing patterns during cell growth and development.

Effects similar to these are also seen in whole plants. Thus the effect of the form in which nitrogen is supplied on the ionic balance and solute content of both roots and shoots has been recognized for many years; it has been discussed recently by Dijkshoorn (1969) and Kirkby (1969). Nitrate-grown plants have a higher total osmotic pressure, a higher potassium-organate content, and a higher sap pH than plants grown on ammonium salts. In addition, differences in the pattern found in different plants and differences within the plant body are found, which reflect differences in the site and extent of nitrate reductase activity in roots and tops. Nitrate reduction is associated with excretion of OH^- in the medium and accumulation of KA in the vacuole, whereas ammonium nutrition is associated with acidification of the medium and much reduced ion accumulation in the vacuole.

Vacuolar salt/sugar substitution in whole plants is also an observation of many years' standing (Prévot and Steward, 1936; Hoagland and Broyer, 1936). Barley plants grown on $CaSO_4$ are low in salt but high in sugar (typically 15 μeq g^{-1} potassium and 60 μmoles g^{-1} reducing sugar); plants grown on potassium salts are high in salt but low in sugar (typically 95 μeq g^{-1} K and 15 μmole g^{-1} sugar). In low salt roots transferred to KCl or NaCl there is rapid accumulation of salt to a high steady level, but this is associated with a fall in sugar level (Pitman et al., 1971). Mott and Steward (1972d) have argued that reversible transitions of this type are probably a normal feature of the interrelations between different organs of the angiosperm body, reflecting the progressive stages of growth, maturity and senescence in its tissues.

The work of Mott and Steward underlines the importance of considering the *solute* relations of plant cells, with ion accumulation as only one aspect of a more general process, namely the maintenance of the large central sap cavity (through its osmotic content), by which the mature plant cell economizes on cytoplasm and spreads it thin. Steward argues that the primary creation of new internal volume during growth should be distinguished from the means by which the vacuolar contents of mature cells are later regulated, and from the responses of mature cells to limited changes in environmental conditions. He deprecates the widespread study of nongrowing systems (chosen because they allow specific well-defined questions to be posed) and is of the view that progress can be made only by the study of plant cells throughout their life history, during their growing phases as well as in their mature state. This last argument could be disputed; it could equally be held likely that the same processes are involved in the maintenance of vacuolar systems as in their initial creation. Thus in mature characean cells vacuoles are maintained, rather than grown. But vacuoles redevelop in nonvacuolate fragments of *Nitella flexilis* obtained by centrifugation and ligation of the cytoplasmic fragment (Hayashi, 1952); this may well be by growth of very small vacuoles remaining after centrifugation, but the capacity to produce a large central salt-filled cavity is clearly still present in the mature cytoplasm. If common mechanisms are involved in the initial growth of vacuoles and the subsequent maintenance of their composition, then the study of systems in which we can frame questions and design experiments relating to the mechanisms will throw light on the more general problem. Steward focuses attention on salt accumulation as a feature of the creation of vacuoles in the cell, and his work suggests that there may be common features in a number of different processes of vacuole formation, leading to accumulation of different solutes in the central cavity. But the translation of these ideas into concrete molecular mechanisms demands detailed study of events in the cytoplasm, and must depend on the experimental feasibility of making such measurements. Initially at least it must rely on the study of the more amenable cells and tissues, even if these are not growing and do not reveal the full range of phenomena involved.

In discussion of the experimental work available we may consider three types of study, that of chloride fluxes in giant algal cells, that of chloride accumulation in higher plant cells of various kinds, and that of the accumulation of potassium organate in higher plant cells.

B. Chloride Fluxes in Giant Algal Cells

1. PLASMALEMMA FLUXES: FRESHWATER SPECIES

A number of characean cells have been used for study, and a great deal of work has been done on *Hydrodictyon africanum*. This work has been reviewed in detail (MacRobbie, 1970a, 1974), and only a brief account of the conclusions will be given here. The characean cells accumulate chloride from dilute solutions (of the order of 1 mM) to vacuolar concentrations of 100–200 mM, against an electrical potential gradient of 100–180 mV, with active influxes of the order of 1–2 pmoles cm^{-2} sec^{-1}; *Hydrodictyon* has similar fluxes and potentials, but a considerably higher surface: volume ratio and maintains a lower chloride concentration of about 40 mM in the vacuole. Estimates of cytoplasmic chloride vary with the cell and the method of obtaining the cytoplasmic sample; they make it clear that the concentration differs considerably in different cytoplasmic compartments. Thus in characean cells the chloroplasts (130–240 mM) have higher chloride concentrations than either the flowing cytoplasm free of chloroplasts (10–87 mM) or the vacuole (100–160 mM) (full references in reviews cited). However, with cytoplasmic potentials of -140 to -170 mV (or more negative), any of these figures for cytoplasmic concentration require there to be active transport of chloride at the plasmalemma, and make it likely that there is also active transport from cytoplasm to vacuole (and at the organelle membranes).

Most work on chloride influxes has been concerned with establishing first the energy supply for the active transport, the agent by which the pump activity is coupled to an energy-yielding metabolic sequence, and second the mechanism of the transfer process. It appears that although chloride transport can be driven either by photosynthesis or by respiration (i.e., by some consequence of these processes), the flux is strongly light-dependent, being up to 10–20-fold higher in the light than in the dark. Three types of coupling have been envisaged by which reaction sequences in the organelles could give rise to ion movements at the plasmalemma—by the provision of two sorts of chemical intermediate, in the form of ATP or reducing power, to be consumed in reactions in the plasmalemma, or by changes in H^+ and OH^- at the plasmalemma as a consequence of energy-dependent H^+ fluxes at the coupling membranes of the organelles (Robertson, 1968). Lüttge (1973) has termed the first two processes "biochemical" linkage, and the last "biophysical" linkage.

In two of the cells studied, *Nitella translucens* and *Hydrodictyon africanum*, there is evidence that the energy source is not ATP. The evidence includes the insensitivity of the chloride influx to uncouplers, and the fact that conditions in which only cyclic photophosphorylation is possible, in far-red light or in the presence of low concentrations of DCMU, do not support chloride influx (MacRobbie, 1965, 1966a; Raven, 1967). The point is best established in *Hydrodictyon* by Raven, both by the use of a wide range of inhibitors (1968, 1969b, 1971) and by the study of light effects (1967, 1969a). His most telling evidence is the red rise in quantum efficiency for potassium influx and sodium efflux, when by contrast the chloride influx, like CO_2 fixation, shows a red drop; the red rise argues that the supply of ATP at the plasmalemma has gone up in the transition to far-red light, and the existence of a red drop is then evidence of a process requiring the operation of both photosystems, or of a product of noncyclic flow. It would therefore appear that some product or consequence of light-driven electron flow other than ATP is responsible for light-dependent chloride uptake in these cells.

However, chloride transport in *Chara corallina* may differ from this pattern, in that it is sensitive to CCCP and to phlorizin and Dio-9 (Coster and Hope, 1968; Smith and West, 1969), and it has been argued that ATP is the energy source for transport. Recent work on a wide range of light-dependent processes in *Chara corallina* suggests that cyclic phosphorylation may not function in this cell, or that its products may not be available outside the chloroplast (Smith and Raven, 1974). If this is so, then we are left with no evidence for or against ATP as the energy source for ion transport in *Chara*, as we are unable to separate the possible products of the light-driven processes of which the cell is capable. But Smith and Raven (1974) showed that even in *Chara* the light-dependence of chloride influx differs from that of phosphate influx or glucose uptake; this suggests that its energy source may differ from that of the other solutes.

2. MECHANISM OF CHLORIDE TRANSPORT

The existence of a degree of linkage between chloride influx and monovalent cation influx is established in *Nitella*, *Hydrodictyon*, *Tolypella*, and often in *Chara* (Raven, 1968; Smith, 1967, 1968; Findlay *et al.*, 1969), although the nature of the linkage is not yet clear. The absence of a large effect of chloride on the membrane potential suggests that electrogenic coupling is not involved, and would argue for chemical coupling, but more elaborate explanations in terms of electrogenic coupling can be produced by making further assumptions. The simplest explanation remains, however, one of salt entry accounting for the major part of the chloride influx, with the short fall between (K + Na) influxes and Cl influx ascribed either to entry of HCl or to an anion exchange process.

There have been two suggestions that the active chloride influx is in fact a secondary consequence of active H^+ extrusion. Spear *et al.* (1969) found that chloride entry in *Nitella clavata* was largely confined to the acid-excreting bands of the cell, and they proposed that release of H^+ by the proton pump to an intramembrane region of restricted accessibility increased the product $[H^+][Cl^-]$ at the transport sites, and allowed the partitioning of molecular HCl inward through the membrane. The degree of acidity required at the pump site depends on the values assumed internally; the vacuolar pH of 5.5 (Hirakawa and Yoshimura, 1964), and chloride of 100 mM would demand a pH of 2.5–3 at the site. Since both H^+ and Cl^- seem to be lower in the cytoplasm than in the vacuole this estimate could be reduced somewhat, provided a subsequent transport process from cytoplasm to vacuole were additionally postulated. There remain, however, difficulties about the diffusion paths to and from the site of transport, to allow H^+ but not HCl to move outward, and Cl^- to move inward to the same site.

The second hypothesis was put forward by Smith (1970) and is a development of older concepts of salt uptake as a double ion exchange process, of K^+ for H^+ and Cl^- for OH^-. Smith suggested that the primary excretion of H^+ leaves the cytoplasm alkaline enough for Cl^-/OH^- exchange to ensue, while K^+ enters down the potential gradient created by electrogenic H^+ extrusion. He adduced a number of observations most easily explained in these terms; these include the inhibition of chloride influx at high external pH, its stimulation by ammonium ions or imidazole or tris buffers (and presumably the inhibition by ammonium ions at high pH), and the fact that pretreatment of *Chara* cells at high pH can put them in a state in which their ability to take up Cl in the dark is higher than that found after pretreatment at low pH (Smith, 1970, 1972). He has argued that the existence of a pH gradient at the plasmalemma, whether generated by H^+ extrusion or by changes in pH of the bathing solution, is the prime determinant of the rate of chloride uptake. The light-dependence of the chloride transport is then taken to imply that the cytoplasmic pH is markedly more alkaline in the light than in the dark, as a consequence of light-stimulated H^+ extrusion at the plasmalemma. The metabolic dependence of the chloride transport would then parallel that of H^+ extrusion, suggesting ways in which the hypothesis might be tested. In cells in which the chloride transport is not ATP-dependent the hypothesis would demand a vectorial redox reaction in the plasmalemma as the driving force for H^+ extrusion, a possibility discussed by Brown *et al.* (1973). Spanswick's conclusion (1974) that the H^+ pump in *Nitella translucens* is ATP-dependent, taken with the work discussed in the previous section suggesting that Cl transport is not dependent on ATP, would, however, argue against a direct link between the two transports. A critical measurement for this hypothesis is the pH of the cytoplasm, since the gradient of OH^- is to provide the driving force for Cl^- entry, but this is a figure not yet available. Smith suggested that a differential of 2 pH units would suffice, but even this might be considered to pose problems; Cl influx takes place readily from pH 7.5, requiring a cytoplasmic pH of 9.5, and some influx is possible from solutions up to pH 9. Since a pH of 9.5 seems unlikely for the bulk cytoplasm, some further means of restricting such alkalinity to the region just inside the plasmalemma would seem to be required if we are to accept the hypothesis. We might also be required to specify that the cytoplasmic chloride outside organelles should be low, and that the main accumulation should be achieved by an active transfer at the tonoplast.

But in fact the effects of ammonium ions, or imidazole buffers, on chloride influx, if they are to be interpreted as Smith has suggested, provide an indication of the value to be assigned to the cytoplasmic pH, since a weak base will only enhance the pH gradient if added to the *acid* side of the membrane, and will reduce the gradient if added to the alkaline side. Smith's demonstration of stimulation of chloride influx by ammonium ions at pH 7.5 or below, but either no effect or inhibition at pH 8 and above, would argue for a cytoplasmic pH between 7.5 and 8, which does not appear to be high enough to drive chloride influx. It might be possible, however, to rescue the hypothesis by making further assumptions. If the H^+ pump works until halted by the energy gradient (i.e., until the chemical reaction is unable to provide the energy to overcome the combination of potential gradient and concentration gradient), then the addition of a very permeable cation outside will reduce the electrical potential required to sustain net cation influx to balance the charge transfer by the pump, allowing a higher pH differential to be maintained. Thus if NH_4^+ is a readily penetrating cation it might be expected to enhance the pH gradient on this basis, irrespective of further effects on the gradient mediated by the equalization of NH_3 across the membrane and $NH_4^+/(NH_3 + H^+)$ equilibration inside.

Indirect indications of feasible values for the pH differential may also be obtained by considering its energetics. We may consider first what might be achieved by a charge-separating ATPase in the plasmalemma, with properties similar to those of the ATPases in coupling membranes. The limiting phosphate potentials for phosphorylation in chloroplasts (Kraayenhof, 1969), mitochondria (Slater, 1971), and membrane fragments of *Azotobacter vinelandii* (Eilermann and Slater, 1970) are all remarkably similar, with values of 17–17.5 kcal mole^{-1}, equivalent to the transfer of 1 electron through about 750 mV. If the primary reaction involves the separation of n electronic charges across the membrane, then the limiting energy barrier for the charge transfer is equivalent to $750/n$ mV. The potential gradient is 140–230 mV, depending on the degree of hyperpolarization, but at 180 mV as commonly observed the residual gradient against which the pump can work must be no greater than $(750/n) - 180$ mV, setting limits to the pH gradient which could be sustained. If we use the figure of $n = 3$ found in chloroplasts (Junge *et al.*, 1970; Witt, 1971), and also suggested by some authors for mitochondria (Skulachev, 1971), then the limit is set at 1.2 pH units. These are similar to the gradients achieved by the ATPase in *Streptococcus faecalis* (Harold and Papineau, 1972; Harold *et al.*, 1970). If we use a figure of $n = 2$, also suggested for mitochondria (Mitchell, 1966) then the limiting pH gradient becomes about 3 pH units. In fact either of these figures suggest that at low external chloride concentrations the system would be working at or beyond the limit; in *Chara* if the cytoplasmic chloride concentration is 10 mM then 1–2 pH units would suffice from external concentrations of 1 mM and 0.1 mM, but in *Nitella* where cytoplasmic chloride has been measured as 65–87 mM (Spanswick and Williams, 1964; Hope *et al.*, 1966), then under some conditions nearer 3 pH units would be required. Arguing on these lines we would predict that the pH at which chloride influx is cut off would depend on concentration (in the absence of any large effects of chloride on membrane potential, or making due allowance for any dependence of the degree of hyperpolarization on chloride outside); this might provide a further test of the hypothesis.

It would seem, however, that neither in *Chara* nor in *Nitella* could this mechanism alone accumulate chloride to a vacuolar concentration of 100–200 mM, and some further transport mechanism between cytoplasm and vacuole would be required; the continued entry of chloride at the plasmalemma would then be dependent on the removal of chloride from the cytoplasm to the vacuole. The hypothesis suggests that study of possible links between H$^+$ transport and Cl$^-$ transport would be valuable, and is consistent with some experimental evidence, but as yet it must be regarded as unproved.

If, however, the energy is supplied to the pump not as ATP but as reducing power, then the energetic problems are a good deal more formidable. Various shuttle systems have been suggested by which NADPH generated inside the chloroplast by light-driven electron flow can produce reducing power in the cytoplasm; these include a triosephosphate/phosphoglycerate shuttle (Latzko and Gibbs, 1969), or a malate/oxaloacetate shuttle (Heldt and Rapley, 1970; Heber and Krause, 1971), but it seems likely that, by whatever indirect means the reducing equivalents are made available from the chloroplast, the end result in the cytoplasm is NADH for subsequent use. If this is to be consumed in a redox chain in the membrane, we might expect the final acceptor to be oxygen, but by way of a chain including NADH, flavoprotein, and cytochrome. Skulachev (1971) has suggested that there may be both ATPase and short dead-end redox chains in all membranes. If so the membrane might be charged up either by ATPase activity or by a properly oriented redox chain of this kind. But while the complete sequence NADH/O$_2$ has a very large associated free energy change, it is unlikely that an individual redox reaction responsible for the charge separation in the membrane has a span of more than about 200 mV. This implies that unless there are several such transmembrane loops in the chain (as in mitochondria or chloroplasts) the redox process could not generate any appreciable pH gradient across the membrane in the face of a potential gradient approaching 200 mV, and could not therefore be expected also to move Cl$^-$ as a secondary consequence of the primary charge separation.

This argument can, in fact, be applied generally to any redox-driven chloride transport; the energy available from any likely single charge separation could overcome the barrier to chloride entry of either the membrane potential or the concentration gradient but not both. Given a favorable concentration gradient for the co- or the counterion, a neutral process of salt entry or anion exchange would lower the energy barrier, or alternatively there might be several redox loops working in concert, and charging the membrane cumulatively.

It should be clear from this section that discussion of the mechanism of chloride entry is at the moment highly speculative. There is evidence which is most simply explained in terms of Smith's hypothesis of linkage to a primary process of H^+ extrusion, but there is an obvious need for some hard experimental facts bearing on the critical conditions of the model and their implications for other cell processes. The degree of parallel behavior of chloride influx and H^+ extrusion needs urgently to be established under a wide range of experimental conditions affecting one or other process; reliable measurements of cytoplasmic pH under the same range of conditions are equally needed. But we need also to be clearer about the degree of chloride accumulation to be achieved by the primary process of entry, and the extent of further accumulation in subsequent transport processes, for example at endoplasmic reticular membranes or the vacuolar membrane. We might also notice that if the plasmalemma has the full combination of electrogenic H^+ extrusion and coupled passive cation entry to achieve charge balance, and both Cl^-/OH^- antiport and Na^+/H^+ antiport to stabilize the internal pH, then the system is both very versatile and very difficult to study, particularly in its open condition with outlet to or input from the vacuole on its inner side.

C. Chloride Accumulation in Higher Plant Cells

1. GENERAL

In vacuolate higher plant cells net salt accumulation from KCl takes place against the concentration gradient and in the face of a negative vacuolar potential of 80–120 mV or often higher; it must therefore involve the active transport of at least the anion, although the end result is the accumulation of both K^+ and Cl^-, usually in equivalent amounts. The main potential is at the plasmalemma (review by Higinbotham, 1973a), and indirect estimates of the chloride content of the cytoplasm in a number of tissues by the method of flux analysis (Pitman, 1963, 1971; Cram, 1968a; Lannoye, 1970) suggest that the primary chloride transport must be at the plasmalemma, although further transport from cytoplasm to vacuole may also be required. Recent direct measurements, using chloride-sensitive microelectrodes (Gerson and Poole, 1972), of the chloride concentration in the cytoplasm of the young, largely nonvacuolate, cells in mung bean root tips (in which vacuoles make up only about 23% of the cell volume rather than the 94% characteristic of mature cells), confirm the conclusion that the influx of chloride at the plasmalemma must be active at both high and low external concentrations.

Studies on the chloride transport have been largely concerned with the effects of external concentration (C_o), the extent of regulation of the influx by the vacuolar content (Q_v), and the effects of inhibitors on plasma-lemma and tonoplast fluxes.

2. Effect of C_o

A great deal of effort has been directed toward the characterization and interpretation of two apparently saturating systems of chloride influx, distinguishable in measurements over a wide range of external concentration—the so-called dual isotherms of salt uptake (Epstein, 1966; Welch and Epstein, 1968, 1969; Laties, 1969). However, it now seems that arguments about the interpretation of influx measurements should be restricted to experimental observations in which well-defined membrane fluxes were measured. As seen in Section I, B, 2 this demands measurements of M_{oc} by short uptake/short wash, i.e., short compared with the half-time for cytoplasmic exchange; as the half-times for removal of extracellular tracer may, in some conditions, be comparable with that for cytoplasmic exchange, this requirement may pose problems, but it remains a prerequisite if the results are to be interpretable. We saw that if cytoplasm and vacuole are in series, then the quantity measured in influx experiments if either uptake or wash are "long" is the quasi-steady state influx to the vacuole, $M_N = M_{oc} \cdot M_{cv}/(M_{cv} + M_{co})$.

It was argued previously (MacRobbie, 1971a) that figures existed in the literature which allowed the concentration dependence of M_{oc}, M_{co}, M_{cv}, and the derived quantity M_N, to be assessed. Such figures are available from flux analysis studies in beet (Pitman, 1963), carrot (Cram, 1968a), and potato (Lannoye, 1970). In excised roots the analysis of efflux curves is complicated by the loss of tracer from the cortical cytoplasm to the solution in two ways, both by transport to the stele and exudation from the cut end and by the more direct path of M_{co}, but this difficulty does not affect the direct comparison of M_{oc} and M_N made in barley roots by Cram and Laties (1971). For the measurement of further fluxes in excised roots, the method of separate collection of tracer lost from the root surface and from the cut end is required (Greenway, 1967; Weigl, 1969, 1971; Pitman, 1971). The difficulty is also overcome by the use of isolated maize root cortical tissue (Cram, 1973a).

It emerges from such comparisons of M_{oc} and M_N at different external concentrations that the apparent saturation of uptake at high concentrations is the result of an increase in the plasmalemma efflux, M_{co}, with little change in M_{cv}; M_{oc} rises steadily with concentration but as the ratio $M_{cv}/(M_{co} + M_{cv})$ falls steadily, the discrepancy between M_{oc} and M_N increases. At low C_o, M_N is close to M_{oc}, since the plasmalemma fluxes are much smaller than the tonoplast fluxes, but at higher concentrations outside M_{oc} is as high or higher than the tonoplast fluxes, and M_N is much closer to M_{cv} than to M_{oc}. (For example, in carrot at 0.5 mM M_{cv} is about 10 $\times$ M_{oc}, but falls to 0.5 $\times$ M_{oc} or less at higher concentrations. In beet in the range 20–40 mM, M_N is about 83–90% of M_{cv}, and in carrot over 25–60 mM about 75% of M_{cv}.) But as C_o increases, both influx and efflux at the plasmalemma increase in parallel, with little or no increase in net influx. This has been shown by Cram (1968a) in carrot tissue and by Weigl (1969) in maize roots, in which the ^{36}Cl efflux from the root surface is stimulated by increasing the external chloride concentration while the transport to the xylem is inhibited. Processes of exchange diffusion at the plasmalemma were suggested, and it is clear that the plasmalemma fluxes do not show the behavior expected of passive diffusion. Thus it appears that, as Welch and Epstein (1968, 1969) claimed, the increased transport at high concentrations is associated with the plasmalemma, and is not a passive movement. However, the appearance of an apparently saturating component at high C_o is a feature of M_N, not of the true influx M_{oc}. It arises because Laties is right in his contention (1969), that at high concentrations the tonoplast flux limits access of tracer to the vacuole, because M_{cv} does not change a great deal as C_o rises. However, as Cram (1973a) pointed out, the fact that the tonoplast imposes a kinetic limitation on transfer of tracer to the vacuole does not imply anything about the mechanism or control of the tonoplast flux.

Most such work has been done on low-salt tissues, either washed storage tissue, or low-salt, high-sugar roots. In roots sugar levels fall as salt accumulation proceeds (Pitman et al., 1971), and we have an instance of the partial replacement of one type of solute by another. A consistent feature of the results is that the final level of salt, in salt-saturated tissue, is independent of external concentration; C_o determines the time necessary to reach salt saturation, but the degree of accumulation achieved seems to be independently regulated. Thus the effect of C_o on fluxes in high salt tissues must be quite different from the effects in low salt tissue discussed in this section, and considerable control must be exerted by the content of the tissue.

3. EFFECTS OF INTERNAL SALT CONTENT, Q_v

The nature of the control exerted by the salt content on the fluxes has been the subject of a number of studies. There are two questions—whether the reduction in net influx as the tissue salt content rises is achieved by reduced influx at the plasmalemma, or by increased efflux at the plasmalemma, or by both, and whether there is control of the tonoplast fluxes also. Again it is important to establish effects of the content Q_v on the

fluxes at specific membranes, rather than on the composite quantity measured as M_N. The further complication in excised roots of the dual nature
of efflux from the cortex is also important, since the effects of Q_v on the
loss M_{co} and on transfer to the stele are likely to be different, and their
implications for control of salt levels in the intact plant are obviously
quite different.

   Jackson and Edwards (1966) found that in barley roots M_{oc} remained
constant as the chloride content rose from 5 μeq gm^{-1} initially to a salt-
saturated level of 80–90 μeq gm^{-1}; the fall in net flux was the result of a
rising efflux. In isolated root cortical tissue from maize also M_{oc} was un-
affected by content, although the long-term influx (M_N, or M_{ov}) was
reduced, particularly at high C_o, implying a reduction in M_{cv} (Cram,
1973a). The independence of final salt level on C_o would then imply that
the efflux is in some way regulated by the influx rate, and is not a simple
function of the internal content. Pitman (1969), by computer simulation
of the behavior of various models of the cell, suggested further that M_{cv}
must be an active flux and that it must also be regulated by the vacuolar
content.

   However, in other studies there are effects on M_{oc} as the tissue fills,
and the approach to salt saturation does not appear to be reached in these
instances solely by rising efflux. Cram and Laties (1971) found that M_{oc}
did fall in barley roots as the salt content rose, and it is argued (Cram,
1973b; Pitman and Cram, 1973) that this fall is not to be attributed to
the concurrent fall in sugar levels. In carrot Cram (1968a, 1969) found
very little change in M_{oc} over the range of content 70–125 μeq gm^{-1}, but
in later work, over a wider range of Q_v, a clear reduction in M_{oc} accompanied
the rise in Q_v as the tissue filled. In the tissue in which M_{oc} remained
constant, there were effects on other fluxes; salt saturation was approached
by increasing efflux M_{co}, but with associated cytoplasmic changes, with
reductions in the flux from cytoplasm to vacuole (M_{cv}), and in the cytoplasmic content as estimated by flux analysis.

   The most comprehensive study is that of Cram (1973b) on the regulation of chloride influx and nitrate influx in carrot and in barley roots.
In both tissues the influx of chloride at the plasmalemma is reduced with
increasing ($Cl^- + NO_3^-$) content of the tissue, and in barley the long-
term influx (M_{ov} or M_N) is reduced more than M_{oc}, suggesting an effect
also on M_{cv}. An interesting feature of the results is that internally chloride
and nitrate are equivalent in their effect on the fluxes, although they are
not equivalent in the primary entry process at the plasmalemma. Cram
suggests that the effect of ($Cl^- + NO_3^-$) content must be either on a
nontransport site or on a common nonselective site at the tonoplast. This
is reminiscent of the equivalence of chloride and bromide in respect of

transfer to the vacuole in *Nitella* cells, in spite of considerable selectivity of chloride over bromide in the entry mechanism (Section IV, E, 2), and might again argue for a process of nonselective salt transport from a cytoplasmic phase to the vacuole. The other interesting feature of the results is that in barley there was no relationship between plasmalemma chloride influx and malate content, although there was some effect of malate content on the flux to the vacuole. In carrot neither flux was significantly correlated with malate content. Thus the regulation appears to be exerted by concentration of $(Cl^- + NO_3^-)$, not by osmotic pressure.

This section has reviewed work which identifies problems rather than solves them. The existence of feedback processes by which the tissue content is controlled seems to be necessary, with regulation of the transfer from cytoplasm to vacuole as the clearest feature to be established, although the mechanism remains unknown. There are, however, some indications from the flux analysis results that the system may be a good deal more complex than the simple model. A curious feature of the results in carrot (Cram, 1968a, 1969) is that as the vacuolar chloride level rises the efflux M_{co} rises, but both M_{cv} and the cytoplasmic content Q_c, as estimated by the simple flux analysis, are found to fall. There are two problems here, first the means by which M_{co} rises while Q_c is falling, and second, the means by which a falling level of salt in the cytoplasm can be reconciled with rising levels in the vacuole. There are osmotic problems here as well as problems of control. Cram (1969) suggested that two cytoplasmic compartments, one responsible for M_{co} and the other for M_{cv}, might be invoked, but if we do so then the simple two-compartment model of the cell is no longer adequate, and the flux and content calculations are no longer valid. Some alternative to indirect analysis by tracer kinetics seems to be required for progress beyond the stage of identifying problems, and it may be that development of techniques of intracellular localization of ions, by autoradiography and electron probe microanalysis, is required for future progress.

D. Accumulation of Potassium Organate in Higher Plant Cells

We have already seen that the vacuolar accumulation of inorganic cations and organic acid anions is important in two conditions, both of which may be regarded as situations of excess cation uptake—by the supply of a poorly absorbed anion such as sulfate, or a metabolizable anion such as nitrate. Such accumulation associated with nitrate reduction is most clearly seen when this process takes place in the leaves, since nitrate reduction in the roots is in part associated with the excretion of OH^- in the medium (Kirkby, 1969). But in short-term experiments, after transfer of barley plants from nitrate-containing solutions to CaSO$_4$, the disappearance of nitrate from the leaves was associated with the accumulation of an equivalent amount of malate (Ben-Zioni et al., 1970).

The further consequences of the two situations of excess cation uptake seem to be equivalent, the synthesis of a particular organic acid anion in the cytoplasm, and its transfer to the vacuole along with the inorganic cation, usually K$^+$. The particular anion synthesized in such conditions depends on the tissue; it is frequently malate (as in barley roots and carrot slices) but oxalate is accumulated in Atriplex leaf tissue (Osmond, 1967, 1968), and other examples are recognized in other tissues. The common feature of the two situations leading to the synthesis and accumulation of potassium organate may be an increase in cytoplasmic pH, either by K$^+$/H$^+$ exchange at the plasmalemma or by consumption of H$^+$ in nitrate reduction, although this is not clearly established. Two problems are involved in the overall process, that of control of the synthesis of organic acid in such circumstances, and that of the mechanism of transfer to the vacuole.

The relations between ion uptake and organic acid metabolism in the tissue are most clearly seen in the comparison of the changes in organic acid content and sap pH in low salt barley roots in K$_2$SO$_4$, in KCl and in CaCl$_2$ (Hiatt, 1967; Hiatt and Hendricks, 1967). In sulfate solutions the accumulation of potassium and organic acid anions in equivalent amounts was associated with the excretion of H$^+$ to the medium, and an increase in the pH of the cell sap. In CaCl$_2$ there was uptake of chloride, with the disappearance of organic acid, although not in equivalent amounts, and the opposite changes in pH—the medium going more alkaline and the sap pH falling. In KCl there was accumulation of KCl with little change in pH of either medium or sap, although in KCl at low pH, or at concentrations below 0.5 mM, the chloride uptake was greater than the potassium uptake and there were smaller changes in organic acid and pH of the kind shown in CaCl$_2$ solutions. It was suggested that the malate level might be controlled by the cell pH, that K$^+$/H$^+$ exchange might increase the cytoplasmic pH and stimulate CO$_2$ fixation into malate, and that in CaCl$_2$ a fall in cytoplasmic pH might stimulate the decarboxylation of malate. These suggestions are consistent with observations on the properties of the relevant enzymes in vitro; it is suggested that the synthesis is achieved by the operation of phosphoenol-pyruvate carboxylase (PEP carboxylase) and malic dehydrogenase, and the decarboxylation by malic enzyme, all associated with the nonparticulate fraction of the cell (Ting and Dugger, 1967; Danner and Ting, 1967). The pH characteristics of the enzymes are appropriate; Davies (1973) quotes figures showing a 7–8-fold increase in the rate of carboxylation by PEP carboxylase from maize, as a result of an increase in pH from 7.2 to 7.6, and a 6-fold decline in the rate of activity of NAD-malic enzyme for an increase in pH from 7 to 8. Davies suggests that a combination of this kind, of a carboxylase which increases in activity and a decarboxylase which decreases in activity with increasing pH, would act as a pH stat and may be responsible for the control of cytoplasmic pH. The relative importance for the regulation of intracellular pH of biochemical mechanisms of this kind, and of processes of H^+ transfer across cell membranes of the kind discussed in the previous section, has not yet been clarified.

It is important to realize, however, that schemes involving malate synthesis in response to excess cation uptake are entirely dependent for continued operation on the removal of malate to the vacuole. The activity of PEP carboxylase is markedly inhibited by both malate and oxaloacetate; Ting (1968) gave values of K_i of 0.8 mM for malate and 0.4 mM for oxaloacetate, compared with a K_m for PEP of about 0.08 mM. These figures may differ under cell conditions, but the conclusion is still likely to be valid, that continued synthesis of malate is possible only if it can be removed to the vacuole.

If a primary electrogenic extrusion of H^+ is to provide the driving force for K^+ entry (as discussed in Section III), then it can also be argued that removal of K^+ to the vacuole is equally necessary for continued operation, and that the process will work in this form only when passive entry of K^+ is possible. At very low external concentrations, if an active potassium pump were induced by low cytoplasmic potassium, then some other means of charge balance would be required. As the hypothesis stands, transfer of both potassium and malate to the vacuole are required; the malate transfer seems likely to be uphill (high vacuolar levels can be achieved, and the tonoplast potential is small), but a salt transfer from cytoplasm to vacuole seems to be the most likely hypothesis.

There is in fact a good deal of evidence that the synthesized malate is transferred to the vacuole. That organic acid synthesis associated with excess cation uptake is a feature of vacuolate rather than nonvacuolate cells was shown by Torii and Laties (1966); the incorporation of $^{14}CO_2$ into malate in proximal sections of barley or maize roots was greatly stimulated in sulfate solutions rather than chloride, whereas there was little salt response in root tips, in which the cells are largely nonvacuolate. They suggested that delivery of potassium malate to the vacuole might stimulate synthesis in the cytoplasm. Jacoby and Laties (1971) showed that root tips, pulse-labeled with $KH^{14}CO_3$, then lost activity fairly rapidly to $CaSO_4$, whereas older portions of the roots retained the label. The amount involved is too large to be only cytoplasmic, and sequestering in the vacuole is indicated. The same is suggested by the efflux kinetics of ^{14}C label from beet tissue, after loading in KCl or in K_2SO_4; tissue labeled in sulfate rather than in chloride had not only much more ^{14}C label, but also a much higher fraction of that label in the slowly exchanging fraction, identified with the vacuole (Osmond and Laties, 1969). Compartmentation of the organic acids of plant tissues was first demonstrated by MacLennan *et al.* (1963), who showed the existence of pools of non-metabolized acids, not readily labeled from acetate-^{14}C, in vacuolate segments of maize roots. Later work by Lips and Beevers (1966a,b) extended these observations, showing that the malate pool labeled from acetate-^{3}H was metabolized readily, whereas labeled malate produced by the fixation of bicarbonate-^{14}C was remote from respiratory activity. Conditions increasing the rate of loss of ^{14}C label were associated with a fall in total malate (pH 5 outside, in the presence of malonate or DNP) but their effects on malate-^{3}H utilization differed (^{3}H release was unchanged at pH 5 compared with pH 7.5, was reduced by malonate but increased by DNP). Malate taken up from the outside solution went to the same pool as that fixed in the tissue, and not to that labeled by exogenous acetate-[3]H. Lips and Beevers suggested that both pools might be cytoplasmic, but the later work by Jacoby and Laties (1971) makes it clear that the bulk of ^{14}C-labeled malate must be vacuolar. Presumably we must envisage also a cytoplasmic pool (labeled by ^{14}C fixation or by malate uptake from outside) from which the vacuolar activity is derived, and which is distinct from the mitochondrial pool labeled from acetate.

Cram and Laties (1974) have made a detailed study, in carrot and in barley roots, of the kinetic behavior of the pools labeled by bicarbonate and malate supplied externally. Three kinetically distinguishable labeled components are found after bicarbonate-^{14}C uptake and fixation, and it is argued that these should be identified with the total pool of HCO_3^-/CO_2 in the tissue, the cytoplasmic malate, and the vacuolar malate. Supply of malate-^{14}C in the solution leads both to vacuolar accumulation of malate, and to the release of [14]CO_2 from a cytoplasmic pool. If the transfer of malate to the vacuole and the release of [14]CO_2 come from the same cytoplasmic pool, its size (with 1 mM malate outside) is estimated as 0.11 μmole gm^{-1}, or about 5 mM in 2% of the tissue volume. While the vacuolar accumulation rate increases steadily as the external concentration increases in the range 1–70 mM, the rate of [14]CO_2 release saturates above about 5 mM. The authors argue that this implies a specific activity of cytoplasmic malate equal to that outside. On the other hand, they argue that during the fixation of bicarbonate the specific activity of cytoplasmic malate remains low (about 0.04–0.08 of that outside is quoted), because of dilution with unlabeled CO_2 produced in respiration. If we are to accept their estimates, then we must additionally postulate that malate outside suppresses the fixation of bicarbonate in the cytoplasm.

The effects of chloride on CO_2 fixation, and on malate fluxes, in carrot and barley have also been studied by Cram (1974). He has pointed out that the question at issue is more likely to be the mechanism by which chloride suppresses carbon fixation, rather than as usually framed, that of the initiation of organic acid synthesis by excess cation uptake. Using labeled malate outside, he found that 10 mM chloride inhibited the influx of malate to the vacuole but stimulated the release of $^{14}CO_2$. Chloride had no effect on bicarbonate influx but a marked effect on the accumulation of fixed carbon as malate in the vacuole. Cram argues that the effects could be explained solely in terms of a suppression of CO_2 fixation by chloride, although the effect of chloride on transfer of exogenous malate to the vacuole might suggest a dual effect, both on production of malate and on its subsequent transfer to the vacuole. In the washout curve the intercept associated with the cytoplasmic malate pool is much reduced in chloride solutions compared with sulfate solutions. Cram converts the intercept to a value for the cytoplasmic content of malate by assuming a cytoplasmic specific activity of 0.22 in sulfate and 0.25 in chloride (from the relative rates of HCO_3^- entry and CO_2 release in respiration), using this to calculate tonoplast fluxes of malate, and then assuming that the rate of production of CO_2 from malate is the same at a given value of tonoplast flux, whether this is the result of endogenous production or uptake from the solution. This results in very high estimates of the cytoplasmic content of malate, of 4.3 μmoles gm^{-1} in sulfate and 0.9 μmoles gm^{-1} in chloride (corresponding to concentrations of 215 mM and 45 mM in 2% tissue volume).

A further result is that the cytoplasmic pools associated with equal fluxes to the vacuole and to CO_2 in the two conditions (of malate uptake, and of CO_2 fixation in chloride), differ by a factor of about 10. Given the properties of PEP carboxylase, it seems unlikely that the pools could be as high, particularly in sulfate when the fixation proceeds at maximum rate, and there seems to be a need to reexamine the assumptions. It may be that the specific activity is a good deal higher than 0.22–0.25 at the site of fixation, but this would imply close structural association between processes of salt entry, of HCO_3^- fixation and of vacuolar transfer, with respiratory CO_2 less readily accessible to the fixation process than transported HCO_3^-. This is difficult, but perhaps not impossible. In spite of the uncertainty over the size of the cytoplasmic pool, Cram's conclusion that chloride inhibits the fixation process seems certain, and the question of mechanism then arises. Cram suggested that the simplest interpretation is that Cl^- uptake involes a Cl^-/OH^- exchange and hence lowers the cytoplasmic pH, reducing the level of HCO_3^- available to PEP carboxylase (whose activity also decreases at lower pH). We have already seen that there are difficulties in explaining chloride uptake as a simple Cl^-/OH^- exchange unless further postulates are added, but the interrelations of inorganic salt accumulation and potassium organate accumulation might be taken to lend support to the hypothesis.

Cram also looked at the effect of salt loading on the malate accumulation rates. The vacuolar accumulation of malate, whether from exogenous malate or from HCO_3^- fixation, was markedly reduced in KCl-loaded tissue; in the first condition the site of action was not clearly identified, but in conditions of carbon fixation, it was established that the effect was on the tonoplast flux. Cram suggests that the most likely common effect is a suppression of malate transfer to the vacuole by high levels of vacuolar chloride, reminiscent of his previous observations (1973b) of reduction in vacuolar transfer of chloride by high vacuolar content of Cl^- or NO_3^-. He argues that this may reflect the existence of common regulatory mechanisms in the processes of vacuolar accumulation of organic anions produced in the cytoplasm and of ions taken up from the solution.

E. Vacuolar Fluxes

1. GENERAL

Two things should be clear from the previous discussion, the importance of processes of salt transfer from cytoplasm to vacuole, and the difficulty of their measurement. Continued accumulation of salt, whether by uptake of both cation and anion from the external solution or by uptake of cation and synthesis of anion, seems to depend on its removal from the cytoplasm by transfer to the vacuole. There is also evidence that this transfer is regulated by the vacuolar content (Cram, 1973b, 1974) and that it reflects *solute* relations rather than *salt* relations (Steward and Mott, 1970). Cram's findings of the equivalence of internal Cl^- and NO_3^- in this regulation, and the effect of Cl^- on the vacuolar transfer of malate have already been discussed; we might also add the observations of inhibition by salt of sugar transport to the vacuole in barley root cells (Pitman *et al.*, 1971).

But in spite of the obvious importance of vacuolar transport processes the experimental difficulties in their measurement are considerable, particularly in higher plant cells where only indirect methods are available. A number of instances have already been discussed in which the method of flux analysis leads to tonoplast fluxes which are unexpected and difficult to explain (or not consistent with the assumptions on which such analysis is based). Another complication is the possibility that the cytoplasm should not be considered as a single kinetic compartment in series with the vacuole, and that fluxes calculated on this basis are incorrect. For these reasons the results obtained by the use of flux analysis, some of which have been previously reviewed (MacRobbie, 1971a), will not be considered here, but only work on vacuolar kinetics in characean cells, with particular reference to chloride accumulation. The possibility of separation of cytoplasm and vacuole should remove some of the uncertainties from the measurements, although we shall see that problems of interpretation remain.

2. CHLORIDE TRANSFER FROM CYTOPLASM TO VACUOLE IN CHARACEAN CELLS

In characean cells, where cytoplasm and vacuole can be separated for analysis, attempts have been made to get information on the movement of ions between cytoplasm and vacuole from the kinetics of appearance in the vacuole of tracer ions from the outside solution. The experiments consist of labeling cells for varying short periods of time, and then counting a length of cell to determine the total tracer uptake (Q_T^*, nmoles cm^{-2}), and a sample of separated vacuolar sap to give the vacuolar content (Q_v^*, also expressed on an area basis in nmoles cm^{-2}). The method has been used on *Nitella translucens* and on *Tolypella intricata* (MacRobbie, 1966b, 1969, 1970b, 1971b, 1973), and on *Chara corallina* (Coster and Hope, 1968; Findlay et al., 1971; Walker and Bostrom, 1973). We may discuss, first, the behavior expected of the simplest models of the cell, second, the experimental observations, and last the interpretation. As the interpretation remains highly speculative and incompletely defined, it is important to separate the experimental facts from subsequent interpretation.

a. Model Behavior. The simplest model is that of a uniformly labeled cytoplasmic phase in series with a much larger vacuolar compartment; in this instance the specific activity in the cytoplasm, and hence the rate of tracer flux to the vacuole, will rise as $(1 - e^{-kt})$, with the rate constant k equal to the ratio of the sum of the fluxes out of the cytoplasm to the cytoplasmic content. After short times of uptake, the fraction of the total cell tracer which is in the cytoplasm will be given by $(1 - e^{-kt})/kt$. At very short times the cytoplasmic fraction will be $(1 - \frac{1}{2}kt)$, or the vacuolar fraction will be $\frac{1}{2}kt$. Thus the vacuolar fraction ($P = Q_v^*/Q_T^*$) will rise linearly with time, and the amount of tracer in the vacuole (Q_v^*) will go up as t^2 (MacRobbie, 1969, 1971a).

The next simplest model is that of two cytoplasmic phases, a large slowly exchanging phase of content Q_c, within a small rapidly exchanging phase, of content Q_r, which receives activity from the solution and delivers it to the vacuole. (Thus Q_c might include the chloroplasts, or other organelles, within a bulk cytoplasmic phase Q_r, or some other arrangement could be envisaged.) In this model we predict that after a very short initial period, in which the specific activity of phase r rises to a quasi-steady level, the vacuolar fraction will take the form: $P = P_o + \gamma t$. The intercept P_o is determined by the relative fluxes out of the rapid phase to the vacuole and to the slow cytoplasmic phase, and the slope γ is related to the exchange of the slow cytoplasmic phase (MacRobbie, 1969, 1973). A vacuolar time course of this form may be an indication that two cytoplasmic phases are involved in the transfer of tracer to the vacuole, and that the specific activity of one of these phases rises very rapidly, while the specific activity of the other is rising linearly with time over the period of measurement. The time course does not, however, define the arrangement of the two phases, and another possible explanation will be discussed later.

b. *Experimental Observations.* Experimentally, the vacuolar time course for ^{36}Cl is of the second form, and P shows both a fast component and a component rising linearly with time. In *Nitella translucens* it was shown by double labeling with ^{36}Cl and ^{42}K that the fast component was a feature of the chloride kinetics but not of the potassium kinetics; this means that the fast component for chloride cannot arise by gross cytoplasmic contamination of the sap sample (MacRobbie, 1969). The existence of a fast component for chloride is also seen in *Chara corallina* (Findlay et al., 1971; Walker and Bostrom, 1973).

Given this time course, the kinetics of the transfer of tracer chloride to the vacuole over short times are characterized by two quantities, the intercept P_o, which is a measure of the fast component, and the subsequent slope γ, a measure of the contribution of the slow component. We then require to establish the behavior of P_o and γ under different influx conditions, and to interpret such behavior.

The fast component P_o seems to be extremely variable. Its average value in *Nitella translucens* was 0.13 in cells collected from the wild, but 0.39 in cells cultured in the laboratory (MacRobbie, 1969). In *Chara corallina* much higher values were observed by Findlay et al. (1971) and by Walker and Bostrom (1973); thus Findlay et al. found that after 10-minute uptake periods 39% of cells had vacuolar fractions of 0–0.35, 22% fell in the range 0.35–0.7, but 35% of cells had values as high as 0.7–1.0; Walker and Bostrom (1973) found P_o equal to 0.5–0.6. These high values contrast with much lower vacuolar fractions found by Coster and Hope (1968) in the same species. They quote values for the exchange constant k calculated on the assumption of a single cytoplasmic phase, but if we calculate back to vacuolar fractions from these derived values we find that, even after 1 hour uptake, values of P come out to be only in the range 0.07–0.25 for cells from one source, and 0.17–0.37 for cells from another, suggesting P_o values lower than these figures.

When the distribution of values of P_o in any single experiment was examined (MacRobbie, 1970b), it was argued that such values were nonrandomly distributed, with a tendency to fall into distinct groups. This conclusion was disputed by Findlay et al. (1971) but reaffirmed by MacRobbie (1971b, 1973). The claim is that, in any given experiment, cells can be divided into groups having the same mean influx but different values of vacuolar fraction, with distinct gaps in the distribution of values of P; it is argued that cells are more frequent in subranges of P with relative values of 0.8–1.2, 1.8–2.2, and 2.8–3.2, and less frequent in the subranges of relative values 1.3–1.7 and 2.3–2.7. That is, there is a clustering around values of vacuolar fraction which fall in the ratio of 1:2:3. The essential point is that the same pattern of deviation from a uniform distribution (or from any single Gaussian distribution) is found in each experiment. While gaps in a distribution are indeed to be expected in experiments on small numbers of cells, the pattern of such gaps should be random. A recurring pattern in different experiments suggests a property of the distribution which has some significance and requires explanation.

The behavior of the slow phase in a range of influx conditions has also been examined, and it too shows unexpected features. This was first seen in earlier work, in which a single cytoplasmic phase was considered, and its rate constant for exchange (k) was calculated from the vacuolar fraction; a clear dependence of k on chloride influx to the cell was seen (MacRobbie, 1966b; Coster and Hope, 1968). That this reflects a dependence of the flux out of the slow cytoplasmic on the influx to the cell is shown in the later work, in which the contributions of the fast and slow phases to the vacuolar transfer were separated (MacRobbie, 1969); it then appears that γ is proportional to the influx to the cell. This is more clearly seen in double labeling experiments using ^{82}Br and ^{36}Cl, and allowing measurement of P_o and γ in the same cell (MacRobbie, 1971b, 1973). Two other properties of the kinetics appear from this work. The first is that there is no discrimination between bromide and chloride within the cell, although the initial influx mechanism has a marked preference for chloride over bromide. But cells labeled for equal times in ^{82}Br and ^{36}Cl have equal values of vacuolar fractions, and this equality holds over a very wide range of relative influx values, in both fast and slow components of vacuolar transfer. The second is that P_o and γ seem to be independent of one another; the same value of γ, and the same dependence on influx, is seen in the groups of cells having different values of P_o.

We may summarize the features of the kinetics which require explanation before discussing possible interpretations. We require a model of the cell that will explain (1) the existence of two kinetic components of vacuolar transfer of chloride, described by the quantities P_o and γ; (2) the tendency for clustering of values of P_o in different cells, around means in the ratios of 1:2:3, with some higher values; (3) the proportionality between the slope γ and the total influx to the cell (M_T); (4) the independence of P_o and γ, seen as the much greater variability of the fast fraction P_o, its clustering in distinct groups, and in the fact that cells in different groups of P_o values have the same value of γ/M_T; (5) the fact that there is no discrimination between chloride and bromide in vacuolar transfer, in spite of a marked discrimination at entry.

c. *Interpretation.* The form of the time course of vacuolar fraction requires two kinetically distinct cytoplasmic components to be involved in the vacuolar transfer over the first 1–2 hours of labeling. The problem is to identify these with distinct structural compartments and to interpret the kinetic properties found. There are two possible ways in which the observed time course might arise, and these may be discussed in turn.

The first is that the two components of vacuolar transfer represent two distinct cytoplasmic compartments; the specific activity in one rises very rapidly (in about a minute or so) to a quasi-steady level, while the other fills more slowly, its specific activity rising linearly with time over the experimental period. The kinetics then establish (i) that the rapid phase is very small, with a content much less than the chloride content of the flowing cytoplasm, (ii) that the slow phase is divided between the stationary and flowing cytoplasm (to explain the large fraction of activity reaching the flowing layers in experiments in which only half the cell is in labeled solution), and (iii) that the flux M_{cr}, from the slow cytoplasmic phase to the rapid cytoplasmic phase, is proportional to the influx to the cell. Identification of the slow phase with the plastids alone is ruled out by (i) and (ii), but the last characteristic is difficult to explain even if the slow phase is enlarged to include also other forms of organelle, including some in the flowing cytoplasm. It is very hard to see how the flux out of a mixed population of organelles, into the bulk cytoplasm around them, could change with influx at the plasmalemma, while the content of the slow phase can change only very little. It was therefore suggested that entry of salt, and transfer to the vacuole, by pinocytotic vesicles was consistent with the observations (MacRobbie, 1969); this would allow exchange between fast and slow compartments to be proportional to influx. The possibility of fusion of such vesicles (if they exist) with the endoplasmic reticulum, and a subsequent budding off from the endoplasmic reticulum of vacuoles which fuse with the central vacuole was also discussed. However the existence of grouping in the fast fraction, in near integral ratios, complicates the interpretation, particularly since this is a feature of P_o, but not of γ, and since the fast and slow phases seem to be independent of one another (MacRobbie, 1971b, 1973). These properties seem to demand that fast and slow components are in parallel rather than in series, and the complexity increases.

The alternative hypothesis is that the vacuolar transfer is primarily concerned with a cytoplasmic phase in which the specific activity is rising linearly with time, but that the fast component represents transfer to the vacuole during cutting, its quantization arising from the occurrence of a variable number of action potentials generated in the cutting and handling. It was originally argued that this was unlikely, for two reasons. The first was that too large a fraction of cytoplasmic chloride was involved (since no change in cytoplasmic content was detected). We could, however, overcome this difficulty by arguing that any transfer during the action potential took place from a cytoplasmic phase containing only a small fraction of the total chloride, but nearly all of the short-term tracer content, say from the cytoplasm minus the plastids. The second argument against the fast fraction as an artifact of cutting concerned the effect of bicarbonate in the solution on the chloride transfer to the vacuole; however, a better-defined time course of the vacuolar chloride transfer in the presence of bicarbonate may be required before this argument can be given full weight. If we argue that the fast fraction does indeed represent the transfer of a fraction of cytoplasmic chloride (or bromide) during action potentials on cutting, then we are left with two properties of the kinetics to explain. These are the link between influx and vacuolar transfer, and the lack of any discrimination between bromide and chloride. These are difficult to explain in terms of single ion transfer across the tonoplast, by a conventional carrier mechanism, but might be expected from the alternative type of mechanism suggested, namely a process of salt transfer to the vacuole by budding of new vacuoles from, say, the endoplasmic reticulum, and their subsequent fusion with the central vacuole. This might well not discriminate between chloride and bromide, and could be imagined to take place at a rate closely linked to the rate of chloride input to the phase.

The kinetics as established so far do not define any single model of the cell and its ion transport systems. The interpretation of the fast component as transfer from a cytoplasmic phase on cutting is in many ways easier to envisage. It explains the variability of P_o, the independence of fast and slow components and the grouping of values of P_o, all of which are difficult to explain if we are concerned with two cytoplasmic phases in parallel, delivering chloride to the vacuole independently of one another. The hope must be that study of ion transfer to the vacuole during the action potential, and of the kinetic characteristics under a wider range of conditions, will allow the speculations to be put on a firmer basis.

In whatever way the fast fraction is to be interpreted, there remain two well-established properties of the slow phase transfer—the link between vacuolar transfer and influx at the plasmalemma, and the non-discrimination between chloride and bromide in the process of transfer to the vacuole. These are both easier to explain in terms of the creation of new vacuolar volume, than in terms of processes of single-ion transport across a preformed tonoplast.

3. CONCLUSIONS

The measured kinetics of vacuolar chloride transfer in characean cells are curious, and may in part represent experimental artifacts. But the characteristic properties which are real and cannot be artifact are still difficult to interpret, and suggest that a change in viewpoint is required, with the stress on transport by a dynamic membrane system rather than a static one.

If the vacuolar accumulation of chloride does involve the creation of new vacuoles then it may be easier to envisage links with other transport processes. If plant cell vacuole is formed from the endoplasmic reticulum then we might expect this to be the site of the processes transferring salt to the vacuole, whether of chloride or of internally synthesized organic anion. It may be that a rise in solute content of the endoplasmic reticulum leads to membrane turnover processes, and the budding-off of vesicles from the endoplasmic reticulum, providing a common control on all vacuolar transfer processes. Links between the vacuolar accumulation of inorganic salts, organic acid anions, and sugars have been discussed in previous sections. The existence of common membrane turnover processes might suggest one way in which such links could be achieved. Perhaps the most pressing problems at the moment concern the nature of the processes of solute transfer to the vacuole and of their interrelations, of their control by solute content, and of the links between vacuolar transfer and entry to the cell.

V. COMPARISON WITH ANIMAL CELLS

The comparative aspects of ion transport in animal cells of various types have been considered by Keynes (1969, 1973), who recognizes five distinct types of ion pump.

Pump I is of major importance in animal cells—the Na/K exchange pump inhibited by ouabain, associated with membrane-bound ATPase activity vectorially sensitive to sodium and potassium concentrations, with a variable coupling ratio and hence of variable electrogenicity. As we have seen this is present in some plant cells in some conditions, but it is not universal; even in cells where it can be present in appropriate conditions its activity may at other times be masked by the operation of other types of ion transport also responsible for sodium or potassium movements.

The relation between the other ion pumps in the Keynes classification and the transport processes in plant cells which have been discussed here is not clear. Pump II is responsible for the net uptake of NaCl from dilute solutions by a range of animal epithelia, including the gills of freshwater fish and crustaceans, anal papillae of insect larvae, the outer surface of frog skin, and the luminal surface of kidney tubules (reviews by Keynes, 1969, 1973; Ussing, 1974). Although the normal function of this pump is the absorption of both Na^+ and Cl^- in equivalent amounts, it can under appropriate conditions transport Na^+ in and Cl^- out, or vice versa. It is therefore considered that the anion and cation transport are independent of one another, and the pump is split into group IIa for sodium uptake and pump IIb for chloride uptake. The suggestion has been made by a number of authors of two independent ion-exchange processes, of Na^+ for H^+ or NH_4^+, depending on tissue or conditions, and of Cl^- for HCO_3^- (or OH^-) (Krogh, 1939; reviews by Garcia Romeu, 1971; Maetz, 1971; Kirschner, 1973; Kristensen, 1973). Similar ion exchanges have been proposed in mammalian systems including the small intestine (review by Schultz et al., 1974). One characteristic of these transports is their sensitivity to acetazolamide (Diamox), presumably because of a role of carbonic anhydrase in maintaining the supply of HCO_3^- for exchange with Cl^-. The relation between pumps IIa and/or IIb and the H^+ extrusion pump and/or chloride influx pump in plant cells is not yet clear, but may indeed be very close. It is therefore interesting that Findenegg (1974) reports differences in Cl uptake in Scenedesmus grown in air and in high CO_2 conditions (in which the level of carbonic anhydrase is abnormally low), and also finds inhibition of Cl uptake by Diamox. On the other hand, Cram (1973b) found no inhibition by Diamox of Cl influx in carrot tissue. The energy source for the ion exchanges in epithelia is not identified; as Keynes (1973) has pointed out, very large pH gradients would be required to overcome the ion gradients against which uptake can occur—a difficulty we have already seen in considering Smith's hypothesis in Chara cells. An energy input may therefore be required. The identification of an ATPase in gastric mucosa and in pancreas, which is stimulated by HCO_3^- and inhibited by thiocyanate, led to suggestions that this enzyme may be responsible for chloride uptake and bicarbonate secretion by the cells (Kasbekar and Durbin, 1965; Simon et al., 1972). In plant studies the proton transport has been considered to be primary, whereas in the animal work the bicarbonate-chloride transport has been given this status, to explain the effects of Diamox and the stimulation of the ATPase by HCO_3^-. However, if a primary proton transport comes to be limited by the gradient it sets up (ΔE or ΔpH or both), and ΔpH is an important part of the total gradient, then the Cl^-/HCO_3^- exchange which dissipates this part of the gradient will influence the rate of proton extrusion and cation uptake; in these circumstances the supply of HCO_3^- for exchange could limit both other ion fluxes and ATPase activity. So it is possible that we are in fact concerned with basically similar ion transport processes in the animal systems and in the plant cell plasmalemma.

Keynes' pump III is also a chloride transport, but is responsible for electrogenic extrusion of Cl^- from animal cells; it is found in oxyntic cells, in the chloride cells of fish gills in sea water, in kidney tubules, and in intestinal mucosa, and it also is sensitive to Diamox. Since pump IIb can achieve net extrusion of chloride given suitably directed gradients, and pump III may be responsible for electrogenic chloride transport into cells in other systems (acinar cells, salivary glands, and intestinal mucosa), their mutual relations are not yet clear. We have seen that the active uptake of chloride in the marine alga *Acetabularia* is strongly electrogenic (Saddler, 1970c), but the transport in plants which is more likely to be akin to pump III (if any is) is the electrogenic Cl^- extrusion in halophytic salt glands of *Limonium* (summary in Hill and Hill, 1973); here it is suggested that electrogenic chloride secretion into long narrow channels in the convoluted cell walls of the secretory cells sets up the driving force for concomitant sodium extrusion.

However, although there may (or may not) be basic similarities in the ion transport processes in the outer cell membranes of animal and plant cells there seem to be very fundamental differences in intracellular processes. The essential feature of the ionic relations of mature plant cells is the transfer of salt of one kind or another to the central vacuole—a process which is absent in animal cells (those with contractile vacuoles for excretion of salt and water excluded). The mechanisms involved in this intracellular storage of solutes, and their relations to the processes of ion entry at the plasmalemma, are perhaps the most interesting problems at the moment, but also those least understood.

REFERENCES

Anderson, W.P. (1972). Ion transport in the cells of higher plant tissues. *Annu. Rev. Plant Physiol.* **23**, 51-72.
Arens, K. (1939), Physiologische Multipolarität der Zelle von *Nitella* während der Photosynthese. *Protoplasma* **33**, 295-300.

Barber, J. (1968a). The influx of potassium into *Chlorella pyrenoidosa*. *Biochim. Biophys. Acta* **163**, 141-149.

Barber, J. (1968b). The efflux of potassium from *Chlorella pyrenoidosa*. *Biochim. Biophys. Acta* **163**, 531-538.

Barber, J., and Shieh, Y.J. (1972). Net and steady-state cation fluxes in *Chlorella pyrenoidosa*. *J. Exp. Bot.* **23**, 627-636.

Ben-Zioni, A., Vaadia, Y., and Lips, S.H. (1970). Correlations between nitrate reduction, protein synthesis and malate accumulation. *Physiol. Plant.* **23**, 1039-1047.

Bentrup, F.W., Gratz, H.J., and Unbehauen, H. (1973). The membrane potential of *Vallisneria* leaf cells: Evidence for light-dependent proton permeability changes. *In* "Ion Transport in Plants" (W.P. Anderson, ed.), pp. 171-182. Academic Press, New York.

Brinckmann, E., and Lüttge, U. (1972). Vorübergehende pH-Änderungen im umgebenden Medium intakter grüner Zellen bei Beleuchtungswechsel. *Z. Naturforsch. B* **27**, 277-284.

Brown, D.F., Ryan, T.E., and Barr, C.E. (1973). The effect of light and darkness in relation to external pH on calculated H⁺ fluxes in *Nitella*. *In* "Ion Transport in Plants" (W.P. Anderson, ed.), pp. 141-152. Academic Press, New York.

Coster, H.G.L., and Hope, A.B. (1968). Ionic relations of *Chara australis*. XI. Chloride fluxes. *Aust. J. Biol. Sci.* **21**, 243-254.

Cram, W.J. (1968a). Compartmentation and exchange of chloride in carrot root tissue. *Biochim. Biophys. Acta* **163**, 339-353.

Cram, W.J. (1968b). The effects of ouabain on sodium and potassium fluxes in excised root tissue of carrot. *J. Exp. Bot.* **19**, 611-616.

Cram, W.J. (1969). Respiration and energy-dependent movements of chloride at plasmalemma and tonoplast of carrot root cells. *Biochim. Biophys. Acta* **173**, 213-222.

Cram, W.J. (1973a). Chloride fluxes in cells of the isolated root cortex of *Zea Mays*. *Aust. J. Biol. Sci.* **26**, 757-779.

Cram, W.J. (1973b). Internal factors regulating nitrate and chloride influx in plant cells. *J. Exp. Bot.* **24**, 328-341.

Cram, W.J. (1974). Effects of Cl⁻ on HCO₃⁻ and malate fluxes and CO₂ fixation in carrot and barley root cells. *J. Exp. Bot.* **25**, 253-268.

Cram, W.J., and Laties, G.G. (1971). The use of short-term and quasi-steady influx in estimating plasmalemma and tonoplast influx in barley root cells at various external and internal chloride concentrations. *Aust. J. Biol. Sci.* **24**, 633-646.

Cram, W.J., and Laties, G.G. (1974). The kinetics of bicarbonate and malate exchange in carrot and barley root cells. *J. Exp. Bot.* **25**, 11-27.

Craven, G.H., Mott, R.L., and Steward, F.C. (1972). Solute accumulation in plant cells. IV. Effects of ammonium ions on growth and solute content. *Ann. Bot. (London)* [N. S.] **36**, 897-914.

Danner, J., and Ting, I. P. (1967). CO₂ metabolism in corn roots. II. Intracellular distribution of enzymes. *Plant Physiol.* **42**, 719-724.

Davies, D.D. (1973). Metabolic control in higher plants. *In* "Biosynthesis and Its Control in Plants" (B.V. Milborrow, ed.), pp. 1-20. Academic Press, New York.

Dijkshoorn, W. (1969). The relation of growth to the chief ionic constituents of the plant. *In* "Ecological Aspects of the Mineral Nutrition of Plants" (I.H. Rorison, ed.), pp. 201-213. Blackwell, Oxford.

Eilermann, L.J.M., and Slater, E.C. (1970). The phosphate potential generated by membrane fragments of *Azotobacter vinelandii*. *Biochim. Biophys. Acta* **216**, 226-228.

Epstein, E. (1966). Dual pattern of ion absorption by plant cells and by plants. *Nature (London)* **212**, 1324-1327.

Epstein, E. (1972). "Mineral Nutrition of Plants: Principles and Perspectives." Wiley, New York.

Etherton, B. (1963). Relationship of cell transmembrane electropotential to potassium and sodium accumulation ratios in oat and pea seedlings. *Plant Physiol.* **38**, 581-585.

Findenegg, G.R. (1974). Beziehungen zwischen Carboanhydraseaktivität und Aufnahme von HCO_3^- und Cl^- bei der Photosynthese von *Scenedesmus obliquus. Planta* **116**, 123-131.

Findlay, G.P., Hope, A.B., Pitman, M.G., Smith, F.A., and Walker, N.A. (1969). Ion fluxes in cells of *Chara corallina. Biochim. Biophys. Acta* **183**, 565-576.

Findlay, G.P., Hope, A.B., and Walker, N.A. (1971). Quantization of a flux ratio in Charophytes? *Biochim. Biophys. Acta* **233**, 155-162.

Garcia-Romeu, F. (1971). Anionic and cationic exchange mechanisms in the skin of anurans, with special reference to Leptodactylidae *in vivo. Phil. Trans. Roy. Soc. London, Ser. B* **262**, 163-174.

Gerson, D.F., and Poole, R.J. (1972). Chloride accumulation by mung bean root tips. A low affinity active transport system at the plasmalemma. *Plant Physiol.* **50**, 603-607.

Gradmann, D., and Bentrup, F.W. (1970). Light-induced membrane potential changes and rectification in *Acetabularia. Naturwissenschaften* **57**, 46-47.

Greenway, H. (1967). Effects of exudation on ion relationships of excised roots. *Physiol. Plant.* **20**, 903-910.

Harold, F.M., and Papineau, D. (1972). Cation transport and electrogenesis by *Streptococcus faecalis.* I. The membrane potential. *J. Membrane Biol.* **8**, 27-44.

Harold, F.M., Baarda, J.R., and Pavlosova, E. (1970). Extrusion of sodium and hydrogen ions as the primary process in potassium ion accumulation by *Streptococcus faecalis. J. Bacteriol.* **101**, 152-159.

Hayashi, T. (1952). Some aspects of behaviour of the protoplasmic streaming in plant cells. *Bot. Mag.* **65**, 765-766.

Heber, U., and Krause, G.H. (1971). Transfer of carbon, phosphate energy and reducing equivalents across the chroplast envelope. *In* "Photosynthesis and Photorespiration" (M.D. Hatch, C.B. Osmond, and R.O. Slatyer, eds.), pp. 218-225. Wiley (Interscience), New York.

Heldt, H.W., and Rapley, L. (1970). Specific transport of inorganic phosphate, 3-phosphoglycerate and dihydroxyacetonephosphate, and of dicarboxylates across the inner membrane of spinach chloroplasts. *FEBS (Fed. Eur. Biochem. Soc.) Lett.* **10**, 143-148.

Hiatt, A.J. (1967). Relationship of cell sap pH to organic acid change during ion uptake. *Plant Physiol.* **42**, 294-298.

Hiatt, A.J., and Hendricks, S.B. (1967). The role of CO_2 fixation in accumulation of ions by barley roots. *Z. Pflanzenphysiol.* **56**, 220-232.

Higinbotham, N. (1973a). The mineral absorption process in plants. *Bot. Rev.* **39**, 15-69.

Higinbotham, N. (1973b). Electropotentials of plant cells. *Annu. Rev. Plant Physiol.* **24**, 25-46.

Higinbotham, N., Etherton, B., and Foster, R.J. (1967). Mineral ion contents and cell transmembrane electropotentials of pea and oat seedling tissue. *Plant Physiol.* **42**, 37-46.

Higinbotham, N., Graves, J.S., and Davis, R.F. (1970). Evidence for an electrogenic ion transport pump in cells of higher plants. *J. Membrane Biol.* **3**, 210-222.

Hill, A.E., and Hill, B.S. (1973). The *Limonium* salt gland: A biophysical and structural study. *Int. Rev. Cytol.* **35**, 299-319.

Hirakawa, S., and Yoshimura, H. (1964). Measurement of intracellular pH in a single cell of *Nitella flexilis* by means of microglass pH electrodes. *Jap. J. Physiol.* **14**, 45-55.

Hoagland, D.R., and Broyer, T.C. (1936). General nature of the process of salt accumulation by roots with description of experimental methods. *Plant Physiol.* **11**, 471-507.

Hope, A.B., Simpson, A., and Walker, N.A. (1966). The efflux of chloride from cells of *Nitella* and *Chara*. *Aust. J. Biol. Sci.* **19**, 355-362.

Hope, A.B., Lüttge, U., and Ball, E. (1972). Photosynthesis and apparent proton fluxes in *Elodea canadensis*. *Z. Pflanzenphysiol.* **68**, 73-81.

Hurd, R.G. (1958). The effect of pH and bicarbonate ions on the uptake of salts by disks of red beet. *J. Exp. Bot.* **9**, 159-174.

Jackson, P.C., and Adams, H.R. (1963). Cation-anion balance during potassium and sodium absorption by barley roots. *J. Gen. Physiol.* **46**, 369-386.

Jackson, P.C., and Edwards, D.G. (1966). Cation effects on chloride fluxes and accumulation levels in barley roots. *J. Gen. Physiol.* **50**, 225-241.

Jacoby, B., and Laties, G.G. (1971). Bicarbonate fixation and malate compartmentation in relation to salt-induced stoichiometric synthesis of organic acid. *Plant Physiol.* **47**, 525-531.

Jefferies, R.L. (1973). The ionic relations of seedlings of the halophyte *Triglochin maritima* L. *In* "Ion Transport in Plants" (W.P. Anderson, ed.), pp. 297-321. Academic Press, New York.

Jeschke, W.D. (1970a). Evidence for a K^+-stimulated Na^+ efflux at the plasmalemma of barley root cells. *Planta* **94**, 240-245.

Jeschke, W.D. (1970b). Lichtabhängige Veränderungen des Membranpotentials bei Blattzellen von *Elodea densa*. *Z. Pflanzenphysiol.* **62**, 158-172.

Jeschke, W.D. (1973). K^+-stimulated Na^+ efflux and selective transport in barley roots. *In* "Ion Transport in Plants" (W.P. Anderson, ed.), pp. 285-296. Academic Press, New York.

Junge, W., Rumberg, B., and Schröder, H. (1970). The necessity of an electric potential difference and its use for photophosphorylation in short flash groups. *Eur. J. Biochem.* **14**, 575-581.

Kasbekar, D.K., and Durbin, R.P. (1965). An adenosine triphosphatase from frog gastric mucosa. *Biochim. Biophys. Acta* **105**, 472-482.

Keynes, R.D. (1969). From frog skin to sheep rumen: A survey of transport of salts and water across multicellular structures. *Quart. Rev. Biophys.* **2**, 177-281.

Keynes, R.D. (1973). Comparative aspects of transport through epithelia. *In* "Transport Mechanisms in Epithelia" (H.H. Ussing and N.A. Thorn, eds.), Alfred Benzer Symp. V, pp. 505-511. Munksgaard, Copenhagen.

Kirkby, E.A. (1969). Ion uptake and ionic balance in plants in relation to the form of nitrogen nutrition. *In* "Ecological Aspects of the Mineral Nutrition of Plants" (I.H. Rorison, ed.), pp. 215-235. Blackwell, Oxford.

Kirschner, L.B. (1973). Electrolyte transport across the body surface of fresh water fish and amphibia. *In* "Transport Mechanisms in Epithelia" (H.H. Ussing and N.A. Thorn, eds.), Alfred Benzer Symp. V, pp. 447-460. Munksgaard, Copenhagen.

Kishimoto, U., and Tazawa, M. (1965). Ionic composition of the cytoplasm of *Nitella flexilis*. *Plant Cell Physiol.* **6**, 507-518.

Kitasato, H. (1968). The influence of H^+ on the membrane potential and ion fluxes of *Nitella*. *J. Gen. Physiol.* **52**, 60-87.

Kraayenhof, R. (1969). 'State 3-State 4 transition' and phosphate potential in 'Class I' spinach chloroplasts. *Biochim. Biophys. Acta* **180**, 213-215.

Kristensen P. (1973). Anion transport in frog skin. *In* "Transport Mechanisms in Epithelia" (H.H. Ussing and N.A. Thorn, eds.), Alfred Benzer Symp. V, pp. 148-156. Munksgaard, Copenhagen.

Krogh, A. (1939). "Osmotic Regulation in Aquatic Animals." Cambridge Univ. Press London and New York.

Lannoye, R.J. (1970). Echange de chlorures et compartimentation dans les tissus charnus de *Solanum tuberosum* L. *Ann. Physiol. Veg. Bruxelles* **15**, 1-30.

Laties, G.G. (1969). Dual mechanisms of salt uptake in relation to compartmentation and long-distance transport. *Annu. Rev. Plant Physiol.* **20**, 89-116.

Latzko, E., and Gibbs, M. (1969). Levels of photosynthetic intermediates in isolated spinach chloroplasts. *Plant Physiol.* **44**, 396-402.

Lips, S.H., and Beevers, H. (1966a). Compartmentation of organic acids in corn roots. I. Differential labeling of 2 malate pools. *Plant Physiol.* **41**, 709-712.

Lips, S.H., and Beevers, H. (1966b). Compartmentation of organic acids in corn roots. II. The cytoplasmic pool of malic acid. *Plant Physiol.* **41**, 713-717.

Lucas, W.J., and Smith, F.A. (1973). The formation of alkaline and acid regions at the surface of *Chara corallina* cells. *J. Exp. Bot.* **24**, 1-14.

Lüttge, U. (1973). Proton and chloride uptake in relation to the development of photosynthetic capacity in greening etiolated barley leaves. *In* "Ion Transport in Plants" (W.P. Anderson, ed.), pp. 205-221. Academic Press, New York.

Lüttge, U., Higinbotham, N., and Pallaghy, C.K. (1972). Electrochemical evidence of specific action of indole acetic acid on membranes in *Mnium* leaves. *Z. Naturforsch. B* **27**, 1239-1242.

MacLennan, D.H., Beevers, H., and Harley, J.L. (1963). Compartmentation of acids in plant tissues. *Biochem. J.* **89**, 316-327.

MacRobbie, E.A.C. (1965). The nature of the coupling between light energy and active ion transport in *Nitella translucens*. *Biochim. Biophys. Acta* **94**, 64-73.

MacRobbie, E.A.C. (1966a). Metabolic effects on ion fluxes in *Nitella translucens*. I. Active fluxes. *Aust. J. Biol. Sci.* **19**, 363-370.

MacRobbie, E.A.C. (1966b). Metabolic effects on ion fluxes in *Nitella translucens*. II. Tonoplast fluxes. *Aust. J. Biol. Sci.* **19**, 371-383.

MacRobbie, E.A.C. (1969). Ion fluxes to the vacuole of *Nitella translucens*. *J. Exp. Bot.* **20**, 236-256.

MacRobbie, E.A.C. (1970a). The active transport of ions in plant cells. *Quart. Rev. Biophys.* **3**, 251-294.

MacRobbie, E.A.C. (1970b). Quantized fluxes of chloride to the vacuole of *Nitella translucens*. *J. Exp. Bot.* **21**, 335-344.

MacRobbie, E.A.C. (1971a). Fluxes and compartmentation in plant cells. *Annu. Rev. Plant Physiol.* **22**, 75-96.

MacRobbie, E.A.C. (1971b). Vacuolar fluxes of chloride and bromide in *Nitella translucens*. *J. Exp. Bot.* **22**, 487-502.

MacRobbie, E.A.C. (1973). Vacuolar ion transport in *Nitella*. *In* "Ion Transport in Plants" (W.P. Anderson, ed.), pp. 431-446. Academic Press, New York.

MacRobbie, E.A.C. (1974). Ion uptake. *In* "Algal Physiology and Biochemistry" (W.D.P. Stewart, ed.), pp. 676–713. Blackwell, Oxford.

Maetz, J. (1971). Fish gills: Mechanism of salt transfer in fresh water and sea water. *Phil. Trans. Roy. Soc. London, Ser. B* **262**, 209-249.

Mitchell, P. (1966). Chemiosmotic coupling in oxidative and photosynthetic phosphorylation. *Biol. Rev. Cambridge Phil. Soc.* **41**, 445-502.

Mott, R.L., and Steward, F.C. (1972a). Solute accumulation in plant cells. I. Reciprocal relations between electrolytes and non-electrolytes. *Ann. Bot. (London)* [N.S.] **36**, 621-639.

Mott, R.L., and Steward, F.C. (1972b). Solute accumulation in plant cells. II. The progressive uptake of non-electrolytes and ions in carrot explants as they grow. *Ann. Bot. (London)* [N.S.] **36**, 641-653.

Mott, R.L., and Steward, F.C. (1972c). Solute accumulation in plant cells. III. Treatments which arrest and restore the course of absorption and growth in cultured carrot explants. *Ann. Bot. (London)* [N.S.] **36**, 655-670.

Mott, R.L., and Steward, F.C. (1972d). Solute accumulation in plant cells. V. An aspect of nutrition and development. *Ann. Bot. (London)* [N.S.] **36**, 915-937.

Osmond, C.B. (1967). Acid metabolism in *Atriplex*. I. Regulation of oxalate synthesis by the apparent excess cation absorption in leaf tissue. *Aust. J. Biol. Sci.* **20**, 575-587.

Osmond, C.B. (1968). Ion absorption in *Atriplex* leaf tissue. I. Absorption by mesophyll cells. *Aust. J. Biol. Sci.* **21**, 1119-1130.

Osmond, C.B., and Laties, G.G. (1969). Compartmentation of malate in relation to ion absorption in beet. *Plant Physiol.* **44**, 7-14.

Pitman, M.G. (1963). The determinations of the salt relations of the cytoplasmic phase in cells of beetroot tissue. *Aust. J. Biol. Sci.* **16**, 647-668.

Pitman, M.G. (1964). The effect of divalent cations on the uptake of salt by beetroot tissue. *J. Exp. Bot.* **15**, 444-456.

Pitman, M.G. (1969). Simulation of Cl^- uptake by low-salt barley roots as a test of models of salt uptake. *Plant Physiol.* **44**, 1417-1427.

Pitman, M.G. (1970). Active H^+ efflux from cells of low-salt barley roots during salt accumulation. *Plant Physiol.* **45**, 787-790.

Pitman, M.G. (1971). Uptake and transport of ions in barley seedlings. I. Estimation of chloride fluxes in cells of excised barley roots. *Aust. J. Biol. Sci.* **24**, 407-421.

Pitman, M.G., and Saddler, H.D.W. (1967). Active sodium and potassium transport in cells of barley roots. *Proc. Nat. Acad. Sci. U.S.* **57**, 44-49.

Pitman, M.G., Mertz, S.M., Graves, J.S., Pierce, W.S., and Higinbotham, N. (1970). Electrical potential differences in cells of barley roots and their relation to ion uptake. *Plant Physiol.* **47**, 76-80.

Pitman, M.G., Mowat, J., and Nair, H. (1971). Interactions of processes for accumulation of salt and sugar in barley plants. *Aust. J. Biol. Sci.* **24**, 619-631.

Pitman, M.G., and Cram, W.J. (1973). Regulation of inorganic ion transport in plants. *In* "Ion Transport in Plants" (W.P. Anderson, ed.), pp. 465-481. Academic Press, New York.

Poole, R.J. (1966). The influence of the intracellular potential on potassium uptake by beetroot tissue. *J. Gen. Physiol.* **49**, 551-563.

Poole, R.J. (1969). Carrier-mediated potassium efflux across the cell membrane of red beet. *Plant Physiol.* **44**, 485-490.

Poole, R.J. (1971). Effect of sodium on potassium fluxes at the cell membrane and vacuole membrane of red beet. *Plant Physiol.* **47**, 731-734.

Poole, R.J. (1973). The H^+ pump in red beet. *In* "Ion Transport in Plants" (W.P. Anderson, ed.), pp. 129-134. Academic Press, New York.

Prévot, P., and Steward, F.C. (1936). Salient features of the root system relative to the problem of salt absorption. *Plant Physiol.* **11**, 509-534.

Raven, J.A. (1967). Light-stimulation of active ion transport in *Hydrodictyon africanum*. *J. Gen. Physiol.* **50**, 1627-1640.

Raven, J.A. (1968). The linkage of light-stimulated Cl influx to K and Na influxes in *Hydrodictyon africanum*. *J. Exp. Bot.* **19**, 233-253.

Raven, J.A. (1969a). Action spectra for photosynthesis and light-stimulated ion transport processes in *Hydrodictyon africanum*. *New Phytol.* **68**, 45-62.

Raven, J.A. (1969b). Effects of inhibitors on photosynthesis and the active influxes of K and Cl in *Hydrodictyon africanum*. *New Phytol.* **68**, 1089-1113.

Raven, J.A. (1971). Inhibitor effects on photosynthesis, respiration and active ion transport in *Hydrodictyon africanum*. *J. Membrane Biol.* **6**, 89-107.

Raven, J.A., and Smith, F.A. (1973). The regulation of intracellular pH as a fundamental biological process. *In* "Ion Transport in Plants" (W.P. Anderson, ed.), pp. 271-278. Academic Press, New York.

Robertson, R.N. (1968). "Protons, Electrons, Phosphorylation and Active Transport." Cambridge Univ. Press, London and New York.

Rothstein, A. (1964). Membrane function and physiological activity of microorganisms. *In* "Cellular Functions of Membrane Transport" (J.F. Hoffman, ed.), p. 23-39. Prentice-Hall, Englewood Cliffs, New Jersey.

Saddler, H.D.W. (1970a). The ionic relations of *Acetabularia mediterranea*. *J. Exp. Bot.* **21**, 345-359.

Saddler, H.D.W. (1970b). Fluxes of sodium and potassium in *Acetabularia mediterranea*. *J. Exp. Bot.* **21**, 605-616.

Saddler, H.D.W. (1970c). The membrane potential of *Acetabularia mediterranea*. *J. Gen. Physiol.* **55**, 802-821.

Schultz, S.G., Frizzell, R.A., and Nellans, H.N. (1974). Ion transport in mammalian small intestine. *Annu. Rev. Physiol.* **36**, 51-91.

Shieh, Y.J., and Barber, J. (1971). Intracellular sodium and potassium concentrations and net cation movements in *Chlorella pyrenoidosa*. *Biochim. Biophys. Acta* **233**, 594-603.

Simon, B., Kinne, R., and Sachs, G. (1972). The presence of a HCO_3^--ATPase in pancreatic tissue. *Biochim. Biophys. Acta* **282**, 293-300.

Skulachev, V.P. (1971). Energy transformations in the respiratory chain. *Curr. Top. Bioenerg.* **4**, 127-190.

Slater, E.C. (1971). The coupling between energy-yielding and energy-utilizing reactions in mitochondria. *Quart. Rev. Biophys.* **4**, 35-71.

Slayman, C.L. (1965). Electrical properties of *Neurospora crassa*. Respiration and the intracellular potential. *J. Gen. Physiol.* **49**, 93-116.

Slayman, C.L. (1970). Movement of ions and electrogenesis in microorganisms. *Amer. Zool.* **10**, 377-392.

Slayman, C.L., and Slayman, C.W. (1968). Net uptake of potassium in *Neurospora*. Exchange for sodium and hydrogen ions. *J. Gen. Physiol.* **52**, 424-443.

Slayman, C.L., Lu, C.Y.-H., and Shane, L. (1970). Correlated changes in membrane potential and ATP concentrations in *Neurospora*. *Nature (London)* **226**, 274-276.

Slayman, C.W. (1970). Net potassium transport in *Neurospora:* Properties of a transport mutant. *Biochim. Biophys. Acta* **211**, 502-512.

Slayman, C.W., and Slayman, C.L. (1970). Potassium transport in *Neurospora*. Evidence for a multisite carrier at high pH. *J. Gen. Physiol.* **55**, 758-786.

Slayman, C.W., and Tatum, E.L. (1964). Potassium transport in *Neurospora*. I. Intracellular sodium and potassium concentrations and cation requirements for growth. *Biochim. Biophys. Acta* **88**, 578-592.

Slayman, C.W., and Tatum, E.L. (1965). Potassium transport in *Neurospora*. II. Measurement of steady state potassium fluxes. *Biochim. Biophys. Acta* **102**, 149-160.

Smith, F.A. (1967). The control of Na uptake into *Nitella translucens*. *J. Exp. Bot.* **18**, 716-731.

Smith, F.A. (1968). Metabolic effects on ion fluxes in *Tolypella intricata*. *J. Exp. Bot.* **19**, 442-451.

Smith, F.A. (1970). The mechanism of chloride transport in Characean cells. *New Phytol.* **69**, 903-917.

Smith, F.A. (1972). Stimulation of chloride transport in *Chara* by external pH changes. *New Phytol.* **71**, 595-601.

Smith, F.A., and Raven, J.A. (1974). Energy-dependent processes in *Chara corallina:* Absence of light stimulation when only photosystem one is operative. *New Phytol.* **73**, 1-12.

Smith, F.A., and West, K.R. (1969). A comparison of the effects of metabolic inhibitors on chloride uptake and photosynthesis in *Chara corallina*. *Aust. J. Biol. Sci.* **22**, 351-363.

Spanswick, R.M. (1972). Evidence for an electrogenic ion pump in *Nitella translucens*. I. The effects of pH, K^+, Na^+, light and temperature on the membrane potential and resistance. *Biochim. Biophys. Acta* **288**, 73-89.

Spanswick, R.M. (1973). Electrogenesis in photosynthetic tissues. *In* "Ion Transport in Plants" (W.P. Anderson, ed.), pp. 113-128. Academic Press, New York.

Spanswick, R.M. (1974). Evidence for an electrogenic ion pump in *Nitella translucens*. II. The control of the light-stimulated component of the membrane potential. *Biochim. Biophys. Acta* **332**, 387-398.

Spanswick, R.M., and Williams, E.J. (1964). Electric potentials and Na, K and Cl concentrations in the vacuole and cytoplasm of *Nitella translucens*. *J. Exp. Bot.* **15**, 193-200.

Spear, D.G., Barr, J.K., and Barr, C.E. (1969). Localization of hydrogen ion and chloride ion fluxes in *Nitella*. *J. Gen. Physiol.* **54**, 397-414.

Steward, F.C., and Mott, R.L. (1970). Cells, solutes and growth: Salt accumulation in plants re-examined. *Int. Rev. Cytol.* **28**, 275-370.

Strunk, T.H. (1971). Correlation between metabolic parameters of transport and vacuolar perfusion results in *Nitella clavata*. *J. Exp. Bot.* **23**, 863-874.

Ting, I.P. (1968). CO_2 metabolism in corn roots. III. Inhibition of P-enol pyruvate carboxylase by L-malate. *Plant Physiol.* **43**, 1919-1924.

Ting, I.P., and Dugger, W.M. (1967). CO_2 metabolism in corn roots. I. Kinetics of carboxylation and decarboxylation. *Plant Physiol.* **42**, 712-718.

Torii, K., and Laties, G.G. (1966). Organic acid synthesis in response to excess cation absorption in vacuolate and non-vacuolate sections of corn and barley roots. *Plant Cell Physiol.* **7**, 395-403.

Ulrich, A. (1941). Metabolism of non-volatile organic acids in excised barley roots as related to cation-anion balance during salt accumulation. *Amer. J. Bot.* **28**, 526-537.

Ussing, H.H. (1974). Transport pathways in biological membranes. *Annu. Rev. Physiol.* **36**, 17-49.

Vredenberg, W.J. (1973). Energy control of ion fluxes in *Nitella* as measured by changes in potential, resistance and current-voltage characteristics of the plasmalemma. *In* "Ion Transport in Plants" (W.P. Anderson, ed.), pp. 153-169. Academic Press, New York.

Vredenberg, W.J., and Tonk, W.J. (1973). Photosynthetic energy control of an electrogenic ion pump at the plasmalemma of *Nitella translucens*. *Biochim. Biophys. Acta* **298**, 354-368.

Walker, N.A., and Bostrom, T.E. (1973). Intercellular movement of chloride in *Chara*— a test of models for chloride influx. *In* "Ion Transport in Plants" (W.P. Anderson, ed.), pp. 447-458. Academic Press, New York.

Walker, N.A., and Hope, A.B. (1969). Membrane fluxes and electrical conductance in Characean cells. *Aust. J. Biol. Sci.* **22**, 1179-1195.

Weigl, J. (1969). Efflux and Transport von Cl^- und Rb^+ in Maiswurzeln. Wirkung von Aussenkonzentration, Ca^{++}, EDTA und IES. *Planta* **84**, 311-323.

Weigl, J. (1971). Diskontinuität des polaren Ionen-efflux durch das Xylem abgeschnittener Mais-wurzeln. *Z. Pflanzenphysiol.* **64**, 77-79.

Welch, R.M., and Epstein, E. (1968). The dual mechanisms of alkali cation absorption by plant cells: their parallel operation across the plasmalemma. *Proc. Nat. Acad. Sci. U.S.* **61**, 447-453.

Welch, R.M., and Epstein, E. (1969). The plasmalemma: Seat of the type 2 mechanisms of ion absorption. *Plant Physiol.* **44**, 301-304.

Witt, H.T. (1971). Coupling of quanta, electrons, fields, ions and phosphorylation in the functional membrane of photosynthesis. Results by pulse spectroscopic methods. *Quart. Rev. Biophys.* **4**, 365-477.

Young, M., and Sims, A.P. (1972). The potassium relations of *Lemna minor* L. I. Potassium uptake and plant growth. *J. Exp. Bot.* **23**, 958-969.

Young, M., and Sims, A.P. (1973). The potassium relations of *Lemna minor* L. II. The mechanism of potassium uptake. *J. Exp. Bot.* **24**, 317-327.

H⁺ Ion Transport and Energy Transduction in Chloroplasts

RICHARD A. DILLEY AND ROBERT T. GIAQUINTA

Department of Biological Sciences
Purdue University
West Lafayette, Indiana

I. INTRODUCTION

The chloroplast, mitochondrial, and bacterial membrane systems carry out energy conversion by a poorly understood series of reactions. Here we will restrict consideration of energy conversion to that occurring in the electron transfer chain and in the mechanism of ATP formation. We will deal in depth only with the chloroplast system, but the generalities of energy conversion may be similar in the three membrane systems.

We know enough about the composition, the gross structural features, and various chemical and redox reactions of membranes to propose several hypotheses as to how the potential energy of electrons in the electron transfer components might be converted partly into ATP formation. Experimental tests of the hypotheses has progressed slowly, and at this writing there are no clear experimental data that rule out a particular hypothesis. A brief discussion of the three most popular hypotheses is in order.

A. The Chemical Intermediate Hypothesis

First formulated by Slater in the 1950s (see Slater's review, 1971), the essence is that redox agents, such as cytochromes, form covalently bonded chemical species with other membrane proteins that then bind phosphate (or ADP) in a configuration with high enough energy content that the reaction $ADP + P_i \rightleftarrows ATP$ goes to the right. Other articles that discuss and support this hypothesis are by Wilson et al. (1973), Erecinska et al. (1973), and Wang (1972). Articles that stress the deficiencies of this viewpoint are by Greville (1969) and Skulachev (1971).

B. The Chemiosmotic Hypothesis

Formulated by Mitchell in the early 1960s (see Mitchell's review, 1966), the main features are:

1. Electron transfer agents are spatially arranged in the membrane so as to vectorially transfer protons, leading to a proton-motive force in the proton electrochemical potential gradient.

2. The proton gradient is hypothesized to be dissipated through a reversible ATPase such that ATP is formed at the expense of the proton gradient. Alternatively, ATP hydrolysis is predicted to give rise to a proton gradient.

Williams (1961) independently formulated a concept of proton and electron separation as the primary steps in energy conversion. Other articles that discuss aspects of the chemiosmotic hypothesis are Jagendorf and Uribe (1966), Greville (1969), Skulachev (1971), Crofts et al. (1971), and Witt (1971).

C. The Mechanochemical, or Membrane Conformation, or Fixed Charge Hypothesis

This hypothesis has been vaguer than the others, as indicated by the three more or less equivalent names given it. The essence is that electron and ion (particularly H^+ ion) transport give rise to changes in fixed charge arrays, changes that store and/or transduce some of the potential energy of the electron transfer chain components so as to favor the reaction $ADP + P_i \rightleftharpoons ATP$ going to the right. Some articles dealing with this viewpoint are Boyer (1965), Boyer et al. (1973), Dilley (1969, 1971), Weiss (1969), Murakami and Packer (1970), Azzone (1972), and Green and Ji (1972).

In this review we will first present a discussion of possible ways for energy storage and exchange that may occur during changes in macromolecular polyelectrolyte conformation. None of the reviews in the bioenergetics area that we are aware of deal with this in any detail. That situation should be corrected, for there are several intriguing aspects of polyelectrolyte behavior as energy transducers that relate reasonably well to what is known to occur in the biological energy-transducing systems. We will then turn to a discussion of the published work on chloroplast ion transport relating to electron flow and phosphorylation and, where possible, compare the observations with various hypotheses.

Detailed discussions of the chemiosmotic and chemical hypotheses abound in the reviews listed above. The mechanochemical or conformational viewpoint has hardly been explored except by Green and colleagues (Green and Ji, 1972). This is probably due in part to the difficulty of dealing with conformational changes on either an experimental or a theoretical basis. In the section below, we will outline some aspects of conformational changes, indicating possible approaches that may, when suitably worked out, provide some quantitative ways to treat energy conversion in polyelectrolyte systems.

D. Free Energy Forms

Without worrying about detailed mechanisms, one may compare various chemical free energy (ΔG) forms that may occur in biological systems, such as chloroplast membranes.

Electron transfer potential

$$D + A \rightarrow D^+ + A^-$$

$$\Delta G^\circ = -nF\Delta E^\circ \tag{1}$$

where n is the valency change, F the Faraday, and $\Delta E°$ the standard redox potential difference of the two couples, and D and A are electron donor and acceptor.

Group transfer potential (cf. Klotz, 1967, Chapter 6)

$$ATP + H_2O \rightarrow ADP + P_i$$

$$\Delta G = \Delta G° + RT \ln \frac{(ADP)(P_i)}{(ATP)(H_2O)} \tag{2}$$

Proton transfer potential:

$$AH + H_2O \rightarrow A^- + H_3O^+$$

$$\Delta G = \Delta G° + RT \ln \frac{(A^-)(H_3O^+)}{(AH)(H_2O)} \tag{3}$$

Electrochemical potential gradients:

membrane

out | in

a_o^+ | a_i^+

$$\Delta G = \Delta G° + RT \ln([a^+]_i/[a^+]_o) + nF\Delta\Psi \tag{4}$$

$\Delta\Psi$ is the electrical potential difference between two phases and $a_{i,o}$ activities (concentrations for dilute solutions) of an ion in the two phases.

The chemical hypothesis attempts to relate electron transfer potential directly to group transfer potential via "high energy" forms of the electron carriers, particularly the b cytochromes (Chance and Williams, 1956; Slater, 1971; Erecinska et al., 1973; Wilson et al., 1973) and cytochrome a_3 (Wilson and Dutton, 1970). Wilson et al. (1973) give a clear discussion of the energetic relationship between the respiratory chain and the ATP synthesizing mechanism from the viewpoint of the chemical hypothesis. Therefore, it need not be repeated in detail here. According to this hypothesis, an osmotically intact membrane system is not required, and Wilson et al. (1973) refer to evidence that "energy transduction" can occur in the absence of an intact membrane.

Such a situation is completely at variance with the chemiosmotic hypothesis (Mitchell, 1966), which requires the development and maintenance of electrochemical potential gradients (Eq. 4) of suitable ions, the flux of which provides the driving force for energy transduction. Further studies will verify whether energy transduction does occur in the absence of an intact membrane. This is an important point, for it is one of the few definitive experimental results that is sufficient to disprove the basic chemiosmotic postulate.

The conformational or mechanochemical hypothesis is not drawn explicitly enough to be on either side of the intact membrane requirement issue. Depending on which energy charging mechanism you wish to postulate, one could build a hypothesis either way. It is appropriate, at this point, to consider the energetic aspects of polyelectrolyte conformation changes.

E. Conformational Changes in Polyelectrolytes

Polymer chemists have shown experimentally and described in thermodynamic formalism that chemical free energy can be converted to mechanical work in a reversible, isothermal system. Katchalsky and Zwick (1955), Kuhn et al. (1960), and Mandelkern (1967) describe various types of mechanochemical energy-transducing systems that are not yet clearly models for the biological energy conversion with which we are concerned here, but the principles involved may be operative in photosynthetic and oxidative phosphorylation. Certainly the structure of all biological membranes includes a good proportion of polyelectrolytes; it is not unreasonable, therefore, to consider that the cardinal membrane functions may utilitize the unique properties of such polymer systems as mechanochemical transducers.

Mandelkern (1967) and Kuhn et al. (1960) describe how the reversible contractile properties of polymers can be affected by changes of the chemical potential of small molecular weight species other than H^+ surrounding it, by pH changes and by changes in the oxidation–reduction potential of the surrounding media. Is it simply coincidence that these three "factors" are intimately involved with biological energy transduction?

Polymers can undergo reversible conformational changes while in an amorphous or liquid state or during a transition from the liquid to the crystalline phase. In the former case, popularly termed rubber elasticity, changes in entropy of the polymer make a major contribution to the exchange of free energy. Guth (1947) has shown that a decrease in entropy of rubber polymers during the first stages of stretching can account for practically all the energization of the stretched material. In the case of charged polyelectrolytes, changes in $[H^+]$ or other co- or counterions or the redox environment can alter the degree of swelling and hence the entropy. This raises the *possibility that, in the chloroplast membrane, changing the chemical and redox milieu of the membrane polyelectrolytes during energization may lead to energy storage and exchange via entropy changes.* Weiss (1969) has discussed the possibility previously. The liquid–crystal transition (Mandelkern, 1967) similarly can occur with large changes in entropy in a reversible, isothermal manner dependent on changes in the ionic species surrounding the polyelectrolyte.

Measuring entropy changes in a complex system can be approached in two ways, one utilizing the temperature dependence of rate constants and the other from knowledge of the statistical probability of polymer configurations. From classical thermodynamic theory, the van't Hoff relation (cf. Bray and White, 1967, p. 94) relates temperature to rate constants or equilibria. The requirements for use are that the equilibrium constant be measurable for the reaction in question, in this case the reversible formation of ATP driven by the redox reactions. The reaction must be near equilibrium to apply classical thermodynamic theory. If such a reaction can be dissected out of the system and its equilibrium constant is measured, $\Delta G°$ is given by:

$$\Delta G° = -RT \ln K \tag{5}$$

where K is the equilibrium constant. The van't Hoff relation is derived from the Gibbs-Helmholtz equation

$$\left(\frac{\partial(\Delta G°/T)}{\partial T}\right)_{P,n_i} = \frac{-\Delta H°}{T^2} \tag{6}$$

where $H°$ is the enthalpy. This can be converted to the van't Hoff relation,

$$d[(\ln K)/dT] = \Delta H/RT^2 \tag{7}$$

which, with suitable rearrangement gives:

$$R\frac{d(\ln K)}{d(1/T)} = -\Delta H° \tag{8}$$

so that a plot of $R \ln K$ vs $1/T$ gives $\Delta H°$ as the slope.

Knowing $\Delta H°$ and $\Delta G°$, one can compute $\Delta S°$ from $\Delta G° = \Delta H° - T\Delta S°$.

The other, less rigorous approach involves using absolute rate theory (Bray and White, 1967) to calculate free energy of activation, $\Delta G\ddagger$, from the rate constants. For a first-order reaction having a rate constant k_f, one assumes that the equilibrium constant $K\ddagger$, for the formation of the activated complex or transition state, is related to the measured rate constant, k_f, by the relation:

$$K\ddagger = (k_f h)/kT \tag{9}$$

where h is Planck's constant, k is the Boltzmann constant, and T the absolute temperature. Using Eq. (5), and this $K\ddagger$ term, the free energy of activation, $\Delta G\ddagger$ is given by:

$$\Delta G\ddagger = -RT \ln[(k_f h)/kT] \tag{10}$$

Plotting $\Delta G\ddagger/T$ vs $1/T$ (from $\Delta G\ddagger = \Delta H\ddagger - T\Delta S\ddagger$) allows $\Delta H\ddagger$ to be calculated from the slope and $\Delta S\ddagger$ from the intercept. Ostroy *et al.* (1966) have used this approach to analyze the transformations occurring in the visual pigments during bleaching of rhodopsin, and have concluded that large positive entropy changes occur following the Meta I → Meta II formation.

Case and Parson (1971) and Callis *et al.* (1973) have used the classical thermodynamic approach in analyzing the thermodynamics of electron transfer in chromatophores of photosynthetic bacteria. They conclude that large negative entropy changes are associated with the reduction of the primary electron acceptor. Could these data be expressing energy storage as "negentropy"? Perhaps so, and further work on this system should verify this by identifying the molecular basis of these entropy changes.

If the conformation of membrane polymers can be better understood and if significant differences in the degree of stretching are shown to occur, then it may be possible to use the statistical theory of polymers (Kuhn and Green, 1946) to calculate entropy changes from the relation:

$$S = k \ln W \qquad (11)$$

where k is the Boltzmann constant and W is the number of possible configurations that a polymer can assume, having different energy levels. A contracted polymer has a lower W than an expanded or stretched polymer, hence ΔS is positive going from a contracted to a stretched conformation.

If energy transduction in the phosphorylation system occurs in a way involving entropy changes of significant magnitude, the examples taken from the work of Ostroy *et al.* (1966) and Case and Parson (1971) indicate that it should be possible to measure such changes. Valid data showing that large-magnitude entropy changes occur during generation and transduction of the high energy state should encourage further serious consideration of the mechanochemical hypothesis.

Entropy changes are not the only factor involved in free-energy changes of a polyelectrolyte system interacting with its surroundings. The electrostatic free energy of a polyelectrolyte molecule changes with the degree of deformation or stretch. Katchalsky (1954) gives an expression for the electrostatic free energy, G_{el}, as:

$$G_{el} = (\nu^2\epsilon^2)/Dh \ln[1 + (6h/\kappa h_o^2)] \qquad (12)$$

where ν is the number of charged groups per macromolecule, ϵ is the charge of an electron, D the dielectric constant of the medium, h is the distance between cross-links in the polymer, or the end-to-end distance; $h_o^2 = NA^2$, where N is the number of statistical elements in the molecule and A the hydrodynamic length of each element, and κ is directly proportional to the square root of the ionic strength. Obviously, changing the number of charges per molecule (e.g., protonating ionized acid groups) would have a profound effect on the electrostatic free energy while changing the length, h, of the molecule will have a lesser but still significant effect and ionic strength changes would have a somewhat smaller effect. No experimental data are available that attempt relating changes in G_{el} to biological energy transduction; but it may be a fruitful approach for future work.

For instance, in chloroplasts under energized conditions, H^+ ions are bound to fixed charge groups and internal K^+ and Mg^{2+} ions are exchanged out into the medium (Dilley, 1964; Dilley and Vernon, 1965; Dilley and Rothstein, 1967; Crofts et al., 1967; Nobel, 1967). The amount of H^+ binding is quite large, from 0.3 to about 1 H^+/Chl, chlorophyll being roughly 0.2 M in the membrane. Accompanying the protonation reaction the membrane undergoes conformational changes as evidenced by light-scattering changes (Dilley and Vernon, 1964; Hind and Jagendorf, 1965; Dilley, 1966, 1971; Deamer et al., 1967; Dilley and Rothstein, 1967; Murakami and Packer, 1971), packed cell volume changes, and electron microscopic data. The work of Dilley (1966), Dilley and Rothstein (1967), and Deamer et al. (1967) clearly showed that a portion of the light-scattering changes were related directly to protonation of fixed charge groups rather than to simple osmotic effects. In other words, membrane protonation induced structural changes in the membrane lipoprotein material, unrelated to osmotic changes (although osmotic changes also can occur under these conditions). Part of the evidence for this came from acid-induced conformational changes (see Fig. 9 of Dilley and Rothstein, 1967). It is interesting that chloroplasts undergoing an acid-to-base transition in the dark can drive ATP formation (see Jagendorf and Uribe, 1966). One can postulate that the energy transduction in this case occurs via H^+-induced conformation changes of the membrane polyelectrolytes. Thus it appears that energization of chloroplast membranes is attended by the kinds of changes expected of a mechanochemical transducer (cf. Katchalsky, 1954), but that is not to say that energy transduction into ATP synthesis does occur through such a mechanism.

A recent paper by Caserta and Cervigni (1973) proposes a piezoelectric transducer model for photophosphorylation that encompasses mechano-chemical and mechanoelectrical energy transduction. This model, to paraphrase the words of its authors, can possibly give a quantitative form to the otherwise vague statements previously made about conformational change models of energy transduction. With the possibility of experimentally detecting piezoelectric resonance, this model may be tested.

Straub (1974) has proposed a hypothesis based on energy transduction via lattice vibrations of the lipoprotein membrane matrix (i.e., generation and propagation of optical or acoustical phonons).

Green and colleagues (Green and Ji, 1972) have put together concepts and facts from diverse approaches into their electromechanochemical (EMC) model. They draw upon views presented by Mitchell (1966), Lumry (1971), Straub (1974), Witt (1971), Skulachev (1971), and many others.

In our opinion, the next few years will see a significant effort being made to test the various parameters associated with mechanochemical and mechanoelectrical properties of the energy-transducing membranes. The Mitchell chemiosmotic hypothesis may well merge with the mechanochemical model, for it is reasonable that protons with high electrochemical potential would be quite reactive with membrane polyelectrolyte binding sites, and proton binding to polyelectrolytes tends to induce polymer conformation changes. It has never been made clear in the extant versions of the Mitchell hypothesis just how the protonmotive force interacts with the membrane components to transduce the alleged gradient energy into ATP synthesis. Until such mechanistic aspects are elaborated, the chemiosmotic view will suffer from a similar vagueness attributed to the chemical hypotheses because of the failure to isolate the "high energy intermediate," or to the mechanochemical model for not identifying precise conformation changes.

The next section will deal with experimental findings on chloroplast H^+ ion transport in relation to electron flow and phosphorylation. We will show that a seemingly disjoint array of measurements relating H^+ fluxes to electron flow can be made reasonably coherent. The larger question, dealing with the mechanism of energy transduction into ATP, is still an elusive issue. No clear experiments with the chloroplast system permit saying either the chemiosmotic, the chemical, or the mechanochemical hypotheses, have been disproved.

II. CHLOROPLAST STRUCTURE AND ION FLUXES

Figure 1 shows a typical higher plant chloroplast that was fixed with glutaraldehyde and osmium tetroxide, embedded in plastic, thin-sectioned and poststained, and then viewed in an electron microscope. The outer membrane is not pertinent to this review, the ion transport and energy transduction events being confined to the thylakoid membranes. Most of the work dealt with here concerns isolated chloroplasts that have had the outer limiting membrane stripped away. The ion transport events are a

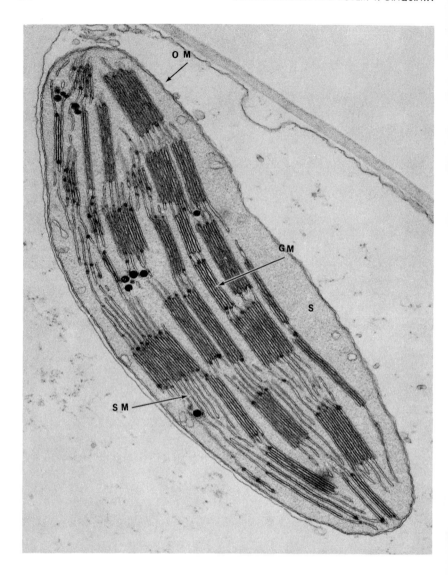

Fig. 1. An electron micrograph of a lettuce (*Lactuca sativa* var. Romaine) leaf chloroplast. Leaf tissue was fixed in glutaraldehyde–osmium tetroxide, embedded in plastic, and thin-sectioned. Poststaining was done with lead citrate and uranyl acetate. The outer, limiting chloroplast envelope, OM, encloses the grana membranes GM; the stroma membranes, SM; and the soluble stroma phase, S. The cell wall is also shown. This preparation was carried through the microscopy procedures by Dr. Robert Fellows. × 24,000.

light (electron transfer) dependent H^+ uptake into the thylakoid osmotic space (first observed by Jagendorf and Hind, 1963) in exchange for K^+ and Mg^{2+} (first shown by Dilley, 1964; Dilley and Vernon, 1965). Nobel (1967, 1969), using a clever, rapid isolation technique, has shown that light-induced K^+ and Mg^{2+} fluxes occur in chloroplasts *in situ*. The H^+ ions taken up are mostly bound to fixed charge buffering groups, the groups being neutralized by the mobile K^+ and Mg^{2+} ions in the "dark state." In some instances, Cl^- may accompany the H^+ ions (Deamer and Packer, 1969), but probably not so as to form an "HCl pool," as some have suggested. Rather, we would suggest that the anion permeability in such cases may exceed the cation (K^+, Mg^{2+}) permeability sufficiently to cause the entry of Cl^- and formation of a "KCl and $MgCl_2$ pool." Otherwise, the acidity inside the chloroplast membrane would drop below the tolerable level. Much remains to be done, however, to clarify relative permeabilities of various ions across the thylakoid membrane.

Before launching into the minutiae of the subject, we should point out that the light-induced Mg^{2+} efflux into, and H^+ out of, the stroma phase have extremely important regulatory effects on the light–dark regulation of the CO_2 fixation enzymes. Jensen and Bassham (1968) Jensen (1971), and Baldry and Coombs (1973) have shown how these factors can exert regulatory control on several enzymes of the Calvin–Benson cycle, particularly on ribulose-5-phosphate kinase, and ribulose-1,5-diphosphate carboxylase. The main theme of this review, however, is the possible involvement of H^+ ions and gradients of H^+ ions in the energy transduction mechanism.

III. ELECTRON TRANSPORT

As a benefit to readers outside this specialized research area, we will briefly review the most widely accepted current viewpoint as to how the chloroplast electron transfer chain is arranged. Figure 2 shows schematically the two photochemical systems, PSI and PSII, and their relationship to the electron transfer agents. A photochemically produced strong oxidant from PSII oxidizes water, the electrons being transferred to the primary acceptor Q, then on through the electron transfer chain, perhaps using cytochrome b_{559}, to plastoquinone (PQ), cytochrome f, plastocyanin (PC), to PSI (P700) another photochemical step, then through the primary acceptor for PSI, ferredoxin (Fd), Fd-NADP⁺ reductase, and thence to NADP⁺. The question of energy transduction encompasses the membrane structure, as it is the determining factor that permits the appropriate interactions of the functional components. Chloroplast membrane struc-

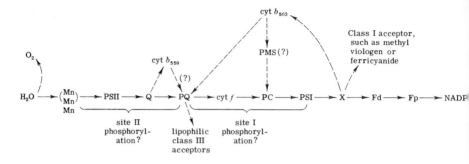

Fig. 2. A popular conceptual representation of electron transport in higher plant chloroplasts. PSI and PSII are the photochemical reaction centers; Q and X the primary electron acceptors for the photoacts; PQ, plastoquinone; cyt, cytochrome, PC, plastocyanin, a copper-containing protein; Fd, ferredoxin; Fp, Fd-NADP⁺ reductase, a flavoprotein; PMS, phenazine methosulfate, an exogenous cyclic phosphorylation cofactor. See Trebst (1974) and Saha et al. (1971) for additional discussion. The question marks indicate aspects that are currently hotly debated or poorly understood, or both.

ture-function relationships are at the same primitive stage that typifies most membrane systems; that is what makes membrane biochemistry so exciting as a research topic.

We know from the work of Howell and Moudrianakis that the coupling factor protein, the ADP kinase (see Arntzen et al., 1969, for references) is located on the external or stroma side of the thylakoid membrane. This laboratory, in conjunction with F. L. Crane, has presented evidence that PSI and PSII are asymmetrically arranged in the thylakoid membrane with PSI on the external portion and PSII on the interior portion (Arntzen et al., 1969). Additional work from this laboratory on this question will be discussed toward the end of the review. Figure 3 shows our diagrammatic representation of two thylakoid membranes, with the particles seen on the cut-away surfaces indicating the freeze-fracture revealed membrane particles. The appearance of the particles in a freeze fracture electron micrograph of a chloroplast is shown in Fig. 4. Such a membrane model has definite features of "sidedness", important in such functions as ion transport. See the recent review by Trebst (1974) on aspects of membrane sidedness and those by Park and Sane (1971) and Arntzen and Briantais (1973) on many other aspects of chloroplast membrane structure.

IV. PROTON FLUXES ACROSS CHLOROPLAST THYLAKOID MEMBRANES

We wish to know (1) the mechanism of H⁺ uptake and efflux; (2) the fate or state of the protons when in the membrane, i.e., bound or free, totally in the inner osmotic compartment or partly in the membrane itself; and (3) how proton fluxes or accumulation interact with the energy transduction mechanism.

A. Mechanism of H⁺ Fluxes

The noncyclic electron flow system. There are three possibilities for H^+ influx that can be visualized: (a) the chemiosmotic $PQH_2 \rightarrow$ cytochrome f redox loop; (b) a redox linked, but separate proton pump or carrier protein; and (c) a change in pK of H^+ binding groups that leads to spontaneous H^+ binding. The H^+/e^- ratio is an important parameter in testing various hypotheses of H^+ uptake. The chemiosmotic hypothesis predicts a stoichiometry of 1 H^+/e^- for a redox loop such as $PQH_2 \rightarrow$ cytochrome f. If water protons from the PSII oxidation step are deposited inside, another H^+ per electron can be accumulated, but it would not be observed as a

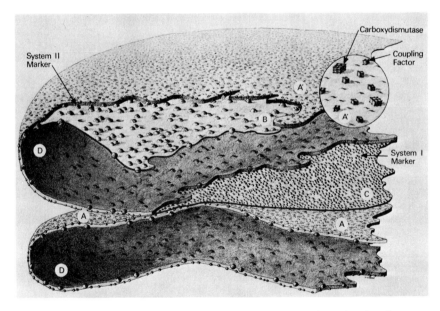

FIG. 3. A schematic representation of two chloroplast grana disks showing the asymmetric internal structure as revealed by the freeze-etch technique. The face labeled B on the inner portion of a single membrane is characterized by widely dispersed 175 Å particles. A detergent-prepared fraction enriched in this 175 Å particle is also enriched in photosystem II photochemical activity. Face C is a layer of 110–120 Å particles that occupies the outer portion of a grana membrane. Detergent-prepared fractions enriched in this particle are correspondingly enriched in photosystem I photochemical activity. From Arntzen *et al.* (1969).

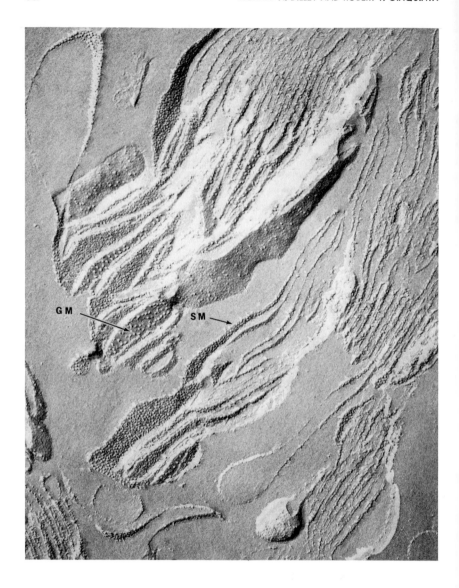

Fig. 4. A freeze-etch micrograph of a corn plant chloroplast (prepared by Dr. Robert Fellows), showing the appearance of the 175 Å particle on the internal face of a grana membrane, GM. To the lower right the grana membranes are shown in cross-sectional view. The stroma membranes, SM, also show particles, but with some differences in size (see Park and Sane, 1971, for a discussion of this). × 39,000.

charge transported from outside to inside, as is the expectation for the $PQH_2 \rightarrow$ cytochrome f loop, the H^+ carrier pump, or the pK change models. However, the water protons, if deposited inside, would contribute to H^+ efflux, and thus would add 1 H^+ per e^- to the measured steady-state H^+/e^- ratios, giving an H^+/e^- of 2. The other two hypotheses have no predictive value regarding H^+/e^- stoichiometries.

The data indicating whether water protons are deposited inside the thylakoid membrane are scant; Fowler (1973, cf. Kok, 1972), Junge and Ausländer (1974), and Gould and Izawa (1973c) reported experiments bearing on this. In these cases, the data support the notion that water protons are deposited inside the loculus, or are bound to the membrane. The first two groups of investigators have shown that agents that increase proton permeability, such as gramicidin and Cl-CCP, have the effect of more quickly releasing the water protons to the outside solution. This is consistent with the possibility that they are initially released *behind* a permeability barrier. Gould and Izawa (1973b) have studied the electron transfer from water to the lipophilic "class III" electron acceptors that are reduced at or before the plastoquinone site (Ouitrakul and Izawa, 1973). Their data can best be explained by assuming at least one-half, and probably all, of the water protons are deposited inside the loculus. The independent experiments of the three groups agree in their conclusions, and we will assume that protons liberated in water oxidation are deposited inside. However, more definitive data are needed to verify this—especially to check the point whether the deposition is all or none. Water oxidation could conceivably occur in the membrane so as to permit partitioning one H^+ to the interior and one to the exterior. As will be seen later, such a circumstance could explain some of our measured H^+/e^- ratios.

B. H^+/e^- Ratios at pH 6

The data in Table I are divided into H^+/e^- values obtained at pH values near 6 and near 8, and where possible the initial and the steady-state H^+/e^- ratios are listed. At pH 6 the initial H^+/e^- ratios vary from 5 to 1.7. Some of the variation, in particular the "superstoichiometry," may be due to technical or measurement artifacts. In the early work, some of these problems were understood and some were not. One that was appreciated but bears mentioning is that at pH 6 there may be a nonlinear electron flow rate, as shown by Crofts (1968), Rumberg et al. (1969), and Bamberger et al. (1973). Spectrophotometric assays for electron flow rates at pH 6 are plagued with light-scattering changes; in fact, that is how one of us (R. A. D.) became interested in the problem of energy-linked membrane

TABLE I

COMPILATION OF H^+/e^- VALUES

pH	Conditions	Initial H^+/e^-	Steady state H^+/e^-	References
6	Chloranil, O_2 and pH electrodes	5	—	Lynn (1967)
6.2	Fe(CN), rapid flow pH electrode	2.2(1.2)[a]–2.6(1.6)	—	Izawa and Hind (1967)
6	Fe(CN), pH electrode	4		Karlish and Avron (1968)
6.8	Fe(CN), NADP⁺ fast pH electrode		2	Schwartz (1968)
6	Fe(CN) glass electrode	2.1	1	Rumberg et al. (1969)
6	NH₄⁺-sensitive electrode and pH electrode	3 by NH₄⁺ uptake		Crofts (1968)
6–8	$H_2O \rightarrow$ DBMIB, a "class III" acceptor, pH electrode	0.4–0.5	0.2–0.25	Gould and Izawa (1973c)
8	Chloranil, O_2 and pH electrodes	1	<1	Lynn (1967)
8.2	MV,[b] pH + O_2 electrodes	1.4–1.9 −ADP 1.0–1.2 +ADP	0.5–0.7 −ADP 0.5 +ADP	Dilley (1970)

pH	Method	Conditions		Reference
8.3	MV, pH and O₂ electrodes	Data from their Fig. 4	$\left.\begin{array}{l}1.00 - \text{Dio-9} \\ 1.20 + \text{Dio-9}\end{array}\right\} - P_i$	Telfer and Evans (1972)
7.4	BV, bromothymol blue for ΔpH, P700 absorption changes for electron flow	2 with BV, etc. 1 with BV + DCMU		Schliephake et al. (1968)
8	Fe(CN) phenol red for ΔpH, or electrodes	$\begin{array}{ll}1 & -\text{val} \quad -\text{ADP} \\ 2 & +\text{val}\end{array}$		Schröder et al. (1972)
8	P705 turnover for e⁻ flow, cresol red for ΔH⁺	$\left.\begin{array}{l}1 \text{ by PQ} \\ 1 \text{ by H}_2\text{O}\end{array}\right\} = 2$	—	Junge and Auslander (1974)
7.8	Fe(CN), glass electrode	$2 \pm 0.5 \quad -\text{ADP}$ $1.0\text{--}2.5 \quad +\text{ADP}$	1.4 +ADP	Dilley (1970)
7.8	MV, pH electrode		0.2 −val 2.0 +val	Karlish and Avron (1971)

[a] The values in parentheses are Izawa and Hind's data before correction for the water oxidation H⁺ release.

[b] MV = methylviologen; BV, benzylviologen; DCMU, dichlorophenyldimethylurea; PQ, plastoquinone; val, valinomycin; DBMIB, dibromothymoquinone; Dio-9.

functions (Dilley and Vernon, 1965). Dual wavelength techniques do not necessarily get rid of this light-scattering artifact (Straub and Lynn, 1965). This problem is not adequately dealt with in the work referred to above, so there is still some uncertainty as to just how much of an electron flow "burst" in fact occurs. Izawa and Hind (1967) avoided this problem by measuring ferricyanide reduction chemically. They have detected no burst in electron flow to ferricyanide, but the problem of assay sensitivity for less than 3-second illumination times seems significant. At pH 8 there is very little burst (Bamberger *et al.*, 1973) with the basal system, and none is detected with the coupled system (Dilley, 1970). To the extent that an electron flow burst may occur at pH 6–7, and given that Clark-type oxygen electrodes respond relatively slowly, H^+/e^- superstoichiometries, such as those reported by Lynn and Brown (1967) using chloranil, could be artifacts.

Another possible source of the superstoichiometry artifact in initial H^+/e^- measurements is the relationship between the lag in water protons getting out of the osmotic compartment and the chemical H^+ binding (in the external phase) to reduced quinones or in the formation of H_2O_2 as reduced methylviologen (MV) is oxidized by oxygen, i.e.

$$2\,MV_{red} + 2\,H^+ + O_2 \rightarrow 2\,MV_{ox} + H_2O_2$$

Junge and Ausländer (1974) deal with this problem, but no one else has. In the first couple of seconds in the light, this chemical reaction will account for 1 H^+/e^- taken up from the media, at pH 8 as well as 6.

To what extent does the $MV_{red} + O_2 + H^+ \rightarrow MV_{ox} + H_2O_2$ reaction proceed after the light is turned off, artificially opposing the decrease in external pH due to $H_i^+ \rightarrow H_o^+$ flux? It is commonly assumed that the turnover of reduced MV with dissolved oxygen is so fast that this reaction is far to the right under the steady-state conditions and hence there is little MV_{red} available to turn over in the light-off period. Recent studies of this show that the reaction time of reduced viologens with O_2 is extremely fast (Farrington *et al.*, 1973).

In the ferricyanide case there are no external H^+ reactions as with MV, but the lag in water protons appearing outside could explain the initial pH rise detected with the ferricyanide or $NADP^+$ systems (Schwartz, 1968; Karlish and Avron, 1968; Dilley, 1970). To the extent that such a problem interferes with calculating true initial H^+/e^- ratios, the H^+/e^- calculations of Karlish and Avron giving 4 H^+/e^- at pH 6 are probably too high, as are the calculations from this lab reported for pH 8 (Dilley, 1970).

If we assume that the water protons are deposited inside the osmotic space and have a $t_{1/2}$ for efflux greater than 4 seconds, then the pH changes in the first couple of seconds of electron transfer to ferricyanide will reflect only the mechanism that transports protons as H^+ (or H_3O^+) into the membrane. Izawa and Hind (1967) show this effect in their Table 1 as the "observed H^+/e^- ratio" (they use H^+/e_2), and the values for the first 3 seconds were 1.2 and 1.6 H^+/e^-. Comparing these data with the steady-state efflux data of Schwartz (1968) for the ferricyanide case ($H^+/e^- = 2$), there is quite good agreement if one assumes that, at pH 6, for each electron one H^+ ion is taken across the membrane and 1 H^+ is deposited inside, owing to water oxidation, giving 2 H^+/e^- for the efflux.

Using the $H_2O \rightarrow NADP^+$ system, Crofts (1968) reported initial H^+/e^- ratios of 2.5 to 3 under conditions of pH 7. He calculated the H^+/e^- ratio measurements coupled to NADP reduction in the presence and in the absence of NH_4Cl. By using an NH_4^+-sensitive electrode [also sensitive to H^+ ion changes as demonstrated by the $-NH_4Cl$ control in Fig. 1 of Crofts (1968)], in addition to a pH electrode, he obtained measurements of H^+ accumulation indirectly by measuring the uptake of NH_4^+ into the osmotic space, the assumed reactions being:

$$NH_{3out} \xrightleftharpoons[\text{across membrane}]{\text{equilibrium}} NH_{3in}$$

$$H^+_{out} \xrightleftharpoons{H^+ \text{ pump}} H^+_{in}$$

$$NH_{3in} + H_i^+ \rightleftharpoons NH^+_{4in}$$

$$NH^+_{4out} \rightleftharpoons NH_{3out} + H^+_{out}$$

Electron transfer to $NADP^+$ results in 1 H^+ being bound to the nicotinamide ring per pair of electrons:

$$NADP^+ + H_2O \rightarrow NADPH + 1 H^+ + \tfrac{1}{2} O_2$$

The initial pH increase upon illumination will reflect this chemical H^+ binding plus any H^+ transport into the membrane, again on the assumption that the water protons are not sensed by the electrode in the first few seconds.

$$\Delta H^+_{obs} = \Delta H^+_{trans. \ in} + NADP_{red}$$

or

$$\Delta H^+_{trans. \ in} = \Delta H^+_{obs} - NADP_{red}$$

where the symbols stand for rates, micromoles of H^+ (or NADP) sec^{-1} (mg Chl)$^{-1}$. The H^+/e^- ratio will be given by the sum of $H^+_{trans. \ in} + NH_4^+$ uptake over $NADP_{red}$ (in μmoles) $\times$ 2, or

$$\frac{H^+}{e^-} = \frac{\Delta H^+_{obs} + NH_4^+ \text{ uptake} - NADPH_{red}}{NADP_{red} \times 2}$$

Crofts used the formula by adding the $NADPH_{red}$ rate in the numerator, on the assumption that the water proton acidification would cancel the protonated NADPH.

$$\frac{H^+}{e^-} = \frac{H_{obs} + NH_4^+ \text{ uptake} + NADP_{red}}{NADP_{red} \times 2}$$

We have recalculated Croft's (1968) data from his Fig. 2, using our assumption about the water protons not being sensed in the external solution in the first few seconds, and we find the H^+/e^- ratios ranging from 0.8 to 1.7, in quite good agreement with the data of Izawa and Hind (1967). In our recalculation, we subtracted from the NH_4^+ uptake values the NH_4^+ electrode rate observed in the absence of added NH_4Cl, assuming that deflection represented the NH_4^+ electrode response to pH changes. That appears to be appropriate in the $-NH_4Cl$ data, which then give an H^+/e^- ratio: $(0.28 + 0 - 0.1)/0.11 \times 2 = 0.18/0.22 = 0.82\ H^+/e^-$.

Interpreted in this way, the data for H^+ transport at pH 6–7 from Izawa and Hind (1967), Schwartz (1968), and Crofts (1968) for both initial and steady-state fluxes are reasonably consistent. The data fit the chemiosmotic hypothesis (shown below) quite well, with the exception that slightly higher than predicted initial influx stoichiometries are found, possibly reflecting proton binding to fixed charges.

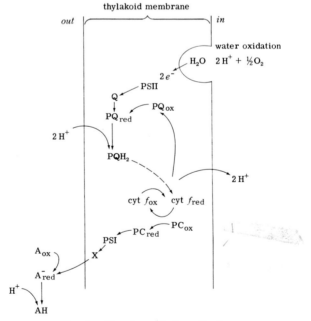

FIG. 5. Chemiosmotic hypothesis.

There is some extra H^+ uptake over that accounted for by the above mechanism (the values of 1.2–1.6 rather than 1). This could be due to a proton binding to membrane fixed charge groups, i.e., a membrane Bohr effect as discussed by Chance et al. (1970). With this possibility in mind, it seems to us that the Mitchell hypothesis best explains the observed proton fluxes at pH 6. Mitchell's (1966) concept postulated water protons deposited inside with the plastoquinone pool carrying 1 H^+/e^-, giving a stoichiometry of 2 H^+/e^- for steady-state conditions. Witt and colleagues (see review, Witt, 1971) have quite ingeniously pursued the testing of this hypothesis, and their work will be discussed at some length below.

Rumberg et al. (1969) supported this interpretation, although their data and the corrections they applied are not very clear. They show a value of 2.1 H^+ transported inward per electron at the initial light on period.

Reinwald et al. (1968) attempted to relate net electron flow to anthraquinone, plastoquinone (PQ) reduction, external pH changes, and membrane electric field formation (515 nm shift). They showed a close correlation up to 20 msec between the rate of PQ reduction, the 515 nm shift, and the pH rise in the medium, taking that as evidence for the PQ carrying H^+ into the membrane. That, however, cannot be accepted as evidence for $PQ^{2-} + 2H_o^+ \rightarrow PQH_2$ unless they account for the pH rise due to the reduced anthraquinone (AQ) chemical reaction with protons that occurs in the medium:

$$AQ^{2-} + 2H_o^+ \rightarrow AQH_2$$

The entire pH_o rise seen in the first 20 msec could have been due to this chemical reaction, which certainly would be expected to keep pace with the electron flux through the PQ pool.

At this time, no *direct* evidence is available that can be fairly interpreted as showing PQ protonation. However, by inference from *indirect* approaches, such as the use of PQ antagonists (Gould and Izawa, 1973b), a consensus is building that accepts PQ as a proton carrier. It is certainly logical in view of the high pK of PQ^{2-}, about 10, and it is a tribute to the work of Mitchell (1966) and Williams (1961) that we should look to quinones for this sort of function. However logical it is to expect PQ^{2-} to form PQH_2, we should not lose sight of the fact that *transmembrane* H^+ transport does not necessarily follow from having PQH_2 in the membrane. We should know more precisely the location of the oxidant for PQH_2 (probably cytochrome f).

C. H^+/e^- Ratios at pH 8

Junge and Ausländer (1974) have measured initial H^+/e^- ratios at pH 8 and interpret their data in complete accord with the scheme shown in

Fig. 5. In their experiments, they compare pH changes, using cresol red as an external pH indicating dye, and electron transfer in PSII and PSI measured by absorption changes at 520 and 705 nm, respectively, as well as by measuring oxygen production. A critical aspect of their data, not clearly explained, is how they arrived at the electron transfer value in the experiments designed to compare H^+/e^- ratios (their Figs. 5 and 6). They apparently estimate PSI activity from P700 bleaching by using an extinction coefficient of $4.2 \times 10^4\, l\, m^{-1}\, cm^{-1}$. This value is questioned by Hiyama and Ke (1972), who measured the extinction coefficient of 6.4×10^4; thus the electron transfer rates may be too high by a factor of 1.55. With this reservation in mind, their data support their conclusions that: (a) protons from water oxidation are trapped behind some permeability barrier (or bound) and not sensed in the external phase in the first couple of seconds, in agreement with the data of Fowler (1973); (b) one H^+ per electron is transferred from the external phase into the membrane phase; (c) A and B result in 2 H^+ ions inside (per electron) that efflux to the external solution; and (d) electron acceptors such as benzylviologen cause the binding of 1 H^+/e^- in the external phase as a result of their being reduced (and reoxidized by O_2 to form H_2O_2).

Our earlier H^+/e^- determinations (Dilley, 1970) under pH 8 conditions fit quite well with the data mentioned above if allowance is made for the time lag in water protons being sensed in the external solution. Using methylviologen (–ADP) we reported initial H^+/e^- ratios that varied from 1.4 to 1.9 (Table I) not adjusted for the water proton aspect (at that time we assumed that the water protons were released immediately and balanced the protons taken up in H_2O_2 formation following oxidation of reduced methylviologen). If we correct the initial H^+ change by attributing one-half the initial pH change to account for uncompensated H_2O_2 formation, the ratio range becomes 0.7 to 0.95. Again, this is in fairly good agreement with the work discussed above.

Schliephake *et al.* (1968) used four PSI electron acceptors (benzylviologen, BV; thymolindophenol, TIP; safranine T, and indigo carmine) and bromothymol blue (BTB) as the pH indicating dye to measure initial H^+/e^- ratios. Electron transfer rates were calculated from the absorbance changes of P700 at 705 nm, also using the questionable extinction coefficient of $4.2 \times 10^4\, l\, m^{-1}\, cm^{-1}$. They concluded that at pH 7.4 the initial H^+/e^- ratio (i.e., H^+ influx per electron) was 2 for any of the four PSI acceptors used; that value becomes 1.3 when the extinction coefficient of Hiyama and Ke is used. Steady-state measurements were not reported. They attributed 1 H^+/e^- as being transferred into the loculus by the PQ shuttle and 1 H^+/e^- released inside as PSII oxidizes water. Their measured ΔpH external apparently reflects both the proton transferred by the PQ and

1 H^+/e^- taken up as electrons from PSI reduce the acceptors that become protonated upon reduction.

Electron transfer from water to lipophilic "class III" acceptors (Ouitrakul and Izawa, 1973), such as dibromothymoquinone (DBMIB), permits study of ion flux associated with that partial, PSII, electron transfer sequence (Gould and Izawa, 1973c). At higher concentrations (25 μM) than needed for inhibition of $H_2O \to MV$ activity, DBMIB is reduced, probably by PQ, and reoxidized by molecular oxygen. The initial H^+/e^- ratio for this system is pH insensitive (from pH 6 to 8) and is close to 0.4–0.5. Parallel measurements at pH 7.5 using $H_2O \to MV$ gave initial, uncorrected H^+/e^- ratios of 1.7. Using the correction as before for the external protons taken up by H_2O_2 or DBMIBH$_2$ formation, the initial H^+/e^- ratios might be about 0.2 for DBMIB and 0.8 for the MV system, the latter in good agreement with the corrected values (our corrections) from other laboratories.

The partial electron transfer chain, $H_2O \to DBMIB$ gives a much lower H^+/e^- ratio, probably because the PQ shuttle is not operating, the electron going to DBMIB either before PQ gets reduced or going from PQ^{-2} to DBMIB before PQ^{2-} can pick up H^+ ions and diffuse to the endogenous oxidizing site. Böhme et al. (1971) and Böhme and Cramer (1972) have concluded that DBMIB inhibits the flux of electrons through the PQ pool. Those studies and the lower H^+/e^- ratios found by Gould and Izawa for the $H_2O \to DBMIB$ system are consistent with PQ normally functioning as a proton transporting shuttle.

Gould and Izawa (1973c) have discussed two mechanisms that could account for initial H^+/e^- ratios of 0.5, but only one is consistent with the maximum steady-state H^+/e^- ratio of 2 found with the MV system, and is shown below:

$$2\ H_2O \to O_2 + 4e^- + 4H^+{}_{in}$$
$$\underline{2DBMIB + 4e^- + 2H^+{}_{in} + 2H^+{}_{out} \to 2DBMIBH_2}$$
$$H_2O + 2DBMIB + 2H^+{}_{out} \to 2DBMIBH_2 + 2H^+{}_{in}$$

They visualize part of the protons needed in DBMIB reduction as coming from the outside and part from the inside—not unreasonable if one pictures the lipophilic DBMIB as being statistically distributed across the membrane, and therefore part of the reduced DBMIB^{2-} may be extruded to the outside, where it picks up protons.

Steady-state H^+/e^- Ratios at pH 8

Turning to steady state H^+/e^- ratios, the agreement between the data from several laboratories and the Mitchell hypothesis in its simplest form breaks down. In the basal (−ADP) system using MV as acceptor, the

H^+/e^- ratios measured in five laboratories are found to be around 1 or much less (Table I), rather than the expected ratio of 2. However, a ratio of 2 is found if energy transfer inhibitors (valinomycin + K^+, Dio-9, DCCD, or synthalin) are present, as first shown by Dilley (1970). Schröder et al. (1972) and Karlish and Avron (1971) have repeated our valinomycin results, and Telfer and Evans (1972) have shown similar Dio-9 data. Karlish and Avron (1971) show the steady state H^+/e^- going from 0.2 (−val) to 2.0 (+val), and Schröder et al. (1971) have shown that val increases the H^+/e^- from 1 to 2. Telfer and Evans (1972) have reported values of 1.00 for H^+/e^- (−Dio-9) and as high as 1.20 (+Dio-9). Telfer and Evans were convinced that the H^+/e^- ratio increase due to Dio-9 was not statistically significant.

The data of Dilley (1970), Karlish and Avron (1971), and Schröder et al. (1972) indicate H^+/e^- increases of 2-fold and greater, not at all consistent with Telfer and Evans' (1972) conclusion that energy transfer inhibitors do not increase the H^+/e^- ratio.

Taken at face value, the pH 8 steady-state H^+/e^- ratio data and the energy transfer inhibitor effects suggest that either (a) the Mitchell "PQ loop" is not operating to carry protons or (b) water oxidation protons, under these conditions, are not deposited within the permeability barrier. Alternative (b) is not consistent with Fowler's (1973) and Junge and Ausländer's (1974) results, which showed that at pH 8 proton permeability-inducing agents greatly increased the rate of appearance of the protons liberated during water oxidation.

Because of the excellent agreement between the Mitchell mechanism for H^+ transport at pH 6 and for the initial light-on conditions at pH 8, simplicity and biological unity lead us to try to find an explanation for the discrepancies mentioned above—i.e., why there is a low steady-state H^+/e^- ratio in the control case, and a significant increase due to energy transfer inhibitors—within the broad framework of the chemiosmotic scheme. Thus, a proton carrier protein, such as was suggested in earlier work from this laboratory (Dilley and Vernon, 1967) does not seem necessary to explain the data. There are several alternatives. (a) One alternative is a change in the H^+ carrying efficiency of PQ induced by energy transfer inhibitors. (b) Another is inhibition of one of two parallel electron transfer chains, an alternative suggested earlier (Dilley, 1970). One of the dual chains was considered to couple electron flow to H^+ uptake and the other was an electron-only chain; the valinomycin (val) or Dio-9 effect was interpreted as inhibiting the electron-only chain, thus increasing the H^+/e^- ratio. This could be considered as one mechanism for operation of alternative (a). Another way of looking at this is to view energy transfer inhibitors as inhibiting the hydrolysis of high-energy intermediates without inhibiting H^+ transport, but that still requires two separate pathways. (c) Energy transfer inhibitors may increase cation (e.g., K^+) permeability, allowing more rapid influx of K^+ in exchange for the effluxing protons, thus allowing H^+ flux to proceed at its "natural" rate, i.e., restrained only by the membrane resistance to H^+ flux rather than by K^+ influx resistance as well. This view was put forth by Schröder et al. (1972). (d) There may be an electric field-driven component of H^+ efflux occurring in the light that does not occur appreciably in the dark. This would "deplete" the H_i^+ pool in the light to an extent proportional to the $\Delta\Psi$, but that extra H^+ efflux would not be detected by slow time resolving pH measurements in the postillumination phase. This view is essentially that of Junge and Witt (1968).

Certainly the effect of val (and low gramicidin concentrations, found by Karlish and Avron (1971) to mimic val effects) is to increase K^+ permeability. However, no evidence indicates that Dio-9 (McCarty et al., 1965), N,N'-dicyclohexyldimethylurea (DCCD) (McCarty and Racker, 1967), or synthalin (Gross et al., 1968) increase cation permeability. The known energy transfer effect is to inhibit coupled electron flow to, or occasionally below, the basal level, with no dependence on a particular ionic environment. On the other hand, val without K^+ does not act as an energy transfer inhibitor (Karlish and Avron, 1971). Therefore it does not act directly on the coupling factor as do some energy transfer inhibitors (e.g., phlorizin, Winget et al., 1969; DCCD, McCarty and Racker, 1967).

If counterion (K^+) influx resistance is the reason why the dark H^+ efflux at pH 8 gives an H^+/e^- ratio of around 1 rather than 2, why is this not found also at pH 6? The membrane permeability is greater at pH 8 than at pH 6 (Dilley and Rothstein, 1967), so the val $+ K^+$ effect should be more pronounced at pH 6, yet it is not; Karlish and Avron (1971) have shown that val $+ K^+$ has little or no effect at pH 6 compared to pH 8.2. If val were simply increasing the K^+ permeability so that in the dark, $H_{out}^+ \to K_{in}^+$ exchange would go faster, there would be no reason for inhibition of the coupled electron flow. Karlish and Avron (1971) were puzzled by the energy transfer inhibition effect of val $+ K^+$, and found no clear explanation for it. Based on the above arguments, we reject alternative (c) as sufficient to explain the data.

A hypothesis consistent with alternative (a), and (b), must explain the following experimental observations: (1) Steady-state H^+/e^- ratios are near 2 at pH 6 and 1 at pH 8 (minus the phosphorylating system). (2) Initial H^+/e^- ratios are very similar at pH 6 and at pH 8, being near 1 to 1.7 H^+/e^-. (3) Energy transfer inhibitors stimulate the pH 8 steady-state H^+/e^- ratio up to near 2 and have no effect on the pH 6 H^+/e^- ratio. (4) Val + K^+ acts as an energy-transfer inhibitor.

A model will be proposed that can account for these observations. Several experimentally testable predictions will follow from the model, and the results of such experiments will be presented. The assumptions needed for this model are as follows.

1. Water oxidation protons are, at least in part, deposited inside (see above).

2. The 515 nm absorption band, elegantly studied by Witt (1971) and co-workers, is caused by an electrochromic shift in pigment absorption produced by an electric field either across the membrane or in the membrane in an unspecified orientation. In our model the electric field need not be transmembrane. Dutton (1971), Jackson and Dutton (1973), Callis et al. (1973), and Chance (1974) have suggested that the photosynthetic membrane electrochromic shifts could well be across certain local regions in the plane of the membrane, rather than across the entire membrane.

3. Plastoquinone is an obligate electron carrier in the electron transfer chain. As shown by Amesz (1973), the evidence in favor of this is quite good.

Hypothesis: Under pH 8 conditions the PQ pool carries electrons but can transfer charge without being protonated. The 515 nm shift-indicated electric field, $\Delta\Psi$, set up in the light, provides an electrophoretic driving force that carries the PQ^{2-} in the steady state from the reduction site to an oxidation site. The time needed for this is less than the time required for aqueous phase protons to react with PQ^{2-}.

At pH 6, the 100-fold greater H^+ concentration, the intrinsically lower electron flow rates, accompanied perhaps by an altered membrane conformation, allows the protonation reaction

$$PQ^{2-} + 2H_o^+ \rightarrow PQH_2 \tag{13}$$

to compete favorably with the electrophoretic reaction

$$PQ^{2-}_{ext} \xrightarrow[\text{in } \vec{E} \text{ field}]{\text{electrophoresis}} PQ^{2-}_{in} \tag{14}$$

The quinone anion electrophoresis model fits the data in the following ways:

1. At an external pH of 6 the electron transfer rate is less than at pH 8

(Bamberger et al., 1973), favoring reaction (13) over (14) in addition to the 100-fold greater H^+ ion concentration.

2. Val + K^+ would increase the steady-state H^+/e^- because it decreases the $\Delta\Psi$ (Witt, 1971; Neumann et al., 1970). This would decrease the $\Delta\Psi$ driving force on PQ^{2-}, favoring the formation of PQH_2.

3. Other energy transfer inhibitors could increase the H^+/e^- ratio by virtue of their braking action on electron transfer, regardless of their actual mode of action. Just by decreasing the net electron flux, they should decrease the $\Delta\Psi$. This favors reaction (13) over reaction (14).

The model predicts that at low light intensities, when $\Delta\Psi$ is reduced, the pH 8 steady-state H^+/e^- would be greater than in high light conditions. We have shown this to be true in the $H_2O \rightarrow MV$ system (Table II). Typical steady-state H^+/e^- ratios near 0.4–0.5 in high light are increased to 0.7–1.0 in low light. Similarly, inhibition of electron transfer by DCMU, inhibits $\Delta\Psi$. In turn this should lead to an increase in the steady-state H^+/e^-. Table II shows this to occur.

Can alternative interpretation (d) equally well explain some of the experimental facts? By that interpretation a different electrical potential driving force on internal H^+ ions would exist in the light compared to the dark. For instance, suppose that a positive inside membrane potential is

TABLE II

EFFECT OF LOW LIGHT INTENSITY AND DCMU ON THE H^+/e^- (STEADY STATE)[a,b]

Expt. No	Conditions	H^+ steady-state rate (μmoles/hr/mg Chl)	Electron transport (μeq/hr/mg Chl)	$H^+/e^-_{(ss)}$
1	High light	88.5	189	0.47
	Low light	78	100	0.78
2	High light	76.5	154	0.50
	Low light	92	94	0.98
3	High light – DCMU	95	215	0.44
	High light + DCMU	28	38	0.74

[a] Chloroplasts were prepared, and electron and proton transport was assayed according to Dilley (1970). The reaction mixture contained in 2 ml: 100 mM KCl, 5 mM MgCl₂, 0.5 mM methylviologen, 0.5 mM sodium azide. The pH of the mixture was 8.0, and the temperature was 15°C. High and low intensity illumination was with heat-filtered white light of 2×10^5 ergs/cm²/sec and 10^4 ergs/cm²/sec, respectively. DCMU concentration was $2 \times 10^{-7} M$.

[b] DCMU, dichlorophenyldimethylurea; Chl, chlorophyll.

set up in the light completely across the membrane, as Junge and Witt
(1968) concluded. Protons on the inside would experience a driving force
$\Delta\tilde{\mu}_H{}^+$, comprising both concentration and electrical components:

$$\Delta\tilde{\mu}_H{}^+ = RT\ln([H_i{}^+]/[H_o{}^+]) + F\Delta\Psi \tag{15}$$

where $\Delta\tilde{\mu}_H{}^+$ is the electrochemical potential gradient for protons.

If we assume for the moment that a transmembrane $\Delta\Psi$ is set up in the
light, then, to that extent, proton efflux in the light would be faster than
in the dark. This would mean that the dark H^+/e^- ratio would not be a
faithful indicator of the steady-state H^+/e^- ratio. Val $+$ K^+, by reducing
the $\Delta\Psi$, would reduce the H^+ efflux in the light and permit a faster K^+–H^+
exchange in the dark, both factors contributing to a greater measured
steady-state H^+/e^- ratio. Other energy transfer inhibitors (Dio-9, synthalin), low light intensity, or pH 6 conditions, by lowering the electron
flow rate and hence the $\Delta\Psi$, could give conditions whereby the $\Delta\Psi$ driving
force on H^+ in the light would be less, allowing the dark H^+ efflux more
closely to match the efflux in the light. Such a situation could be tested by
carefully measuring the $\Delta\Psi$ and H^+ efflux rates under the diverse conditions.

An attractive feature of this alternative is that it can account for the
general finding that initial H^+/e^- ratios are always greater than steady-
state ratios, the latter always being measured in the dark following the
light period.

The important point relevant to this, and still not established to our
satisfaction, is the geometry and magnitude of the $\Delta\Psi$ indicated by the
515 nm shift. Witt and colleagues assumed that the photochemical charge
separation (rise time 20 nsec) produces a transmembrane $\Delta\Psi$, although it
could equally well be a local intramembrane electric field, as discussed
earlier.

The membrane potential has not been quantitatively measured by the
Berlin group– they assumed, with no evidence, that the 515 nm shift
reflects charge separation across the membrane. Another questionable assumption is the assignment of a dielectric constant of 2 for the environment
across which the electric field is impressed (see p. 422 of Witt, 1971). This
assumption is the basis for calibrating the absorption change, whereby a
single turnover flash gives rise to an alleged 50 mV transmembrane potential, and a steady state potential near 100 mV. Attempts to measure
the transmembrane potential in chloroplasts have all resulted in values for
the steady state near or less than 10 mV, quite a serious discrepancy with
the theoretically-based $\Delta\Psi$ assumed by Witt and colleagues (Schroder
et al., 1971; Rottenberg *et al.*, 1972; and Vredenberg and Tonk, 1974).
The first two groups measured distributions of ions between the suspension
phase and the membranes and calculated the $\Delta\Psi$ from the Nernst equation,

while Vredenberg and Tonk used a microelectrode technique to penetrate the grana membrane of the large chloroplasts of *Peperomia metallica* and read the potential directly.

Water oxidation on the inside would be an electrogenic source of positive-inside potential. If so, it should be mentioned in passing that one would expect a component of $\Delta\Psi$ (i.e., the 515 nm shift) to be in phase with the flash-induced H_2O oxidation and H^+ production as measured by Fowler (1973). Such a correlation has not yet been reported, but if Fowler (1973) is correct, i.e., if the H^+ release is not maximal until the third or fourth flash, then a similar flash-dependence of the 515 nm shift should be found, perhaps superimposed on a component of the 515 nm shift due to primary charge separation. If such data are not found, it would either call into question the conclusion of Witt (1971) that the 515 nm shift indicates transmembrane potential, or it would suggest that the water protons are not released into the loculus space.

Both the quinone anion electrophoresis and the field-driven H^+ efflux models can quite well explain the H^+/e^- data. Perhaps further experiments could rule out one alternative. It is also possible, and in fact highly probable that both phenomena occur. If the $\Delta\Psi$ is transmembrane, the protons on the inside will have an increased electrochemical potential, and if the $\Delta\Psi$ is across the lipid region where the PQ is located, the anion will be in the field and tend to electrophorese.

The implications of the quinone anion electrophoresis model are quite clear regarding the expected H^+/ATP ratios. They would be decreased considerably below the ratios predicted for the case where the H^+/e^- is two. The chemiosmotic hypothesis requires a stoichiometric H^+ efflux per ATP produced, the value most discussed being 2 H^+/ATP. A value of 1 H^+/ATP has also been put forth (Mitchell, 1966), but values less than 1 are not acceptable in the chemiosmotic view.

The mechanochemical hypothesis is vaguer yet, there being no predictions that an H^+ efflux must be linked as a driving force to ATP synthesis. If macromolecular conformation changes are energized by some combination of protonation and redox-driven fixed charge alterations, perhaps the stoichiometric involvement of H^+ per ATP could be less than 1.

The issue of H^+/e^- stoichiometry in relation to hypotheses for energy transduction has been discussed before on the basis of initial H^+/e^- ratio data (Dilley, 1970, 1971). In those instances it was noted that the coupled system (ADP + arsenate) gave H^+/e^- ratios consistently less than the basal ($-ADP$) conditions. At that time, it was recognized that such data are inconsistent with the chemiosmotic view, but consistent with the chemical intermediate hypothesis, or with the fixed charge or mechano-chemical concept. It was proposed that dual electron flow pathways, as mentioned above, could be involved, one responsible for H^+ transport and the other just for electron flow, both being required to generate the high energy state. Unless the field-driven H^+ efflux explanation (alternative d above) provides the reason for these "substoichiometries," these data are at variance with the chemiosmotic hypothesis.

D. Cyclic Electron Flow-Linked H^+ Fluxes

The cyclic electron flow system cannot readily be compared to H^+ fluxes by H^+/e^- ratios since there is no easy way to measure electron flow rates. Quantum yield measurements have been used instead. Two laboratories have measured quantum yield of H^+ flux, and agreement is quite good between them. Dilley and Vernon (1967) measured about $5H^+/h\nu$ (initial light-on phase) for wavelengths greater than 700 nm with three different electron transfer cofactors at pH 6 (PYO, DCIP, and trimethyl-hydroquinone, $TMQH_2$). The technique employed a glass electrode. Heath (1972) used bromocresol purple as an external pH indicator at pH 6 and found quantum yields of 6.7 ± 0.4 $H^+/h\nu$ above 700 nm for the initial light-on phase and about 3.2 $H^+/h\nu$ for the dark H^+ efflux. At pH 7.8, Dilley (1970) measured quantum yields of 3 $H^+/h\nu$ in the initial light-on phase cyclic system.

The above data have been interpreted as contradicting the Mitchell chemiosmotic hypothesis. Before proceeding, we should ask whether unrecognized artifacts could elevate the quantum yields. One artifact is that reduced forms of PMS and PYO, $TMQH_2$, and $DPIPH_2$ are all likely to permeate readily across the thylakoid membrane (Lynn, 1967; Izawa, 1970). If the PYO is oxidized by PSI on the inside as suggested by Nelson *et al.* (1972) and Hauska *et al.* (1973) and reduced at the outer membrane surface, one H^+ per electron could be taken from the external phase and deposited in the osmotic space. The redox chemistry of PYO and PMS is discussed in more detail below. If we apply these corrections to the initial and steady-state $H^+/h\nu$ ratios we find the values for H^+ transported by the endogenous cyclic H^+ transport mechanism to be 4–5.7 $H^+/h\nu$ (initial) and 2.2 $H^+/h\nu$ (steady state) for pH 6 and about 2 $H^+/h\nu$ (initial) for pH 7.8. The corrected steady-state quantum yields are at wavelengths beyond 700 nm, so PSII water oxidation is not a source of protons, since the quantum yield for PSII activity is very poor at these wavelengths. Sauer and Park (1965) and Sun and Sauer (1971) find quantum yields around 0.1 $e^-/h\nu$ for $H_2O \rightarrow$ dichlorophenolindophenol (DCIP) in the 700–710 nm region.

If detailed mechanisms are ignored, there is a certain consistency between the noncyclic and cyclic systems regarding the maximum capacity to accumulate protons; both give close to 2 H^+/e^- when the various chemical reaction aspects are corrected for. In the noncyclic system, 1 H^+/e^- seems to derive from H_2O protons deposited on the inside, and the PQ pool is a logical candidate to transport 1 H^+/e^-. However, if the electrophoretic PQ^{2-} motion hypothesis is correct, this may not occur in steady-state conditions at pH 8. The cyclic system deprived of the internal source of protons (water oxidation) may compensate by translocating 2 (as a minimum) H^+ per electron (or quantum). The noncyclic system appears to translocate only one H^+/e^- from out to in, the water protons making an equal contribution to the internal H^+ pool.

The superstoichiometry of 4–5 $H^+/h\nu$ at pH 6 is not understood. The membrane Bohr effect mentioned above is a popular but as yet unproved explanation.

The endogenous cyclic electron flow and phosphorylation system raises an interesting question in comparison to the cyclic system using PYO or PMS. Both oxidized PYO and PMS are cations at pH values around 5–8, and the reduced species are neutral.

PMS, oxidized   PMS, reduced

PYO, oxidized   PYO, reduced

The reduced forms of these cofactors, if oxidized at an internal site (as mentioned above), would give rise to protons inside, as well as contribute to an inside positive potential. If both internal protons and a positive inside electrical potential are necessary for phosphorylation, then it is easy to see why these compounds are such good cyclic cofactors. What about cyclic phosphorylation catalyzed by ferredoxin (Arnon et al., 1967; Forti and Zanetti, 1969)? In the model favored by Arnon and co-workers, reduced ferredoxin directly reduces a component of the cyclic system; but the iron-protein, being water soluble and not a proton carrier, cannot carry protons across the membrane as do reduced PYO and PMS. If the ferredoxin system catalyzes cyclic ATP formation, why will the sulfonated forms of PYO, PMS, and DCIP (Hauska *et al.*, 1973) not catalyze cyclic phosphorylation? Perhaps the lower redox potential of ferredoxin (Fd) in the cyclic system permits it to reduce an electron carrier that cannot be reduced by PYO, PMS, or DPIP. If so, we must expect that electron flux in this system goes through a site(s) capable of translocating protons, if H^+ accumulation is necessary for ATP formation, as we assume here. The cytochrome $b_{563} \to PQ$ step is a possibility but this yields only 1 H^+/e^-. The cyclic mechanism proposed by Forti and Zanetti (1969) is

$$PSI \to Fd \to Fd:NADP \text{ reductase (a flavoprotein)} \to cytochrome f \to PSI$$

It just as easily could be

$$PSI \to Fd \to Fd:NADP \text{ reductase } \xrightarrow{\overset{H_o^+ \quad H_i^+}{\smile\frown}} cytochrome \, b_{563}$$

$$\to PQ \xrightarrow{\overset{H_o^+ \quad H_i^+}{\smile\frown}} cytochrome \, f \to PSI.$$

In this model, two H^+ translocating loops are possible; the Fd:NADP reductase flavin ($FAD + 2e^- + 2H^+ \to FADH_2$) could reduce an electron acceptor such that the two H^+ ions are released to the inner side of the membrane and the $PQH_2 \to$ cytochrome f couple could function to yield the second H^+ translocated per electron. To determine whether this attractive and reasonable hypothesis is valid will require further study, quantum yield measurements being the technique of choice.

It is obvious that a proton carrier protein is a less reasonable model for translocating protons. The noncyclic driven H^+ fluxes are easily explained by the Mitchell loop model, and we can predict, from arguments of biological unity, that similar Mitchell-type redox loops probably operate in the cyclic system. If experiments show that this is so, we should accept the redox loop feature of the chemiosmotic hypothesis as correct.

One aspect of the chemiosmotic hypothesis that is neither as objective nor as quantitative as the redox loop idea is that dealing with how electrochemical potential gradients might *interact with or operate on* the membrane to transduce energy. For instance, the *molecular nature* of quinones and cytochromes embedded in the membrane led Mitchell to the redox loop model for H^+ transport; however, the chemiosmotic hypothesis does not define the molecular nature of the membrane nor any of its constituents in relation to the electrochemical potential gradients, largely because so little is known about the molecular state of membrane constituents. In what follows, we shall discuss some aspects of proton gradients and energy transduction. However, we are still very far from knowing how the membrane really interacts with proton gradients.

V. THE PROTON GRADIENT AND ENERGY TRANSDUCTION

This important parameter must be measured under a variety of energy conversion circumstances to test various hypotheses relating proton gradients to energy transduction. One measurement involves the effect of phosphorylating conditions on the steady-state H^+ accumulation assayed by the external pH changes (Schwartz, 1968; Dilley and Shavit, 1968; Dilley, 1970, 1971; Karlish and Avron, 1968; Schröder et al., 1972; Gould and Izawa, 1973c), and the other utilizes the distribution of weak amines between the external medium and the membrane phase as an indicator of pH differences between the external and internal phases or of amine binding to fixed charges (Kraayenhof, 1970; Kraayenhof et al., 1971; Gaensslen and McCarty, 1971; Rottenberg et al., 1972; Schuldiner et al., 1972a; Rottenberg and Grunwald, 1972; Kraayenhof and Fiolet, 1973).

Schwartz (1968) used a pH electrode and Schröder et al. (1971) used phenol red to measure spectrophotometrically the extent of H^+ efflux in the dark in the presence and in the absence of ADP. The data of both groups indicate that under ATP synthesizing conditions there was about one-half as much pH drop in the dark as under basal electron flow conditions. However, any postillumination ATP synthesis would tend to compensate for an equivalent amount of H^+ efflux. Schröder et al. (1972) stated that this is no problem. However, the low postillumination ATP yield (Hind and Jagendorf, 1963; Izawa, 1970) found by adding ADP and P_i to chloroplasts in the dark after a period of light may differ markedly from the yield in the dark when the cofactors had been present in the light phase. Another difficulty with the data of Schröder et al. (1972) is the use of 2×10^{-4} M imidazole. Imidazole acts as an internal H^+ buffer (Lynn, 1968) and may produce artifacts that in no way reflect the true relationship between H^+ fluxes and ATP synthesis. Phenol red does not appear to be free of spectral shifts due to differential dye binding. This difficulty plagues the use of indicator dyes (Cost and Frenkel, 1967; Jackson and Crofts, 1969; Mitchell et al., 1968; Montal and Gitler, 1973). Therefore before use each dye must be studied carefully to make sure that it does not give rise to binding artifacts.

The ADP + arsenate system has been used by Karlish and Avron (1968), Dilley and Shavit (1968), Dilley (1970, 1971), and Gould and Izawa (1973c) to study H^+ fluxes in a "coupled" system that does not exhibit a continual pH rise resulting from ATP formation. It has been reported that the H^+ uptake is greater in the plus ADP case compared to the minus ADP case (Dilley, 1971). However, an artifact in this system is the undetermined steady-state amount of ADP-arsenate. This compound provides a proportional increase in pH, as shown by measuring pH changes in digitonin subchloroplast particles prepared by the Nelson et al. (1970) procedure (R. A. Dilley, unpublished observations). The preparation exhibited no detectable ΔpH in the absence of ADP (AsO_4 and $MgCl_2$ present), but an easily measurable ΔpH when ADP was added. It is therefore probable that the positive ΔpH reported for the ADP + AsO_4 case in regular chloroplasts (Dilley, 1971) is partly artifactual.

Gould and Izawa (1973c) used the $H_2O \rightarrow DBMIB$ system and found clear evidence that the ADP + AsO_4 "coupled" reaction decreases the steady-state gradient. The rates in their system are a good deal lower than those reported in our earlier results (Dilley, 1970, 1971), so that the positive ΔpH due to $ADP-AsO_4$ was probably less. However, the actual decrease in net ΔpH due to coupling could be even greater. The advantage of Gould and Izawa's system is that coupling conditions have no effect on the electron flow rate. This is not the case in the usual $H_2O \rightarrow MV$ or $Fe(CN)_6^{3-}$ situation.

On balance, the evidence regarding the extent of external pH change in a basal compared to a coupled system favors the view that there is less net H^+ accumulation in the coupled system. Proponents of the Mitchell chemiosmotic hypothesis would interpret this as due to utilization of the gradient in driving phosphorylation. This may indeed be true. However, as mentioned above, a significant decrease in the initial H^+/e^- ratio under coupled, compared to basal, conditions (Dilley, 1970, 1971) could result in the observed decrease of net proton accumulation.

What are the data for measurements alleged to measure the internal H^+ pool? Rottenberg et al. (1971, 1972), Gaensslen and McCarty (1971), Schuldiner et al. (1972a), and Rottenberg and Grunwald (1972) have developed the techniques of measuring the distribution of weak amines either by methylamine-^{14}C, NH_4^+ transport using a cation electrode, or a fluorescence quenching of atebrin or 9-aminoacridine. The principle involves application of the Henderson–Hasselbach equation for the dissociation of a weak acid or base. With the assumption that the uncharged form of the base is in equilibrium across the membrane, and since the external pH, the pK_b, the amount of RNH_3^+ + RNH_2 within the membrane osmotic space, and the volume of the osmotic space are known, the pH within the osmotic space may be calculated. Wadell and Butler (1959) developed this technique for determining intercellular pH using a weak acid, and Dilley and Rothstein (1967) used it to study the Donnan distribution of protons in chloroplasts.

Gaensslen and McCarty (1971) and Rottenberg *et al.* (1972) have determined the osmotic water space and the free space, two values that are critical in calculating the pH. Both groups found the amine-^{14}C distribution in the light indicated a ΔpH of about 2.2–2.4 pH units when the pH_o was 8 (Table III).

Using an NH_4^+-sensitive electrode, light-induced NH_4^+ accumulation can be measured (Crofts, 1968), and Rottenberg and Grunwald (1972) found a light-dependent ΔpH of about 3.0–3.4 by this technique (Table II).

Schuldiner *et al.* (1972a) used a fluorescence quenching technique to measure uptake of atebrin and various acridine dyes. Again a value of about 3–3.4 ΔpH, pH_o8) was found. Kraayenhof and Fiolet (1973) are critical of this technique, believing that the light-energized membranes may *bind* the dyes to newly exposed negatively charged groups, rather than that the dyes become distributed across the membrane in response to a transmembrane pH gradient. Their evidence is quite good that fluorescence quenching occurs as a result of external binding; for instance, atebrin attached to the protein lysozyme still shows light-dependent fluorescence quenching. Interestingly, the lysozyme atebrin complex was a good uncoupler, an unexpected finding since it has become something of a dogma in bioenergetics that weak amines uncouple via their penetration into the osmotic space where they dissipate the pH gradient (Crofts, 1968).

The Kraayenhof and Fiolet results indicate that the methylamine-^{14}C technique for measuring ΔpH could also reflect binding of the amine to the negatively charged fixed groups rather than a pH differential. In that case, there should be competition between acridine binding and NH_4^+ binding. If, however, both indicators behave as Rottenberg *et al.* (1972) and Schuldiner *et al.* (1972a) assume, then at the low concentrations of NH_4Cl and acridine dye used (below uncoupling concentrations) each compound would distribute noncompetitively across the membrane in response to a pH gradient, i.e., independently of one another. These alternatives can be tested experimentally.

Under some conditions, amines appear to cross the membrane into the osmotic space (Deamer and Packer, 1969; Gaensslen and McCarty, 1971). According to Gaensslen and McCarty (1971) chloride is taken up at a ratio of nearly 1 chloride to 1 amine when the Cl⁻ concentration exceeds 50 mM. In the simple case, Cl⁻ ion would not be taken into the osmotic space unless the amine were there. In other words, the external amine that is bound to newly exposed fixed negative charge sites would not induce Cl⁻ uptake (or binding). Conceivably, a more complex rearrangement of fixed charges could occur such that a positive fixed charge is generated on the inside phase as a negative charge is generated on the outside. This would

TABLE III

pH Gradient across Chloroplast Membranes under Various Conditions

pH$_o$	Conditions		$\Delta pH \left(\dfrac{pH_i}{pH_i}\right)$	Technique used	Reference
8	Dark	no cofactor	0.52	MeNH$_2$-^{14}C centrifuged to a pellet	Rottenberg et al. (1972)
			1.97		
8	Light	PYO^a $\begin{cases}-\text{ADP}\\ \text{methylviologen}\end{cases}$	2.18	MeNH$_2$-^{14}C	Rottenberg et al. (1972)
			2.25		
8	Light, PYO −ADP		3.0–3.4	9-NH$_2$-acridine fluorescence quenching	Schuldiner et al. (1972a)
8	Light, PYO −ADP		3.0–3.4	NH$_4^+$ electrode	Rottenberg and Grunwald (1972)
8	Light, PYO $\begin{cases}-\text{ADP}\\ +\text{ADP}\\ -\text{ADP}\end{cases}$		3.81	9-NH$_2$-acridine fluorescence quenching	Pick et al. (1973)
			3.41		
			3.83		
8	Light, PYO $\begin{cases}+\text{ADP}\\ +\text{ADP} + \text{D10-9}\end{cases}$		3.37	9-NH$_2$-acridine fluorescence quenching	Pick et al. (1973)
			3.95		
8	Light, Fe(CN)$_6^{-3}$ $\begin{cases}-\text{ADP}\\ +\text{ADP}\end{cases}$		2.8	pH$_o$ effect on P700 turnover	Rumberg and Siggel (1969)
			2.6		
8	Light, PYO −ADP		2.4	Ethylamine-^{14}C centrifuged through silicone layer	Gaensslen and McCarty (1971)

^{a} PYO, pyocyanine.

provide attraction for Cl^- ions to enter without the amine entering. There is no evidence to support this, but it could explain the amine and Cl^- data. Could it be that both external amine binding and transmembrane distribution of amines occur? There is strong evidence that fixed negative charges undergo changes during energization cycles (Dilley and Vernon, 1965; Dilley, 1966; Dilley and Rothstein, 1967; Deamer et al., 1967; Murakami and Packer, 1971; Gross et al., 1969; Heath, 1973). The external membrane surface has a net negative charge. This is shown by the fact that the membranes bind polycations (Dilley and Platt, 1968; Brand et al.) and that chloroplasts electrophorese toward the positive electrode (Nobel and Mel, 1966). The inside phase of chloroplast membranes also has a net negative charge, neutralized in the resting state by the counterions Mg^{2+} and K^+. Addition of Triton X-100, valinomycin, or nigericin to a darkened chloroplast suspension releases K^+ into the medium (Shavit et al., 1968). The light-induced H^+ uptake occurs with a nearly stoichiometric exchange of Mg^{2+} and K^+ equivalents, and 95% or more of the H^+ ions taken up are bound, consistent with negative charges losing Mg^{2+} and K^+ counterions and binding H^+ (Dilley and Vernon, 1965). Amines could interact with the fixed negative charge groups on either side or on both sides of the membrane.

One way to accomodate the data showing internal accumulation of amines with the Kraayenhof and Fiolet external binding interpretation is to picture the initial energy dependent protonation of fixed charge groups (R_1 in the drawing on p. 86) occurring in the membrane phase on polyelectrolytes that have acridine dye binding sites (R_2 in the drawing on p. 86) that are only generated (or exposed) in the energized state. If the protonated $-R_1H$ group could be in acid-base equilibrium with the interior osmotic compartment, that would account for internal proton accumulation (and amine uptake), while providing external acridine dye binding groups. The two phenomena, external acridine dye binding and internal proton accumulation, would be closely linked through the reversible formation and decay processes. The more simple concept of H^+ accumulation into the osmotic space with binding to buffering groups there does not offer a model as mechanistically satisfying.

Rumberg and Siggel (1969) estimated the light-induced ΔpH from effects of external pH on P700 reduction rate, arriving at a value of 2.8 pH units. As pointed out by Bamberger et al. (1973), the technique may suffer from the problem that the electron transfer rate is not a simple function of either pH_i or pH_o, but depends on some other unidentified factor. However, their data, obtained prior to other ΔpH data, are in agreement with the later values.

Given the uncertainties concerning just what the fluorescent dye quench-

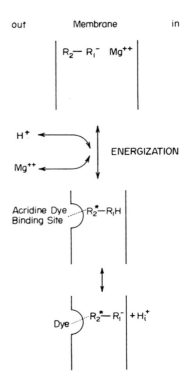

ing or the amine distributions indicate, how do energy transduction states affect the indicators? Phosphorylating conditions decrease the steady-state ΔpH by about 0.5 pH unit (or decrease the dye binding), as shown by Pick *et al.* (1973), who used the 9-aminoacridine fluorescence technique. Energy transfer inhibitors give the expected response, restoring the quenching to the same value as that found with ADP absent. Crofts (1968) found that ATP synthesis was competitive with light dependent NH_4^+ uptake (or binding).

These results would be predicted by either the chemiosmotic or the chemical intermediate hypotheses. In the chemiosmotic hypothesis, the energy required for esterifying ADP comes at the expense of the H^+ gradient that would be dissipated, and in the latter, the phosphorylating system uses the energy competitively. The mechanochemical hypothesis does not predict how energy-transducing conditions affect the H^+ gradient.

THE H^+/ATP RATIO

The above discussion raises the issue of the H^+ efflux per ATP under steady-state conditions. Table IV shows data taken from the literature.

TABLE IV

H$^+$ Efflux/ATP Ratios

Measuring technique	H$^+$ efflux/ATP	References
Glass electrode ΔpH, comparing steady-state $+\Delta$pH (ATP formation) to $-\Delta$pH during light off	2	Schwartz (1968)
Calculated from the decay of the 515 nm shift.	3	Junge et al. (1970)
Glass electrode ΔpH; energy transfer inhibitors D10-9 or phlorizin added	3	Schröder et al. (1972)
H$_2$O → DBMIB^a system, glass electrode comparing H$^+$/e$^-$ and e$^-$/ATP	Near 2	Gould and Izawa (1973c)

^{a} DBMIB, = dibromothymoquinone.

Schwartz (1968), measuring the decrease in pH upon darkening a phosphorylating system, found an H$^+$/ATP ratio of 2. He assumed that the initial decay rate, if added to the steady-state rate of pH increase (caused by the reaction ADP + P$_i$ + H$^+$ → ATP) would give an accurate measure of the actual H$^+$ efflux that occurs under phosphorylating conditions. That assumption is valid only if it is first assumed that phosphorylation can be energized by the H$^+$ gradient. That assumption is reasonable in view of the close dependence of postillumination (X$_E$) ATP synthesis (Hind and Jagendorf, 1963; Izawa, 1970; Girault et al., 1969) and acid-base (Uribe and Jagendorf, 1967) ATP synthesis on the proton accumulation. The kinetics of the onset and the decay of the high-energy state as measured by X$_E$ ATP yield are virtually identical to the onset and decay kinetics of the proton fluxes. Such a correlation is consistent with the proton gradient being directly involved in energization of the membrane. What the mechanism may be whereby the protons interact with the membrane is not clear. It should not be assumed that this rules out some manner of involvement of b type cytochromes or other redox components. However, net electron flow stoichiometric with ATP yield in the acid-base phosphorylation is unlikely in view of experiments by Miles and Jagendorf (1970).

Girault et al. (1969) and Izawa (1970) found that postillumination ATP synthesis occurred with an observed stoichiometry of 1 ATP per 4 or 5 H$^+$ ions accumulated in the light phase. Izawa corrected for efficiency of X$_E$ capture and estimated that 1 ATP per about 2 H$^+$ is the probable stoichiometric ratio.

Nishizaki and Jagendorf (1971) and Nishizaki (1972, 1973) studied the acid-base ATP formation in relation to H^+ fluxes with the aid of a stopped-flow apparatus. These elegant studies resulted in estimating that the maximum ATP yield in the acid-base transition is about 1 ATP per 2 H^+ ions.

Gould and Izawa (1973c) found H^+/ATP ratios near 2.0–2.5 in the partial electron transfer chain $H_2O \rightarrow DBMIB$, with the experimentally determined initial H^+/e^- ratio of 0.4–0.5 and an ATP/e_2 ratio of 0.4. Trebst and Reimer (1973) have measured similar ATP/e_2 ratios using class III lipophilic electron acceptors, such as phenylenediamine. The two groups have a quite different interpretation as regards the interaction of the electron transfer chain with the energy transduction mechanism.

Junge et al. (1970) indirectly calculated an H^+/ATP ratio of 3 from comparison of the ATP yield and the decay kinetics of the 515 nm field indicating absorption change. This computation rests on several assumptions needed to be able to correlate the H^+ flux with the 515 nm shift. Consequently this value is no more than a rough estimate.

Schröder et al. (1972) reported an H^+/ATP ratio of 3. Their technique involved measuring ATP production and H^+ efflux by phenol red dye absorption changes and may not take into account possible dye binding artifacts (see above). The phosphorylating mixture contained the usual components, plus 2×10^{-4} M imidazole and 10^{-7} M valinomycin, compounds that, as discussed in an earlier section, may alter H^+ uptake and flux characteristics. Nonetheless, if one recalls that steady-state H^+/e^- ratios at pH 8 are near or less than 1, and if one uses experimental values of ATP/e^- between 0.5 to 0.65 (Ouitrakul and Izawa, 1973), ATP/e_2 near 1, one gets an H^+/ATP ratio of 2 to 1.5. In more recent H^+/e^- measurements (Table II) at pH 8, we have consistently found steady-state H^+/e^- ratios near 0.5 for the basal system. If this value and an ATP/e^- ratio of 0.5 to 0.65 are used, the H^+/ATP ratios equal 1 to 0.77. Until the problem of estimating the true steady-state H^+/e^- ratio is solved, there will be uncertainty concerning the H^+/ATP ratio.

VI. INTERPRETATIONS OF H^+/ATP RATIO DATA

Can a mechanistic significance be attributed to an H^+/ATP ratio? To answer this we need to know what the effluxing protons interact with. One obvious candidate is the coupling factor (CF), protein, or membrane constituents closely associated with CF_1.

McCarty and Racker (1967) showed that removal of the coupling factor increases H^+ permeability, whereas reconstitution restores the lower permeability state (cf. Lynn and Straub, 1969, for further work). Telfer and Evans (1972) showed that compounds which interact with the coupling factor, Dio-9, ADP, and ATP, can significantly alter membrane H^+ permeability. These effects could be at a level other than the coupling factor, but it is more likely that they reflect an interaction of the coupling factor with protons, or with a permeability channel for protons. McCarty and Racker (1967) and Uribe (1972b) have shown from studies with carbodiimide reagents that CF_1 interacts closely, but in an unknown manner, with H^+ efflux. The lipid-soluble carbodiimide, dicylohexylcarbodiimide, is an energy transfer inhibitor when applied to intact membranes and serves to reduce the H^+ leakiness induced by EDTA removal of CF_1. Obviously, it must interact with membrane chemical groups (probably carboxyl) other than those of CF_1 itself. If this were not the case, it would not restore a H^+ permeability barrier produced by removal of CF_1. The nature of CF_1-membrane interactions is not understood, but the molecular structure(s) involved in that association is likely to constitute a channel for proton efflux. What happens in terms of energy exchange or in terms of polyelectrolyte-H^+ interactions (and other ions or fixed charges) cannot yet be specified.

Another correlation between proton fluxes and CF_1 is the observation that ATP hydrolysis by the membrane-bound CF_1 results in protons being transported into the membrane. This was observed in chloroplasts by Carmeli (1970). The chloroplast membrane, treated with dithiols and a period of light, hydrolyzed ATP in a subsequent dark period. This in turn resulted in a proton gradient quite similar to that generated by light-driven electron transfer. About 2 H^+ were accumulated per ATP hydrolyzed, in accord with the ATPase type II proposed by Mitchell (1966). Packer and Marchant's (1964) experiments with ATPase-supported light-scattering changes first suggested an ATPase-generated H^+ gradient. Crofts (1968) and Gaensslen and McCarty (1971) found that ATP hydrolysis supported amine uptake and explained it on the basis of an ATPase-driven H^+ uptake; the latter report showed that about 2 H^+ per ATP hydrolyzed were accumulated.

The coupling factor protein, acting as an ATPase or as an ADP kinase, may interact directly with H^+ ions. Alternatively, there may be a spatial separation of the CF_1 from the site(s) that interact(s) with the protons. Conformational changes occur in the CF_1 protein subsequent to light energization, as shown by Ryrie and Jagendorf (1971) and McCarty and Fagan (1973). Uncouplers abolish the changes. The significance of the conformational change(s) is not understood; it could be a manifestation of direct interaction of CF_1 with protons, but further work is necessary to clarify this. Regardless of whether the coupling factor itself or an ancillary membrane component is responsible for the interactions with protons that lead to their uptake in the ATPase mode, the fact remains that the energized state has associated with it a proton gradient, whether activated by electron transfer or by the ATPase.

VII. ARE THERE DIFFERENT SITES OF ATP SYNTHESIS?

The answer to this question will be a key factor in elucidating the energy transduction mechanism. The chemical intermediate hypothesis explicitly postulates direct interaction of a redox carrier with the phosphorylation steps (Slater, 1971). In the chemiosmotic postulate, no such direct involvement is considered. The gradient is thought to have been formed by the redox loops, while its dissipation through the coupling mechanism drives ATP formation independent of the redox components. (Mitchell, 1966). The mechanochemical hypothesis is not explicit as to whether conformational changes in b-type cytochromes may be involved in energy transduction. The peculiar nature of b-type cytochromes is intriguing and indicates that these redox components are likely more than just electron transfer agents. Chance and Williams (1956), Slater and colleagues (Slater, 1971), Erecinska et al. (1973), and Wilson et al. (1973) discussed mitochondrial b cytochromes with a view toward their involvement in energy transduction. Fan and Cramer (1970) have shown that the chloroplast cytochrome b_{559} has a pH-dependent redox potential.

Chloroplast studies in several laboratories have implicated two or more "sites" of ATP synthesis. Izawa and Good (1968) based their support for two sites mainly on their observation that P/e_2 ratios are greater than 1. Neumann et al. (1971) observed that electron donation from $DCIPH_2$ seemed to occur at two different sites on the electron transfer chain, one site permitting electrons to flow through a rate-limiting, internal, pH-controlled phosphorylating site and a second coupling site that did not have those characteristics. Black (1967) and Laber and Black (1969) found that heptane-treated chloroplasts had two levels (rates) of ATP formation—one that was sensitive to the heptane treatment, the other much less so. They therefore suggested there may be two functionally different ATP formation sites. More recent work from the Izawa and Good laboratory, using electron acceptors with differing lipid solubility, is interpreted as evidence for two ATP formation sites having different characteristics (Saha *et al.*, 1971; Ouitrakul and Izawa, 1973; Gould and Izawa, 1973a,b,c; Ort and Izawa, 1973a,b; Gould and Ort, 1973). The main observation is that lipophilic acceptors (class III) such as 2,5-dimethylquinone and oxidized p-phenylenediamine accept electrons from PSII (water) prior to the rate-limiting plastoquinone step. The phosphorylation associated with this system has a P/e_2 of near 0.4–0.6, which is pH insensitive from pH 6 to 8.5, (although the rate of ATP formation is much less at pH 6 than at pH 8, just as the full chain phosphorylation) and the electron transfer rate (photosynthetic control) are not affected by phosphorylating conditions. These properties differ from those of the "normal" electron flow sequence such as water to NADP⁺, MV, or ferricyanide (class I acceptors), where P/e_2 is near 1–1.3 and very pH dependent and there exists strong photosynthetic control. Izawa and colleagues have attributed the differences to two phosphorylating sites that have different molecular mechanisms. They believe that the site II energy conserving mechanism is associated with an irreversible step, either photochemical or a step between the photochemistry and the water splitting step (Gould and Ort, 1973). Ort and Izawa (1973a) have shown that the site II coupling step is not dependent on water oxidation per se, since inactivation of water oxidation by hydroxylamine treatment allows PSII artificial electron donation to the PSII and PSI system (MV reduction) to exhibit the same P/e_2 as the water → MV system.

Trebst and Reimer (1973, 1974) have studied electron transfer and phosphorylation in the H_2O → phenylenediamine or dimethylquinone system and find a data pattern similar to that of Izawa and colleagues. However, they interpret their data to indicate no difference in the phosphorylation mechanism, just different degrees of proton accumulation with the different electron transfer sequences. They believe that class III acceptors (phenylenediamine, DBMIB, dimethylquinone) prohibit the PQ pool from transporting 1 $H⁺/e⁻$ into the osmotic space. This reduces the proton gradient by about half (they attribute 1 $H⁺/e⁻$ inside due to water oxidation, as do Gould and Izawa, 1973c); this in turn reduces the P/e_2 ratio by about half.

VIII. MEMBRANE POTENTIAL AND THE ENERGIZED STATE

If the chemical free energy gradient of protons across the membrane is important as part of the energy transduction mechanism, then one should be able to vary it experimentally by effects on ions other than H⁺. This has been done by several groups. McCarty (1970) was the first to show that valinomycin-induced K^+ influx stimulates the postillumination ATP yield in subchloroplast particles. Schuldiner et al. (1972b) showed a similar effect in the postillumination ATP synthesis in intact chloroplasts. Uribe and Li (1973) found that valinomycin + K^+ stimulated the ATP yield in an acid-base transition. It is not a K^+-specific effect, since Na^+ and the transport antibiotic nonactin, which increases Na^+ permeability, also gave the effect (Schuldiner et al., 1972b). Two important points emerge from these studies: first, that some internal proton accumulation is necessary to get significant ATP synthesis; and second, a K^+ diffusion potential gradient from outside to inside stimulates ATP yield and the reverse direction K^+ gradient decreased ATP yield. A valinomycin-induced K^+ gradient out $\rightarrow$ in makes the electrical potential more positive inside, while a reverse K^+ gradient leads to a more negative (or less positive) value inside electrical potential. The electrochemical potential gradient of protons inside will respond according to:

$$\Delta \tilde{\mu}_{H_{in}^+} = RT \ln([H_{in}^+]/[H_{out}^+]) + F\Delta\Psi$$

The $K^+\Delta\Psi$, when positive inside, will increase the electrochemical activity of the internal H^+ ions. It is the H^+ potential gradient that appears to be the critical factor, not the electrical gradient per se.

IX. ON THE MECHANISM OF ENERGY TRANSDUCTION

The search goes on for a single mechanism of energy transduction. In fact there may be several discrete steps, each with different characteristics. In the chloroplast system, the excellent work of Jagendorf and colleagues (Hind and Jagendorf, 1963; Jagendorf and Uribe, 1966) has shown that the energization mechanism can be easily separated from the final steps of ATP formation. Energization may be driven by electron transfer in the light or by an acid–base transition in the dark, both taking place in the absence of the phosphorylating cofactors. The energized state can then drive the esterification of ADP. Both stages are characterized by conversion of energy between at least two different forms. Energy transduction from oxidation-reduction potential energy into the energized state seems far better understood than the utilization of the energized state for net ATP synthesis. The Mitchell chemiosmotic hypothesis elegantly deduces, from the general nature of membranes and the known chemical properties of the redox components a model for energy storage in electrochemical potential gradients. The membrane asymmetry necessary for this model has not been worked out in detail, but the work to date supports asymmetric positioning of PSI on the outside and PSII *mainly* on the inside (Arntzen *et al.*, 1969; Dilley *et al.*, 1972). The chemical nature of quinones and flavins makes them ideal proton carriers when they are coupled appropriately to electron carriers in such an asymmetric membrane. The evidence from many studies generally supports the redox-loop mechanism for proton gradient generation with H^+/e^- stoichiometries consistent with the PQ-cytochrome f couple translocating 1 H^+/e^-. Although less well documented, the studies reported to date indicate that protons are released to the inside in the water oxidation step. In this review, we introduced the concept that the PQ pool may not always carry protons. This could account for anomalies in the observed steady-state H^+/e^- ratios. Such a possibility need not detract from the generality of energization via a proton-motive force, but may simply be a necessary control mechanism to avoid overprotonating the interior and/or coupling electron flow too tightly to obligate proton transfer.

If one accepts for now that the accumulation of protons inside the grana lamellae membrane can readily be explained by the properties of the membrane and the redox agents in it, what is the evidence that the energized state is closely tied to a proton motive force? Three lines of evidence discussed earlier are summarized here. (1) When the proton electrochemical potential gradient, which is more positive inside, is increased by means of an entering cation, such as K^+ or Na^+, the ATP yield in acid–base and postillumination ATP synthesis increases as well, as both of these modes depend on a proton gradient. (2) The ATPase function drives H^+ accumulation. (3) A dissipation of the H^+ gradient is correlated with ATP synthesis. Perhaps alternative explanations can be offered. At present these correlations appear to be consistent with the energized state being a state in which free energy is "stored," at least in part, in an electrochemical potential gradient.

Recent experiments of Racker and Stoeckenius (1974), show that ATP formation can be reconstituted in a system comprised of soybean phospholipid vesicles, mitochondrial coupling factor (F_1), the oligomycin sensitivity conferring factor, mitochondrial hydrophobic protein, and bacteriorhodopsin. The bacteriorhodopsin in the vesicle membrane causes protons to be taken up when the vesicles are illuminated, and the other factors allow ATP formation to proceed. ATP formation is uncoupler sensitive. This suggests that phosphorylation is dependent on proton uptake. These interesting experiments constitute further evidence that proton gradients are intimately involved with energy transduction. The question of the mechanism of linking a proton gradient to ATP formation is unanswered; added hydrophobic mitochondrial proteins may be required to interact with the coupling factor and the proton gradient via an energy transduction cycle involving a conformational change.

At this point *all* current hypotheses are equally vague as to how the energized state might be transduced into ATP or other forms of chemical or mechanochemical energy. Given charge separation and electrochemical potential gradients, it is fascinating to speculate that there exists mechano-chemical energy transduction in a polyelectrolyte membrane immersed in such gradients. Because of its nature, a polyelectrolyte must change its shape and/or its energy content when subjected to a change in ion concentrations (particularly H^+ ion) and a change in electric fields. Unless the charges of membrane polyelectrolytes are completely insulated from the phases experiencing electrochemical potential gradients—an unlikely situation—shape and energy changes will take place in the membrane constituents (see the Introduction). Since the mid-1960s (Dilley, 1966, 1969, 1971), we have held that such conformational changes are at the heart of energy transduction. This does not constitute an explicit mechanistic hypothesis, but rather provides a point of view that leads us to look for polyelectrolyte characteristics that may provide molecular mechanisms for transducing energy from free energy of proton gradients and redox gradients into energy-requiring processes, such as net ATP synthesis. We are presently attempting chemical modification of fixed-charge functional groups to perturb polyelectrolyte interactions in the same sense that protein chemists have used this tool to study molecular interactions in proteins.

X. A MOLECULAR APPROACH FOR MEASURING CONFORMATIONAL CHANGES

We have been able to document (Giaquinta *et al.*, 1973, 1974a) what appears to be a conformational change that occurs upon energization of PSII electron transfer. In turn, this results in the unmasking of diazonium and carbodiimide reactive groups, concomitant with inhibition of water oxidation. The data of Table V indicate the PSII dependent inhibition of water oxidation by diazonium benzene sulfonate (DABS). The DABS treatment is given in light or dark, the unreacted reagent is washed away by repeated centrifugation, and the resuspended chloroplasts are then assayed for electron transfer activity. Table VI shows that PSII-dependent extra DABS incorporation is about four times greater than that which occurs in the dark. PSI cyclic electron flow with PMS, sufficient to activate normal H^+ uptake, did not lead to the extra DABS binding. This finding is somewhat startling, since an active conformational change is thought to occur with the cyclic system (Dilley and Vernon, 1964; Deamer *et al.*, 1967). The PSII-dependent DABS incorporation is not affected by un-

TABLE V

EFFECT OF DIAZONIUM BENZENE SULFONATE (DABS)
TREATMENT[a] ON ELECTRON TRANSPORT

Assay	% Control	DABS Treatment (% inhibition)	
		Light	Dark
1. $H_2O \rightarrow MV^{b,g}$	100^d	60	0
2. $H_2O \rightarrow DCIP^c$	100^e	60	0
3. $DCIPH_2 \rightarrow MV$ (PSI)^c	100^f	0	0

[a] The DABS treatment (2 mM) lasted 1 minute in a reaction mixture consisting of: 100 mM NaCl, 50 mM K_2HPO_4 pH 7.2, with 1 mg of chlorophyll per milliliter. Light exposure was through a red glass filter (Corning No. 2403) giving an intensity of 10^6 ergs cm^{-2} sec^{-1}. The chloroplasts were washed free of the reagents by repeated centrifugations, resuspended in 0.4 M sorbitol, and assayed for electron transfer activity. From Giaquinta et al. (1973).

[b] $H_2O \rightarrow MV$ assay contained: 100 mM KCl, 5 mM $MgCl_2$, 5×10^{-4} M methylviologen, 5×10^{-4} M NaN$_3$, 20 mM Tricine (pH 8.2) and 20 μg of chlorophyll per milliliter in a total volume of 2 ml.

[c] $DCIPH_2 \rightarrow MV$ assay mixture was similar with the addition of 50 μM DCIP, 5×10^{-4} M ascorbate, and 10 μM DCMU. Reaction mixtures were at 15°C and illuminated with heat-filtered white light (2×10^5 ergs/cm^2/sec). $H_2O \rightarrow DCIP$ assay contained: 10 mM Tris-Cl (pH 7.6), 10 mM KCl, 40 μM DCIP, 3 mM NH$_4$Cl and 13 μg of chlorophyll per milliliter in a total volume of 3 ml. The cuvette was illuminated with red light (2×10^5 ergs/cm^2/sec) and the reduction of DCIP was followed at 590 nm.

[d] Control rate 300 μeq(hr·mg Chl)$^{-1}$

[e] Control rate 242 μeq(hr·mg Chl)$^{-1}$

[f] Control rate 2400 μeq(hr·mg Chl)$^{-1}$

[g] MV, methylviologen; DCIP, dichloroindophenol; PSI, photosystem I; Chl, chlorophyll; DCMU, dichlorophenyldimethylurea.

couplers. This suggests that the effect is not simply related to H$^+$ accumulation, as is energization for postillumination ATP synthesis (Giaquinta et al., 1974a). Moreover, the effect is also independent of water oxidation activity per se, since Tris-treated chloroplasts that have lost

TABLE VI

EFFECT OF DIAZONIUM BENZENE SULFONATE (DABS) ON PSII ACTIVITY
AND DAB ^{35}S BINDING[a,b]

Condition during DABS treatment	PSII activity (μeq(hr·mg Chl)$^{-1}$) $H_2O \rightarrow DCIP$	DAB^{35}S binding (μmoles DABS $\times$ 10^{-2} mg Chl)
Control (no DABS)	225	—
Light (DABS)	88	31.0
Dark (DABS)	215	7.0
Light + DCMU (DABS)	255	7.7
Light + DCMU + PMS (DABS)	220	8.2
Light + nigericin (DABS)	95	36.0
Light + methylviologen (DABS)	85	28.0

[a] Chloroplasts (1 mg Chl/ml) in the presence of either 20 μM DCMU, 60 μM PMS, 10^{-5} M nigericin or 6 $\times$ 10^{-4} methylviologen were light-energized for 30 seconds and then reacted with 2 mM DAB^{35}S as described in Table V. $H_2O \rightarrow DCIP$ assay as in Table V. The chloroplasts were assayed for light-dependent H$^+$ uptake prior to and during the DABS treatment. It was found that the normal activity of H$^+$ uptake occurred during all treatments except nigericin; i.e., the PMS + DCMU treatment did not suffer a lack of electrons for mediating a cyclic flow. From Giaquinta et al. (1973).

[b] PSII, photosystem II; DCIP, dichlorophenolindophenol; Chl, chlorophyll; DCMU, dichlorophenyldimethylurea; PMS, phenazine methosulfate.

their water oxidation capacity can be induced to bind extra DABS in the light, provided an alternate PSII electron donor, manganese, is used to restore electron flux through PSII (Table VII) (cf note about Mn^{2+} oxidation in footnote).

An acid-base transition in the presence of DABS leads to water oxidation inhibition and increased DABS binding, shown in Table VIII. An un-coupler (nigericin) had no influence on these DABS effects. This again indicates a phenomenon not dependent on proton impermeability.

We interpret the DABS inhibition of water oxidation, concomitant with extra DABS covalently binding to membrane functional groups, as due to an electron transfer-dependent conformational change in membrane proteins associated with PSII. While there is DABS binding to CF$_1$ (Giaquinta et al., 1974b), our experiments show that the CF$_1$ DABS binding is not related closely to the DABS effects described above (Giaquinta et al., 1974a). Hence, the conformational changes attributed to PSII are not related to the energy-dependent conformational changes that occur in CF$_1$, as reported by Ryrie and Jagendorf (1971) and McCarty and Fagan (1973). The acid–base transition is known to give conformational changes by virtue of the protonation–deprotonation effect on membrane

TABLE VII

BINDING OF ³⁵S-LABELED DIAZONIUM BENZENE SULFONATE (DABS) TO TRIS-TREATED
CHLOROPLAST MEMBRANES DURING Mn²⁺ DONATION[a]

Control	Relative electron transport prior to DABS addition	DABS-³⁵S incorporation (cpm/mg Chl × 10³)
Treatment		
Light (DABS)	100%	220
Dark (DABS)	—	80
Tris-treated		
Light −Mn (DABS)	0	150
Light +Mn (DABS)	107%	240
Dark −Mn (DABS)	—	130
Dark +Mn (DABS)	—	120

[a] Chloroplasts were Tris-treated. Electron transport (H_2O or Mn^{2+} → methylviologen) was measured prior to ³⁵S-DABS addition in 100 mM NaCl, 0.5 mM MV, 0.5 mM azide, ±3 mM MnCl₂. Chloroplast mixtures were illuminated with white light during the DABS reaction. One micromole of DABS = 1.25×10^7 cpm. From Giaquinta et al. (1974a). It should be noted that Mn^{2+} oxidation releases H⁺ ions. Our viewpoint is that H⁺ ions released upon water or alternate donor oxidation are required to give the extra DABS binding.

TABLE VIII

EFFECT OF DARK ACID–BASE AND BASE–ACID TRANSITION ON DIAZONIUM BENZENE
SULFONATE (DABS) INHIBITION OF DCIP[a] REDUCTION AND BINDING[b]

Treatment	Rate (μeq/hr/mg Chl)		DABS-³⁵S incorporation (cpm/mg Chl × 10³)
Control (no DABS)			
Transition			
pH 4.5 → 4.5	88		—
pH 4.5 → 8.5	104		—
pH 8.5 → 8.5	82		—
pH 8.5 → 4.5	97		—
DABS present		(+nigericin)	
pH 4.5 → 4.5	99	126	490
pH 4.5 → 8.5	35	25	900
pH 4.5 → 8.5 + DPC	109	—	—
pH 8.5 → 8.5	78	—	352
pH 8.5 → 4.5	28	—	740

[a] DCIP, dichloroindophenol.
[b] The acid–base and base–acid transitions were performed in the dark. When nigericin (5–10 μM) was present in the reaction mixture, 50 mM KCl was included. From Giaquinta et al. (1974a).

ionizable groups (Dilley and Rothstein, 1967; Deamer *et al.*, 1967; Murakami and Packer, 1970; Ryrei and Jagendorf, 1971). Just how a pH-induced conformational change may relate to the postulated PSII electron flow-dependent conformational change is not clear, but such a relationship is suggested by the DABS effects.

While this manuscript was in preparation, a publication came to our attention (Hirose *et al.*, 1973) showing that mitochondrial membranes in the energized state have a greater amount of $-NH_2$ groups exposed for reaction with acrolein than nonenergized membranes. Those authors interpret their results as indicating that a conformational change occurs in the membrane upon energization, an analogous situation to that we find for chloroplasts.

Energy-linked conformational changes are not a new discovery, but the chemical modification approach may provide a molecular tool of immense value in elucidating conformational changes and the underlying molecular events that give rise to them.

Recent experiments have extended these findings by further dissecting the electron flow in PSII with the use of silicomolybdate as an electron acceptor (from water oxidation) that is reduced prior to the DCMU inhibition site (Giaquinta *et al.*, 1974c). This PSII partial reaction is not coupled to ATP formation, nor is there measurable proton accumulation (measured by amine uptake, Giaquinta and Dilley, 1974). The silicomolybdate is not an uncoupler or a proton transporter since it does not inhibit the post-illumination ATP formation (X_E) as do uncouplers. The water $\rightarrow$ silicomolybdate ($+DCMU$) partial reaction does not potentiate the extra DABS binding. We hypothesize that electron flow through the region Q to plastoquinone—perhaps involving cyt *b* 559—is linked to a conformational change that in some way is required to permit (or cause) protons from water oxidation to be deposited within the membrane. The internal deposition of protons released upon water oxidation is, of course, a possible energization step for site II phosphorylation. The relationship between the redox-linked conformational change and PSII water oxidation–proton deposition is not yet clearly understood, but there are fascinating possibilities that are being actively pursued.

Boyer *et al.* (1973) have proposed the interesting hypothesis that the energy-requiring step in oxidative phosphorylation may serve to release preformed ATP . . . ATP that is formed at the catalytic site with limited or no energy imput. Their data are based on mass spectrometer assays of oxygen exchange. They find that $P_i \rightleftharpoons H_2O$ exchange is less sensitive to uncouplers than $P_i \rightleftharpoons ATP$ or $ATP \rightleftharpoons H_2O$ exchange and that the $P_i \rightleftharpoons$ water exchange is dependent on ADP. They explain these data, and the finding of a small amount of ATP (bound) $\rightleftharpoons P_i$ exchange in the presence of uncouplers, by their postulate of an energy-dependent conformational change being required to release ATP from the catalytic site. As pointed out by Lumry (1971), Green and Ji (1972), and Boyer (1974), energy transduction via conformation coupling includes the possibility of energy exchange in amounts less than the total required to synthesize 1 ATP. Increments of energy may be put into macromolecules in the form of electrostatic free energy and entropy changes driven by redox, electric field, and pH changes.

New approaches are needed to measure more precisely changes in macromolecular conformation. Until we can experimentally monitor conformational changes, little progress will be made in critically testing the mechanochemical energy transduction hypothesis.

ACKNOWLEDGMENTS

The authors thank Dr. Bruce Selman for stimulating discussions of viewpoints expressed in this review, and Mrs. Barbara Anderson for excellent technical assistance. This work was carried under NSF Grant GB-30998, NIH Grant 5R01GM19595, and an NIH Career Development Award to R. A. D.

REFERENCES

Abrahamson, E.W., and Ostroy, S.E. (1967). The photochemical and macromolecular aspects of vision. *Prog. Biophys. Mol. Biol.* **17**, 181–215.

Amesz, J. (1973). The function of plastoquinone in photosynthetic electron transport. *Biochim. Biophys. Acta* **301**, 35–51.

Arnon, D.I., Tsujmoto, H.Y., and McSwain, B. (1967). Ferredoxin and photosynthetic phosphorylation. *Nature (London)* **214**, 562–566.

Arntzen, C.J., and Briantais, J.M. (1973). *In* "Bioenergetics of Photosynthesis" (Govindjee, ed.). Academic Press, New York.

Arntzen, C.J., Dilley, R.A., and Crane, F.L. (1969). A comparison of chloroplast membrane surfaces by freeze-etch and negative staining techniques; and ultrastructural characterization of membrane fractions obtained from digitonin-treated spinach chloroplasts. *J. Cell Biol.* **43**, 16–31.

Azzone, G.F. (1972). Oxidative phosphorylation, a history of unsuccessful attempts: Is it only an experimental problem? *J. Bioenerg.* **3**, 95–104.

Baldry, C.W., and Coombs, J. (1973). Regulation of photosynthetic carbon metabolism by pH and Mg⁺⁺. *Z. Pflanzenphysiol.* **69**, 213–216.

Bamberger, E.S., Rottenberg, H., and Avron, M. (1973). Internal pH, ΔpH and the kinetics of electron transport in chloroplasts. *Eur. J. Biochem.* **34**, 557–563.

Banks, B.E.C., and Vernon, C.A. (1970). Reassessment of the role of ATP *in vivo*. *J. Theor. Biol.* **29**, 301–326.

Black, C.C. (1967). Evidence supporting the theory of two sites of photophosphorylation in green plants. *Biochem. Biophys. Res. Commun.* **28**, 985–990.

Böhme, H., and Cramer, W. (1972). The role of cytochrome b₅₆₃ in cyclic electron transport: Evidence for an energy coupling site in the pathway of cytochrome b₅₆₃ oxidation in spinach chloroplasts. *Biochim. Biophys. Acta* **283**, 302–315.

Böhme, H., Reimer, S., and Trebst, A. (1971). The effect of dibromothymoquinone, an antogonist of plastoquinone, on non cyclic and cyclic electron flow systems in isolated chloroplasts. *Z. Naturforsch. B* **26**, 341–352.

Boyer, P.D. (1965). *In* "Oxidases and Related Redox Systems" (T.E. King, H. S. Mason, and M. Morrison, eds.), pp. 994–1030. Wiley, New York.

Boyer, P.D., Cross, R.L., and Momsen, W. (1973). A new concept for energy coupling in oxidative phosphorylation. *Proc. Nat. Acad. Sci. U.S.* **70**, 2837–2839.

Brand, J., Baszynki, T., Crane, F.L., and Krogmann, D.W. (1972). Selective inhibition of photosynthetic reactions by polycations. *J. Biol. Chem.* **247**, 2814–2819.

Bray, H.G., and White, K. (1967). "Kinetics and Thermodynamics in Biochemistry." Academic Press, New York.

Callis, J.B., Parson, W.W., and Gouterman, M.P. (1973). Fast changes of enthalpy and volume on flash excitation of Chromatium chromatophores. *Biochim. Biophys. Acta* **267**, 348–362.

Carmeli, C. (1970). Proton translocation induced by ATPase activity in chloroplasts. *FEBS (Fed. Eur. Biochem. Soc.) Lett.* **7**, 297–300.

Case, G.D., and Parson, W.W. (1971). Thermodynamics of the primary and secondary photochemical reactions in Chromatium. *Biochim. Biophys. Acta* **253**, 187–202.

Caserta, G., and Cervingi, T. (1973). A piezolectric transducer model for phosphorylation in photosynthetic membranes. *J. Theor. Biol.* **41**, 127–142.

Chance, B. (1974). Carotenoid and Merocyanine probes in chromatophore membranes. (In press.)

Chance, B., and Montal, M. (1971). Ion-translocation in energy conserving membrane systems. *Curr. Top. Membranes Transp.* **2**, 99–156.

Chance, B., and Williams, G.R. (1956). The respiratory chain and oxidative phosphorylation. *Advan. Enzymol.* **17**, 65–105.

Chance, B., Crofts, A.R., Nishimura, M., and Price, B. (1970). Fast membrane H^+ binding in the light activated state of chromatium chromatophores. *Eur. J. Biochem.* **13**, 364–374.

Cost, K., and Frenkel, A.W. (1967). Light induced interactions of *R. rubrum* chromatophores with bromothymol blue. *Biochemistry* **6**, 663–670.

Crofts, A.R. (1968). Ammonium uptake by chloroplasts and the high energy state. *In* "Regulatory Functions of Biological Membranes" (J. Jarnefelt, ed.), pp. 247–263. Elsevier, Amsterdam.

Crofts, A.R., Deamer, D.W., and Packer, L. (1967). Mechanisms of light induced structural change in chloroplasts. II. The role of ion movements in volume changes. *Biochim. Biophys. Acta* **131**, 97–118.

Crofts, A.R., Jackson, J.B., Evans, E.H., and Cogdell, R.J. (1971). The high-energy state in chloroplasts and chromatophores. *Proc. Int. Cong. Photosyn., 2nd*, pp. 873–902.

Deamer, D.W., and Packer, L. (1969). Light dependent anion transport in isolated spinach chloroplasts. *Biochim. Biophys. Acta* **172**, 539–543.

Deamer, D.W., Crofts, A.R., and Packer, L. (1967). Mechanisms of light induced structural changes in cloroplasts. *Biochim. Biophys. Acta* **131**, 81–96.

Dilley, R.A. (1964). Light-induced potassium efflux from spinach chloroplasts. *Biochem. Biophys. Res. Commun.* **17**, 716–722.

Dilley, R.A. (1966). Ion and water transport processes in spinach chloroplasts. *Brookhaven Symp. Biol.* **19**, 258–280.

Dilley, R.A. (1969). Evidence for the requirement of H^+ ion transport in photophosphorylation by spinach chloroplasts. *Prog. Photosyn. Res.* **3**, 1354–1360.

Dilley, R.A. (1970). The effect of various energy conversion states of chloroplasts on proton and electron transport. *Arch. Biochem. Biophys.* **137**, 270–283.

Dilley, R.A. (1971). Coupling of ion and electron transport in chloroplasts. *Curr. Top. Bioenerg.* **4**, 237–271.

Dilley, R.A., and Platt, J.S. (1968). Effect of polylysine on energy linked chloroplast reactions. *Biochemistry* **7**, 338–346.

Dilley, R.A., and Rothstein, A. (1967). Chloroplast membrane characteristics. *Biochim. Biophys. Acta* **135**, 427–443.

Dilley, R.A., and Shavit, N. (1968). On the relationship of H+ transport to photophosphorylation in spinach chloroplasts. *Biochim. Biophys. Acta* **162**, 86–96.

Dilley, R.A., and Vernon, L.P. (1964). Changes in light absorption and light scattering properties of spinach chloroplasts upon illumination. Relationship to photophosphorylation. *Biochemistry* **3**, 817–824.

Dilley, R.A., and Vernon, L.P. (1965). Ion and water transport processes related to the light-dependent shrinkage of chloroplasts. *Arch. Biochem. Biophys.* **111**, 365–375.

Dilley, R.A., and Vernon, L.P. (1967). Quantum requirement of the light-induced proton uptake by spinach chloroplasts. *Proc. Nat. Acad. Sci. U.S.* **57**, 395–400.

Dilley, R.A., Peters, G.A., and Shaw, E.R. (1972). A test of the binary chloroplast membrane hypothesis by using a non-penetrating chemical probe, *p*-(diazonium)-benzenesulfonic acid. *J. Membrane Biol.* **8**, 163–180.

Dutton, P.L. (1971). *Biochim. Biophys. Acta* **226**, 63–70.

Erecinska, M., Wagner, M., and Chance, B. (1973). Kinetics of cytochromes *b*. *Curr. Top. Bioenerg.* **5**, 267–303.

Fan, H.N., and Cramer, W.A. (1970). The redox potential of cytochromes b-559 and b-563 in spinach chloroplasts. *Biochim. Biophys. Acta* **216**, 200–207.

Farrington, J.A., Ebert, M., Land, E. J., and Fletcher, K. (1973). Bipyridylium quarternary salts and related compounds. *Biochim. Biophys. Acta* **314**, 372–380.

Forti, G., and Zanetti, G. (1969). The electron pathway of cyclic photophosphorylation. *Prog. Photosyn. Res.* **3**, 1213–1216.

Fowler, C.F. (1973). Proton evolution during the photooxidation of water in photosynthesis. *Biophys. J.* **13**, 64a.

Gaensslen, R.E., and McCarty, R.E. (1971). Amine uptake in chloroplasts. *Arch. Biochem. Biophys.* **147**, 55–65.

Giaquinta, R.T., Dilley, R.A., Anderson, B.J., and Horton, P. (1974). A chloroplast membrane conformational change activated by electron transport between the region of photosystem II and plastoquinone. *Bioenergetics* **6**, 167–177.

Giaquinta, R.T., Dilley, R.A., and Anderson, B.J. (1973). Light potentiation of photosynthetic oxygen evolution inhibition by water soluble chemical modifiers. *Biochem. Biophys. Res. Commun.* **52**, 1410–1417.

Giaquinta, R.T., Dilley, R.A., Selman, B.R., and Anderson, B.J. (1974a). Chemical modification studies of chloroplast membranes. III. Water oxidation inhibition by diazonium benzene sulfonic acid. *Arch. Biochem. Biophys.* **162**, 200–209.

Giaquinta, R.T., Selman, B.R., Dilley, R.A., and Anderson, B.J. (1974b). Chemical modification of chloroplast membranes. V. Diazonium inhibition of coupling factor activity. *J. Biol. Chem.* **249**, 2873–2878.

Giaquinta, R.T., Dilley, R.A., Crane, F.L., and Barr, R. (1974c). Photophosphorylation not coupled to DCMU-insensitive Photosystem II oxygen evolution. *Biochem. Biophys. Res. Commun.* **39**, 985.

Girault, G., Tyszkiewicz, E., and Galmiche, J. (1969). Photophosphorylation and light-induced increase of pH. *Prog. Photosyn. Res.* **3**, 1347–1353.

Good, N.E. (1962). Uncoupling of the Hill reaction from photophosphorylation by anions. *Arch. Biochem. Biophys.* **96**, 653–661.

Gould, J.M., and Izawa, S. (1973a). Studies on the energy coupling sites of photophosphorylation. I. Separation of site I and site II by partial reactions of the chloroplast electron transport chain. *Biochim. Biophys. Acta* **314**, 211–223.

Gould, J.M., and Izawa, S. (1973b). Photosystem II electron transport and phosphorylation with dibromothymoquinone as electron acceptor. *Eur. J. Biochem.* **37**, 185–192.

Gould, J.M., and Izawa, S. (1973c). Studies on the energy coupling sites of photophosphorylation. IV. The relation of proton fluxes to the electron transport and ATP formation associated with photosystem II. *Biochim. Biophys. Acta* **333**, 509–524.

Gould, J.M., and Ort, D.R. (1973). Studies on the energy coupling sites of photophosphorylation. III. The different effects of methylamine and ADP plus Pi on electron transport through coupling sites I and II in isolated chloroplasts. *Biochim. Biophys. Acta* **325**, 157–166.

Green, D.E., and Ji, S. (1972). The electromechanochemical model of mitochondrial structure and function. *J. Bioenerg.* **3**, 159–202.

Greville, G.D. (1969). A scrutiny of Mitchell Chemiosmotic hypothesis of respiratory chain and photosynthetic phosphorylation. *Curr. Top. Bioenerg.* **3**, 1–78.

Gross, E., Shavit, N., and San Pietro, A. (1968). Synthalin: An inhibitor of energy transfer in chloroplasts. *Arch. Biochem. Biophys.* **127**, 224–228.

Gross, E., Dilley, R.A., and San Pietro, A. (1969). Control of electron flow by cations. *Arch. Biochem. Biophys.* **134**, 450–467.

Grünhagen, H.H., and Witt, H.T. (1970). Primary ionic events in the functional membrane of photosynthesis. *Z. Naturforsch. B* **25**, 373–386.

Guth, E. (1947). Muscular contraction and rubberlike elasticity. *Ann. N. Y. Acad. Sci.* **47**, 715–766.

Hauska, G. (1972). The Hill reaction of chloroplasts action spectra and quantum requirements. *FEBS (Fed. Eur. Biochem. Soc.) Lett.* **28**, 217–221.

Hauska, G., Trebst, A., and Draper, W. (1973). Quinoid compounds as artificial carriers in cyclic photophosphorylation and photoreductions by photosystem I. *Biochim. Biophys. Acta* **305**, 632–641.

Heath, R.L. (1972). Light requirements for proton movement by isolated chloroplasts as measured by the bromocresol purple indicator. *Biochim. Biophys. Acta* **256**, 645–655.

Heath, R.L. (1973). Ethyl red as a probe into the mechanism of light driven proton translocation by isolated chloroplasts. I. The spectral shift of ethyl red and membrane conformational changes. *Biochim. Biophys. Acta* **292**, 444–458.

Hind, G., and Jagendorf, A.T. (1963). Separation of light and dark stages in photophosphorylation. *Proc. Nat. Acad. Sci. U.S.* **49**, 715–722.

Hind, G., and Jagendorf, A.T. (1965). Effect of uncouplers on the conformational and high energy states in chloroplasts. *J. Biol. Chem.* **240**, 3202–3209.

Hirose, S., Tamawa, Y., Iida, K., and Inada, Y. (1973). Energy-linked conformational change of the mitochondrial membrane measured with the $-NH_2$ modifying Reagent, acrolein. *Biochim. Biophys. Acta* **305**, 52–58.

Hiyama, T., and Ke, B. (1972). Difference spectra and extinction coefficient of P 700. *Biochim. Biophys. Acta* **267**, 160–171.

Izawa, S. (1970). The relation of post-illumination ATP formation capacity (X_E) to H^+ accumulation in chloroplasts. *Biochim. Biophys. Acta* **223**, 165–173.

Izawa, S., and Good, N. (1968). The stoichiometries relation of phosphorylation to electron transport in isolated chloroplasts. *Biochim. Biophys. Acta* 162, 380–391.

Izawa, S., and Hind, G. (1967). The kinetics of the pH rise in illuminated chloroplast suspensions. *Biochim. Biophys. Acta* 143, 377–390.

Jackson, J.B., and Crofts, A.R. (1969). Bromothymol blue and bromocresol purple as indicators of pH changes in chromatophores of *R. rubrum*. *Eur. J. Biochem.* 10, 226–234.

Jackson, J. B., and Dutton, P.L. (1973). The kinetic and redox potentiometric resolution of the carotenoid shift in *R. spheroides* chromatophores. *Biochim. Biophys. Acta* 325, 102–113.

Jagendorf, A.T., and Hind, G. (1963). Studies on the mechanism of photophosphorylation. *Nat. Acad. Sci.—Nat. Res. Counc., Publ.* 1145, 599–610.

Jagendorf, A.T., and Uribe, E.G. (1966). Photophosphorylation and the chemiosmotic hypothesis. *Brookhaven Symp. Biol.* 11, 215–245.

Jensen, R.G. (1971). Activation of CO₂ fixation in isolated spinach chloroplasts. *Biochim. Biophys. Acta* 234, 360–370.

Jensen, T.R., and Bassham, J.A. (1968). Light activation of the carboxylation reaction. *Biochim. Biophys. Acta* 153, 227–234.

Junge, W., and Ausländer, W. (1974). The electric generator in photosynthesis of green plants. I. Vectorial and protolytic properties of the electron transport chain. *Biochim. Biophys. Acta* 333, 59–70.

Junge, W., and Witt, H.T. (1968). On the ion transport system of photosynthesis-investigations on a molecular level. *Z. Naturforsch. B* 23, 244–254.

Junge, W., Rumberg, B., and Schröder, H. (1970). The necessity of an electric potential difference and its use for photophosphorylation in short flash groups. *Eur. J. Biochem.* 14, 575–581.

Karlish, S.J.D., and Avron, M. (1968). Analysis of light induced proton. uptake in isolated chloroplasts. *Biochim. Biophys. Acta* 153, 878–888.

Karlish, S.J.D., and Avron, M. (1971). Energy transfer inhibition and ion movements in isolated chloroplasts. *Eur. J. Biochem.* 20, 51–57.

Katchalsky, A. (1954). Polyelectrolyte gels. *Prog. Biophys. Mol. Biol.* 4, 1–59.

Katchalsky, A., and Zwick, M. (1955). Mechanochemistry and ion exchange. *J. Polym. Sci.* 16, 221–234.

Klotz, I.M. (1967). "Energy Changes in Biochemical Reactions." Academic Press, New York.

Kok, B. (1972). Photoacts and electron transport. *Proc. Int. Congr. Photobiol., 6th, 1972.*

Kraayenhof, R. (1970). Quenching of uncoupler fluorescence in relation on the energized state in chloroplasts. *FEBS (Fed. Eur. Biochem. Soc.) Lett.* 6, 161–166.

Kraayenhof, R. (1973). Energy dependent binding of acridine probes to chloroplast membranes. *Proc. Int. Cong. Biochem., 9th, 1973,* 4g5.

Kraayenhof, R., and Fiolet, J.W.T. (1973). On the interaction of 9-amino substituted acridine probes with energy conserving membranes. *Biochim. Biophys. Acta* 13, 355–364.

Kraayenhof, R., Katan, M.B., and Grunwald, T. (1971). The effect of temperature on energy linked functions in chloroplasts. *FEBS (Fed Eur. Biochem. Soc.) Lett.* 19, 5–10.

Kuhn, W., and Green, F. (1946). Statistical behavior of the single chain molecule and its relation to the statistical behavior of assemblies consisting of many chain molecules. *J. Polym. Sci.* 1, 183–212.

Kuhn, W., Ramel, A., Walters, D.H., Ebner, G., and Kuhn, H. J. (1960). The production of mechanical energy from different forms of chemical energy with homogenous and cross striated high polymer systems. *Fortschr. Hochpolym.-Forsch.* 1, 540–592.

Laber, L.J., and Black, C.C. (1969). Site specific uncoupling of photosynthetic phosphorylation in spinach chloroplasts. *J. Biol. Chem.* 244, 3463–3467.

Liberman, E.A., and Skulachev, V.P. (1970). Conversion of biomembrane produced energy into electric form. IV. General discussion. *Biochim. Biophys. Acta* 216, 30–37.

Lumry, R. (1971). Some fundamental problems in the physical chemistry of protein behavior. *In* "Electron and Coupled Energy Transfer in Biological Systems" (T. King and M. Klingenger, eds.), pp. 1–54. Dekker, New York.

Lynn, W.S. (1967). Inhibition of photophosphorylation by phenazine methosulfate. *J. Biol. Chem.* 242, 2186–2191.

Lynn, W.S. (1968). H^+ and electron poising and photophosphorylation in chloroplasts. *Biochemistry* 7, 3811–3820.

Lynn, W.S., and Brown, R. (1967). Competition between phosphate and protons on phosphorylation and cation exchange in chloroplasts. *J. Biol. Chem.* 242, 426–432.

Lynn, W.S., and Straub, K.D. (1969). Isolation and purification of a protein from chloroplasts required for phosphorylation and H^+ uptake. *Biochemistry* 8, 4789–4973.

McCarty, R.E. (1970). The stimulation of post-illumination ATP synthesis by valinomycin. *FEBS (Fed. Eur. Biochem. Soc.) Lett.* 9, 313–316.

McCarty, R.E., and Fagan, J. (1973). Light-stimulated incorporation of N-ethylmaleimide into coupling factor 1 in spinach chloroplasts. *Biochemistry* 12, 1503–1512.

McCarty, R.E., and Racker, E. (1967). Partial resolution of the enzymes catalysing photophosphorylation. The inhibition and stimulation of photophosphorylation by *N,N'*-dicyclohexylcarbodimide. *J. Biol. Chem.* 242, 3435–3439.

McCarty, R.E., Guillory, R.J., and Racker, E. (1965). Dio-9, an inhibitor of coupled electron transport and phosphorylation in chloroplasts. *J. Biol. Chem.* 240, PC4822–PC4823.

McCarty, R.E., Fuhrman, J.S., and Tsuchiya (1971). Effects of adenine nucleotides on H^+ ion transport in chloroplasts. *Proc. Nat. Acad. Sci. U.S.* 68, 2522–2526.

Mandelkern, L. (1967). Some fundamental molecular mechanisms of contractility in fibrous macromolecules. *J. Gen. Physiol.* 50, 29–60.

Miles, D., and Jagendorf, A.T. (1970). Evaluation of electron transport as the basis of ATP synthesis after acid-base transition by chloroplasts. *Biochemistry* 9, 429–434.

Mitchell, P. (1966). Chemiosmotic coupling in oxidative and photosynthetic phosphorylation. *Biol. Rev. Cambridge Phil. Soc.* 41, 445–540.

Mitchell, P. (1972). Chemiosmotic coupling in energy transduction: A logical development of biochemical knowledge. *J. Bioenerg.* 3, 5–24.

Mitchell, P., Moyle, J., and Smith, L. (1968). Bromthymol blue as a pH indicator in mitochondrial suspensions. *Eur. J. Biochem.* 4, 9–20.

Montal, M., and Gitler, C. (1973). Surface potential and energy-coupling in bioenergy conserving membrane systems. *J. Bioenerg.* 4, 363–382.

Murakami, S., and Packer, L. (1970). Protonation and chloroplast membrane structure. *J. Cell Biol.* 47, 332–351.

Nelson, N., Dreschsler, Z., and Neumann, J. (1970). Photophosphorylation in digitonin subchloroplast particles, absence of a light induced pH shift. *J. Biol. Chem.* 245, 143–151.

Nelson, N., Nelson, H., and Racker, E. (1972). Photoreaction of FMN-tricine and its participation in photophosphorylation. *Photochem. Photobiol.* 16, 481–490.

Neumann, J., Ke, B., and Dilley, R.A. (1970). The relation of the 515 nm absorbance change to ATP formation in chloroplasts and digitonin subchloroplast particles. *Plant Physiol.* **46**, 86–92.

Neumann, J., Arntzen, C. J., and Dilley, R.A. (1971). Two sites for ATP formation in photosynthetic electron transport mediated by photosystem I. Evidence from digitonin subchloroplast particles. *Biochemistry* **10**, 866–873.

Nishizaki, Y. (1972). Two phase kinetics of proton release from chloroplasts by acid-base transition. *Biochim. Biophys. Acta* **275**, 177–181.

Nishizaki, Y. (1973). Kinetics of ATP formation and proton efflux by acid-base transition in chloroplasts. *Biochim. Biophys. Acta* **314**, 312–319.

Nishizaki, Y., and Jagendorf, A.T. (1971). Kinetics of acid efflux from chloroplasts following the acid-base transition. *Biochim. Biophys. Acta* **226**, 172–186.

Nobel, P.S. (1967). Relation of swelling and photophosphorylation to light induced ion uptake by chloroplasts *in vitro*. *Biochim. Biophys. Acta* **131**, 127–140.

Nobel, P.S. (1969). Light induced changes in the ionic content of chloroplasts in *Pisum sativa*. *Biochim. Biophys. Acta* **172**, 134–143.

Nobel, P.S., and Mel, H.C. (1966). Electrophoretic studies of light-induced charge in spinach chloroplasts. *Arch. Biochem. Biophys.* **113**, 695–702.

Ort, D.R., and Izawa, S. (1973a). Studies on the energy coupling sites of photophosphorylation. II. Treatment of chloroplasts with NH$_2$OH plus EDTA to inhibit water oxidation while maintaining energy coupling efficiencies *Plant Physiol.* **52**, 595–600.

Ort, D.R., and Izawa, S. (1973b). Studies on the energy coupling sites of photophosphorylation V. Phosphorylation efficiencies (P/e$_2$) associated with aerobic photooxidation of artificial electron donors. *Plant Physiol.* **53**, 370–376.

Ostroy, S.E., Erhardt, F., and Abrahamson, E.W. (1966). Protein configuration changes in the photolysis of rhodopsin. *Biochim. Biophys. Acta* **112**, 265–277.

Ouitrakul, R., and Izawa, S. (1973). Electron transport and photophosphorylation in chloroplasts as a function of the electron acceptor. II. Acceptor-specific inhibition by KCN. *Biochim. Biophys. Acta* **305**, 105–118.

Packer, L., and Marchant, R. (1964). Action of ATP on chloroplast structure. *J. Biol. Chem.* **239**, 2061–2069.

Park, R.B., and Sane, P. V. (1971). Distribution of function and structure in chloroplast lamellae. *Annu. Rev. Plant Physiol.* **22**, 395–430.

Pick, U., Rottenberg, H., and Avron, M. (1973). Effect of phosphorylation on the size of the proton gradient across chloroplast membranes. *FEBS (Fed. Eur. Biochem. Soc.) Lett.* **32**, 91–94.

Racker, E., and Stoeckenius, W. (1974). *J. Biol. Chem.* **249**, 662.

Reinwald, E., Stiehl, H.H., and Rumberg, B. (1968). Correlation between plastoquinone reduction, field formation and proton translocation in photosynthesis. *Z. Naturforsch. B* **23**, 1616–1617.

Rottenberg, H., and Grunwald, T. (1972). Determination of ΔpH in chloroplasts. 3. Ammonium uptake as a measure of ΔpH in chloroplasts and sub-chloroplast particles. *Eur. J. Biochem.* **25**, 71–77.

Rottenberg, H., Grunwald, T., and Avron, M. (1971). Direct determination of ΔpH in chloroplasts, and its relation to the mechanism of photo induced reactions. *FEBS (Fed. Eur. Biochem. Soc.) Lett.* **13**, 41–44.

Rottenberg, H., Grunwald, T., and Avron, M. (1972). Determination of ΔpH in chloroplasts. 1. Distribution of [C-14] methylamine. *Eur. J. Biochem.* **25**, 54–63.

Rumberg, B., and Siggel, U. (1969). pH changes in the inner phase of thylakoids during photosynthesis. *Naturwissenschaften* **56**, 130–132.

Rumberg, B., Reinwald, E. Schröder, H., and Siggel, U. (1969). Correlations between electron transfer, proton translocation and phosphorylation in chloroplasts. *Prog. Photosyn. Res.* **3**, 1374–1382.

Ryrie, I. J., and Jagendorf, A.T. (1971). An energy linked conformational change in the coupling factor protein in chloroplasts. *J. Biol. Chem.* **246**, 3771–3773.

Saha, S., Ouitrakul, R., Izawa, S., and Good, N.E. (1971). Electron transport and phosphorylation in chloroplasts as a function of the electron acceptor. *J. Biol. Chem.* **246**, 3204–3209.

Sauer, K., and Park, R. S. (1965). The Hill reaction of chloroplasts. Action spectra and quantum requirements. *Biochemistry* **4**, 2791–2798.

Schliephake, W., Junge, W., and Witt, H.T. (1968). Correlation between field formation, proton translocation, and the light reactions in photosynthesis. *Z. Naturforsch.* **B 23**, 1571–1578.

Schröder, H., Muhle, H., and Rumberg, B. (1972). Relationship between ion transport phenomena and phosphorylation in chloroplasts. *Proc. Int. Congr. Photosyn. Res., 2nd, 1971* Vol. II, pp. 919–930.

Schuldiner, S., Rottenberg, H., and Avron, M. (1972a). Determination of ΔpH in chloroplasts. *Eur. J. Biochem.* **25**, 64–70.

Schuldiner, S., Rottenberg, H., and Avron, M. (1972b). Membrane potential as a driving force for ATP synthesis in chloroplasts. *FEBS (Fed. Eur. Biochem. Soc.) Lett.* **28**, 173–176.

Schwartz, M. (1968). Light induced proton gradient links electron transport and phosphorylation. *Nature (London)* **219**, 915–919.

Shavit, N., Dilley, R.A., and San Pietro, A. (1968). Ion translocation in isolated chloroplasts. Uncoupling of photophosphorylation and translocation of K^+ and H^+ ions induced by nigericin. *Biochemistry* **7**, 2356–2363.

Skulachev, V.P. (1971). Energy transformations in the respiratory chain. *Curr. Top. Bioenerg.* **4**, 127–190.

Skulachev, V.P. (1972). Solution of the problem of energy coupling in terms of chemiosmotic theory. *J. Bioenerg.* **3**, 25–38.

Slater, E.C. (1971). The coupling between energy yielding and energy utilizing reactions in mitochondria. *Quart. Rev. Biophys.* **4**, 35–71.

Straub, K.D. (1974). A solid state theory of oxidative phosphorylation. *J. Theor. Biol.* **44**, 191–206.

Straub, K.D., and Lynn, W.S. (1965). Turbidity effects in dual wavelength spectrophotometry. *Biochim. Biophys. Acta* **94**, 304–306.

Sun, A.S.K., and Sauer, K. (1971). Pigment systems and electron transport in chloroplasts. 1. Quantum requirements for the two light reactions in spinach chloroplasts. *Biochim. Biophys. Acta* **234**, 399–414.

Telfer, A., and Evans, M.C.W. (1972). Evidence for chemiosmotic coupling of electron transport to ATP synthesis in spinach chloroplasts. *Biochim. Biophys. Acta* **256**, 625–637.

Trebst, A. (1974). Energy conservation in photosynthetic electron transport of chloroplasts. *Annu. Rev. Plant Physiol.* **25**, 423–458.

Trebst, A., and Harth, E. (1970). On a new inhibitor lf photosynthetic electron transport in isolated chloroplasts. *Z. Naturforsch.* **B 25**, 1157–1159.

Trebst, A., and Reiner, S. (1973). Properties of photoreduction by photosystem II in chloroplasts. An energy-conserving step in the photoreduction of benzoquinones by photosystem II in the presence of dibromothymoquinone. *Biochim. Biophys. Acta* **305**, 129–139.

Trebst, A., and Reimer, S. (1974). Properties of photoreductions by photosystem II in chloroplasts. III. The effect of uncouplers on phenylenediamine shuttles across the membrane in the presence of dibromothymoquinone. *Biochim. Biophys. Acta* **325**, 546–557.

Uribe, E.G. (1972a). Interaction of an artificially imposed membrane potential with the ATP synthesizing system of spinach chloroplasts. *Plant Physiol.* **49**, Suppl., 9.

Uribe, E.G. (1972b). The interaction of N,N'-dicyclohexylcarbodiimide with the energy conservation systems of the spinach chloroplast. *Biochemistry* **11**, 4228–4234.

Uribe, E.G., and Jagendorf, A.T. (1967). On the localization of organic acids an in acid-induced ATP synthesis. *Plant Physiol.* **42**, 697–705.

Uribe, E.G., and Li, B.C.Y. (1973). Stimulation and inhibition of membrane dependent ATP synthesis in chloroplasts by artifically induced K⁺ gradients. *J. Bioenerg.* **4**, 435–444.

Vredenberg, W. J., and Tonk, W.J.M. (1974). Analysis of light-induced electrical membrane signals in intact chloroplasts. *3rd Int. Congr. Photosyn. Res., Rehovot, Israel* (in press).

Wadell, W.J., and Butler, T.C. (1959). Calculation of intracellular pH from the distribution of 5,5-dimethyl-2,4-oxazolidinedione (DMO). *J. Clin. Invest.* **38**, 720–731.

Wang, J.H. (1972). On the coupling of electron transport to phosphorylation. *J. Bioenerg.* **3**, 105–114.

Weiss, D.E. (1969). Energy transducing reactions in biological mambranes. 1. Energy transduction in ion- and electron-exchange polymers. *Aust. J. Biol. Sci.* **22**, 1337–1354.

Williams, R.J.P. (1961). Possible functions of chains of catalysts. *J. Theor. Biol.* **1**, 1–17.

Wilson, D.F., and Dutton, P.L. (1970). The oxidation-reduction potentials of cytochromes a and a₃ in intact rat liver mitochondria. *Arch. Biochem. Biophys.* **136**, 583–592.

Wilson, D.F., Dutton, P.L., and Wagner, M. (1973). Energy transducing components in mitochondrial respiration. *Curr. Top. Bioenerg.* **5**, 234–265.

Winget, G.D., Izawa, S., and Good, N.E. (1969). The inhibition of photophosphorylation by phlorizin and closely related compounds. *Biochemistry* **8**, 2067–2078.

Witt, H.T. (1971). Coupling of quanta, electrons fields, ions and phosphorylation in the functional membrane of photosynthesis. *Quart. Rev. Biophys.* **4**, 365–477.

Wolff, C., Buchwald, H.E., Rüppel, H., Witt, K., and Witt, H.T. (1969). Rise time of the light induced electrical field across the function membrane of photosynthesis *Z. Naturforsch. B* **24**, 1038–1041.

The Present State of the Carrier Hypothesis

PAUL G. LeFEVRE

Department of Physiology and Biophysics, Health Sciences Center
State University of New York at Stony Brook
Stony Brook, New York

I. EVOLUTION OF THE QUESTION

A. Delimitation of the "Carrier" Concept

For twenty years or more, the most widely accepted framework for interpretation of the many observed complexities in biological permeability processes has been the postulated interaction of the permeant solutes with specific *membrane carriers*. In the broadest usage of this term (Wyssbrod *et al.*, 1971), a "carrier" may be *any* constituent of a biological barrier (usually a cell's plasma membrane) that exhibits the property of *associating transiently* with a "passenger" (or "substrate") molecule or ion when it impinges on the membrane from either adjacent compartment, and thereby in some way allowing the passenger to penetrate through the barrier and emerge into the compartment on the other side at a rate significantly greater than would prevail were it not for this intermolecular association. For much of the discussion in this chapter, this inclusive usage will be appropriate. However, it will be essential also to deal with the concept of "carrier" in its specifically restricted sense, as a mobile "ferryboat" entity (Ussing, 1952), in effect *solubilizing* its passenger within the membrane, *the passenger–carrier complex traversing the barrier as such* (either by simple diffusion or by some similarly first-order activation process). When this more rigorous definition is intended in the course of this chapter, the restriction will be apparent from a context of such terms as "mobile carrier," "true carrier," or "ferry system" or by explicit identification with a particular model.

The most widely cited ("conventional" or "classical") model is the restricted symmetrical one which Widdas (1952) suggested originally for the mediation of monosaccharide transfer through the sheep placenta and human erythrocyte membranes. It is included as a special case of the more general mobile-carrier picture given by Britton (1964) and by Regen and Morgan (1964), as illustrated in Fig. 1; the restrictions defining the classical Widdas model are listed in the figure legend.

It will be the thesis of this chapter that the present status of the "true-carrier" concept as the basis of operation of any natural system for biological transport is quite precarious indeed, but that (in spite of ardent pronouncements to the contrary) no seriously tenable alternative to the proposition of carrier mediation in the *general* sense has been offered to account for a variety of observed "permeability" phenomena. Moreover, in bacterial systems a number of membrane constituents that would appear to meet the criteria for carrier function in this broad sense have now in fact been identified (Section I, E, 3 below; and Boos, 1974). The apparent failings of the classical mobile-carrier picture, however, have become especially acute in the very system to which it had been most satisfactorily applied: the translocation of simple sugars through human red blood cell membranes. The detailed kinetics of this system documented in recent years have been taken by many workers in the field as impossible to reconcile with the presumption of any sort of translational mobility in the passenger–carrier complex; and there is widespread doubt that the simple shuttle concept will ultimately prove to be applicable to *any* natural system. In fact, one prominent recent review (Lieb and Stein, 1972a)

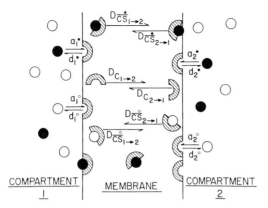

FIG. 1. Generalized model for mobile-carrier operation. Here, as in Figs. 3, 4, and 5, black and white balls represent two species of substrate (*passengers*) unable as such to enter the membrane phase, but capable of associating (process a) with or dissociating (process d) from specific *reactive sites* (marked by hatching) if these sites are suitably disposed at a membrane interface. In this instance, sites are "true" *carriers* free to move within the membrane, and the transmembranal step is presented as an actual migration, involving mobility or diffusion constants (Ds), but this may take the form of any relocation or reorientation that provides a shift in the compartment to which a given carrier has access. Full definition of a single passenger's transport behavior requires specification of eight constants, of which two ($D_{C_{1\to2}}$ and $D_{C_{2\to1}}$) are common to all, and six (a_1, d_1, a_2, d_2, $D_{\overline{CS}_{1\to2}}$, and $D_{\overline{CS}_{2\to1}}$) are substrate-specific (as marked by · or °). The *classical Widdas model* imposes three simplifying assumptions: (1) the system is symmetrical: $a_1 = a_2$; $d_1 = d_2$; $D_{C_{1\to2}} = D_{C_{2\to1}}$; and $D_{\overline{CS}_{1\to2}} = D_{\overline{CS}_{2\to1}}$; (2) the carrier movement is not altered by complexing with substrate: $D_C = D_{\overline{CS}}$ (so that *all* Ds are equal, and consequently $C + \overline{CS} = C_T$, a constant, at each interface); (3) the transmembranal process is rate-limiting, so that the a and d reactions are taken as at equilibrium; thus, at each interface,

$$\frac{\overline{C\dot{S}}}{C_T} = \frac{[\dot{S}]/K^{\cdot}}{1 + [\dot{S}]/K^{\cdot} + [\overset{\circ}{S}]/K^{\circ}}$$

where the brackets refer to the substrate concentrations in the adjacent compartments, and $K^{\cdot}$ and K° are respective dissociation constants for $\overline{C\dot{S}}$ and $\overline{C\overset{\circ}{S}}$. In this case, the necessary parameters for characterizing a single substrate's behavior are reduced to two: the product, $D \cdot C_T$ (a *fixed* $V_{\max}$ for the system), and the K defining the *specific* affinity for that passenger species.

begins with what amounts to a formal funeral announcement for the true-carrier hypothesis.

It is difficult to pin down precisely the historical origins of the carrier postulate; its essence is implicit in early discussions of "physiological permeability" (mainly in connection with intestinal absorption) dating back at least to the turn of the century (Höber, 1899), but the frame of reference here is the translocation through an entire epithelium, and the carrier was evidently considered to be a *cytoplasmic* ingredient. The original definitive proposition of a *membrane* carrier mechanism is generally attributed to Osterhout (1933, 1935) or to Lundegårdh (1935), in connection with cellular penetration by electrolytes. Little specific application of this concept was then made until the late 1940s; but the ensuing quarter-century has seen an ever-increasing use of this basic assumption in the analysis of biological translocation data, with the emergence of "transport physiology" as a major sector of active biomedical exploration and the subject of many outstanding symposia, monographs, and reviews. Among these, special attention to development of the implications of carrier theory is given by Rosenberg and Wilbrandt (1955), Christensen (1960), Wilbrandt and Rosenberg (1961), Stein (1967a), Mitchell (1967), Kotyk and Janáček (1970), and Wyssbrod *et al.* (1971); the last two of these references are particularly recommended to any readers who may wish to examine in some depth the various quantitative relationships expected in carrier kinetics, while avoiding the unduly abstruse formal presentations frequently encountered in this field. An exceptionally readable condensation of the complexities in current developments is that of Kotyk (1973).

B. Common Addenda to the Basic Concept

The possible *enzymic catalysis* of the carrier-passenger association in some systems has occasionally been proposed (Rosenberg and Wilbrandt, 1952, 1955; Koch, 1964) and is eminently compatible with both the high sensitivities to established enzyme inhibitors and the high substrate structural specificities (even *stereo*specificities) commonly observed in biological transport systems. In the case of nutrilite accumulations in bacteria, in fact, such a class of enzymes has been formally recognized by the coining of the unfortunate term, permease (Cohen and Monod, 1957). However, in some instances (Winkler and Wilson, 1966; Kepes, 1969), the "permease" has been viewed as itself subserving the carrier function, rather than merely catalyzing the interaction (Kepes, 1960; Koch, 1964; Peterson *et al.*, 1967).

A mechanistically differing but functionally parallel type of enzyme action designated by Mitchell and Moyle (1958) as *group translocation* appears to underlie some transport phenomena. Here, the passenger may be pictured as either (a) a molecule which through enzymic intervention is delivered into the trans[1]-compartment not as such, but in covalent linkage

---

[1] Throughout the chapter, the prefixes cis- and trans- will denote, respectively, the side of the barrier from which the movement in question originates, and the side toward which it is directed.

with an acceptor ingredient, or (b) a submolecular chemical group which is liberated from a donor molecule on the *cis*-side to an acceptor on the trans-side. Such a mechanism has in recent years been characterized in some detail (Roseman, 1972), as the *phosphotransferase system* underlying the superficially concentrative uptake of various sugars by a number of species of anaerobic bacteria. Again the enzyme (here phosphorylative) may either itself constitute the membrane-spanning entity, or (as concluded by Gachelin, 1970) be coupled at the inner membrane surface with an antecedent translocation step. Meister's (1973) proposal of dipeptide formation by a membranal γ-glutamyl transpeptidase as the basis for active amino acid transport might also formally be classed as a group translocation; but in Meister's picture the free substrate is soon regenerated on the *trans*-side by a γ-glutamyl cyclotransferase.

Enzymic catalysis may also become indirectly associated with transport processes in the form of a coupling (not necessarily obligatory) between the initial attack on a metabolizable solute by a hydrolytic enzyme at the cell surface and the carrier-mediated delivery into the cytoplasm of the products of the hydrolysis. Thus a number of disaccharides are evidently brought into various yeast cells in the form of their hexose components, by such a sequence of events (Sols and de la Fuente, 1961). A similar arrangement for the intestinal absorption of disaccharides (offering a "kinetic advantage" to the hexoses delivered to the transport mechanism by way of the disaccharidases) has been described by Crane (1966, 1968). Also, presentation of amino acids to the gut mucosa in the form of dipeptides has been said to allow their more rapid appearance in the bloodstream (Smyth, 1972; Matthews, 1972).

In many instances the strongest apparent argument for the involvement of a carrier entity in a biological translocation may be the demonstration of an "uphill" movement from a lower to a higher chemical or electrochemical potential (or *active transport*, as defined by Rosenberg, 1948). This feature, however, is in no way a necessary consequence of a carrier operation; and certainly its *absence* in any particular system tells the investigator nothing at all as to whether the translocation may be mediated by a membrane carrier. Numerous attempts have been made to formulate a definitive operational criterion for "active transport" that will allow an unambiguous thermodynamic categorization (Ussing, 1949, 1954; Rosenberg, 1948, 1954; Patlak, 1956; Scheer, 1958; Jardetzky and Snell, 1960; Kedem, 1961). For the purposes of the present chapter, it will suffice to reserve the term loosely for systems capable of establishing and maintaining an electrochemical gradient of the passenger species across the barrier concerned. (In the author's publications prior to 1955, the equilibrative red-cell sugar transfer was designated as "active transport" to indicate the membrane's specific participation; but general acceptance of a more restrictive usage now makes this retroactively misleading.) Such "uphill pumping" clearly must entail a linkage of some step in the carrier operation, directly or indirectly, with an exergonic reaction capable of energetically underwriting the osmotic work performed. A number of significant distinctions in the behavior expected, according to which step is the seat of the activation-coupling, have been developed by Jacquez (1961, 1964; Jacquez and Sherman, 1965). The "active" nature of the transport may thus be viewed as an auxiliary refinement or accessory mechanism, superimposed on the carrier apparatus which provides the basic *equilibrative* transport (Danielli's *facilitated diffusion*, 1954). In this vein, Rosenberg and Wilbrandt (1963) have properly stressed that nearly all the qualitative distinctions from passive diffusion behavior which characterize an active transport are seen also in the merely equilibrative carrier-mediated systems. Winkler and Wilson (1966) in fact present a convincing set of arguments that the facilitated diffusion of galactosides which survives in metabolically poisoned bacteria utilizes the same carrier apparatus that mediates the *active* transport of these same substances in normal cells. There is thus no need to postulate any fundamental *mechanistic* distinction between facilitated diffusions and active transports (Wong, 1965); the major deviations from diffusive behavior (the subject of the following section) accompany the *carrier* function, *not the energizing* of the transport.

C. Criteria for Postulation of Mediated Transport: Significance of Each Class of Evidence

1. Classical Kinetic Signs of Saturation

Since there must be both a finite carrier population-density on a membrane surface and a finite rate of turnover (or of regeneration of the carrier, if it is presumed to be actually consumed in the transfer process), it is clear that the movement of the passenger by this mechanism cannot exceed some limiting value. This characteristic may express itself in diverse ways, according to the experimental situation.

a. Nonlinearity with Gradient. Whenever the substrate concentration is sufficiently high on the cis-side to bring about occupation of a sizable fraction of the membrane sites, and sufficiently low on the trans-side to give almost total dissociation, the net translocation rate will begin to approach the maximal attainable figure; further increments in the substrate gradient will produce progressively lesser proportional increments in the transfer. This contrasts sharply with the essentially unlimited linearity of rate with gradient expected normally in a diffusive system. However, a number of theoretical analyses have agreed that if a diffusion is constrained to occur only within passageways with cross-sectional dimensions not substantially larger than the diffusing particles (and particularly if so constrictive that only a single-file migration can occur), a saturation behavior may become apparent even though no specific intermolecular association with the substance of the barrier is involved. Zierler (1961) has even demonstrated this in a gross mechanical model (randomly shaken steel balls finding their way through a rather small hole in a plastic enclosure); the basis for this is presumed to be the increasing significance, at higher particle concentrations, of diverting collisions in the immediate vicinity of the escape aperture. Also, Läuger (1973) has emphasized how electrostatic considerations might limit the occupancy of a membrane pore by ions, so as to give rise to saturation phenomena. Accordingly, in concluding from the observation of such saturation in a biological system that a translocation must have involved reaction with a membrane ingredient in limited supply, at least a mild reservation must always be kept in mind.

In any case, even if specific passenger-binding sites have definitely been established in a given case, the occurrence of saturation kinetics still provides no direct evidence that a "true carrier" is *shuttling* the passenger through the barrier. Stein and Danielli (1956) discuss conceivable arrangements whereby *fixed* adsorption sites, bordering narrow aqueous channels or pores, could give rise to saturation phenomena. The detailed theory for appearance of saturation in "single-file diffusion" along two-site-long pores operating in this manner has been developed by Heckmann (1965a,b). Moreover, the several recently proposed "noncarrier" models for facilitated diffusion (to be discussed in Section II, D) would all show saturation behavior under essentially the same circumstances as would apply to a shuttling carrier system; this is not surprising, since all these schemas share the postulate of a limited supply of specific adsorption sites.

It is even *less* indicative of the nature of the translocation mechanism if a given system *fails* to exhibit saturation kinetics: this leaves open the question not only of involvement of a carrier, but even whether it may be pumping "uphill." Nonlinearity of the rate of transfer as a function of the applied potential gradient will seldom be apparent until perhaps a third of the available sites are occupied, and nearly complete occupancy is required for a fully convincing display of saturation. Consequently, if the "affinity" in a carrier-substrate association is not sufficiently high in relation to the substrate concentrations at which the experimental tests are performed, a potentially saturable system will show no signs of this characteristic. The apparent affinity is in fact so low in some systems well established as facilitated diffusions that solubility limitations alone would preclude ever achieving a saturating level of substrate; an example of this is fructose entry into human erythrocytes (LeFevre and Davies, 1951). The particular pattern expected in the dependence of transport rate on the standing gradient varies substantially, according to which of several general classes of interaction with carrier or enzyme (or both) is presumed to occur; Rosenberg and Wilbrandt (1955) provide a convenient formal analysis of this.

Experimental data relating transfer rates to the concentrations of passenger presented at the two sides of a membrane commonly serve as the basis for estimation of maximal transport rates and other "operational constants" which, *within the framework of a particular working model*, are taken as expressing "affinities" or other kinetic characteristics of the system. If, as with the "classical" mobile-carrier model (legend of Fig. 1), a rapid, reversible passenger–carrier complex formation be presumed, followed by a relatively slow translocation step, an obvious formal parallel with Michaelis–Menten enzyme operation is presented. Under these circumstances, the familiar Lineweaver–Burk and other established procedures for enzymological data-analysis may rationally be applied in studies of *unidirectional* transport fluxes. This common practice in no way implies the presumption that the transport is necessarily enzyme-catalyzed, although the derived operational constants are often designated by the symbols V_{max} and K_m borrowed from enzymology. Often lost sight of, however, is the illegitimacy of this type of analysis for *net* transfer data under circumstances where the forward flux does *not* vastly exceed the back-flux; here, *each* of the oppositely directed fluxes may well conform to the Michaelis–Menten pattern, but a more complex analysis is required if it is only the *difference* between these two terms (i.e., the net transfer) that is being measured.

b. Competition for Transfer. Just as self-(or homo-)saturation appears upon addition of excessive levels of one species of passenger, the addition of a second species utilizing the same transport mechanism may reveal a hetero-saturation—the rate of translocation of either species (perhaps both) suffering from the presence of the other ingredient. Such inter-substrate *competitive inhibition* in a mediated transport, having its basis again in the finite population and turnover rate of the reacting sites, is subject to all the same reservations and restrictions of interpretation noted in the preceding section.

Moreover, not every instance of transport inhibition observed upon addition of a homolog, isomer, or other structural relative of the passenger is necessarily to be attributed to a *competitive* situation. In developing this type of evidence of transport mediation, it is expedient wherever feasible to vary systematically the concentrations of the two presumed competitors so as to establish whether in fact the inhibition follows systematically a pattern consistent with a competitive interaction. Unless the situation is complicated by the concurrent function of two or more such mechanisms operating in parallel, the comparative prowess among a series of substrates in such competitive interplay should parallel the affinity sequence calculable from quantitative study of the respective homosaturations. In fact, it is generally possible to derive apparent affinity constants from the competition behavior, for direct comparison with those based on the saturation kinetics. Such comparison may prove to be especially crucial in evaluating the relative merits of alternative hypotheses regarding the molecular mechanisms of the translocation process (e.g., as in Sections II, B, 1, II, C, 3, and II, D).

c. High Unidirectional Fluxes. It is obvious that with any substrate that shows a limiting transfer rate at reasonable ranges of concentration, a symmetrical facilitated-diffusion mechanism may readily generate a state in which the opposed unidirectional fluxes are nearly equal long before the final equilibrium is approached. Thus, with an initially "zero-trans" situation, such as upon mixing a glucose solution with a sugar-free suspension of erythrocytes (LeFevre, 1948), a relatively rapid initial movement may soon dwindle to a mere trickle while a substantial difference still remains between the substrate concentrations in the two compartments. The reality of the bilateral flux-saturation basis for this phenomenon was demonstrated in this same system by LeFevre and McGinniss (1960), who observed that ^{14}C-glucose *tracer* equilibration proceeded about two orders of magnitude more rapidly than *net* glucose equilibration in the same cell suspensions. (Appropriate analysis of such measurements in terms of a given carrier model will also generally provide an independent means of estimating the system's hypothetical parameters, but the method is exceedingly cumbersome for general use for this purpose.) Like the other corollaries of saturation discussed above, this high ratio of flux to net movement will not be seen in a carrier-mediated system unless the affinity for the passenger is sufficiently high in relation to the experimental concentrations; thus, tests in human erythrocytes with the low-affinity monosaccharide D-ribose, paralleling those with glucose just described, showed no such high relative tracer fluxes (LeFevre, 1963).

It is possible that in some instances a similar apparent flux excess may arise from the parallel operation of a *separate* carrier capable of shuttling *only when occupied* (thus mediating only equal and opposite exchange fluxes, making no contribution to *net* transport). This variant of carrier behavior has been designated as *exchange diffusion* (Levi and Ussing, 1948); unfortunately this convenient label has tended in many minds to hypostatize the process as a distinctive mechanism in itself, obscuring the fact that *any* carrier system (whether subject to the above shuttling restriction in its original absolute form, or only partially, or not at all) will operate mainly as an exchange diffusion whenever the sites at both interfaces are highly saturated. Thus, dealing again with the passage of simple sugars through erythrocyte membranes, Lacko and Burger (1961, 1963, 1964) have repeatedly maintained a dichotomous view of the carrier mediation, holding that an exchange-diffusion mechanism, distinct from the net-transfer system, is available to certain of the sugars, but not to others. Although this interpretation may be equally consonant with the observations, Occam's law of parsimony would suggest conceptualization simply in terms of a unitary carrier apparatus with differing reactivities among the several substrates. As Wilbrandt (1959) has emphasized, the flux ratio in a carrier-mediated flow may assume *any value between unity and the activity ratio* of the substrate in the two pools, according to the degree of saturation obtaining.

Rosenberg and Wilbrandt (1957) had concluded that the flux-ratio characteristics of any transport mediated by a lattice of *fixed* membrane sites would be essentially like those of a free diffusion. Thus the type of behavior reported by LeFevre and McGinniss (1960) has been generally taken as sound evidence that a true (mobile) carrier is involved. However, qualitatively similar kinetics are predicted for some more recent models (to be taken up in Section II, D) in which the reacting sites cannot strictly be classed as true carriers: their individual access to an extramembranal compartment is limited to one side only. There has unfortunately been extensive confusion in the review literature in this field, between two distinctly different phenomena relating the flow of a labeled substrate to the composition of the trans-compartment: the *simple saturation phenomenon* just discussed, and the *true trans-effect* which Levine et al. (1965) have appropriately designated as "substrate-facilitated transport of the carrier" (Section I, C, 2, a). The latter phenomenon, unlike the bilateral flux-saturation situation, requires that the translocation step proper be more easily initiated by an occupied carrier (complex) than by an empty carrier. This fundamental distinction must be borne in mind even more carefully in connection with the following section dealing with "countertransport"; the confusion of this simple competitive-inhibition phenomenon with the genuine trans-effect has been particularly widespread, both in print and in public discussions.

d. Counterflow (Countertransport). In his original proposal of the classical mobile-carrier model, Widdas (1952) pointed out that, in spite of the fact that no linkage to any driving metabolic reaction was postulated, such a system might be expected to generate *transient* "uphill" movements of a passenger species under certain circumstances. Whenever a second passenger species is appropriately distributed asymmetrically across the membrane so that its competitive displacement of the primary substrate is substantially greater on one side than on the other, the respective occupancies by the primary species will no longer be set by the simple relation to its own concentrations in the two compartments, but by the equation given under point (3) in the legend of Fig. 1. Thus a transmembranal gradient for the complex may develop in the absence of an external concentration difference for the substrate in question, or even in the direction opposite to the external concentration difference; so long as this condition persists, there will be an uphill *counterflow* (*countertransport*), equivalent to one variety of Mitchell's (1962) "*antiport*." Experimental confirmation of this expectation from the model was first observed in sugar transfer in rabbit erythrocytes (Park *et al.*, 1956) and human erythrocytes (Rosenberg and Wilbrandt, 1957). Similar phenomena had previously been reported by Mitchell (1954) in bacterial cells, but not dealt with in these terms.

The counterflow phenomenon exemplifies the powering of an appreciable uphill transfer of one ingredient by the dissipation of a gradient for a second ingredient, the process continuing only so long as the latter gradient persists. This is one of several types of behavior that have made it difficult to formulate a universally acceptable, rigorous definition of "active transport." The counterflow phenomenon, more than any other, has been widely cited as a definitive criterion of a mobile-carrier mechanism; but this is a proper conclusion only if the latter term is taken to encompass *all* types of mechanism in which the interacting sites undergo any *transition* (change of state, orientation, conformation, etc.) *associated with a shift in accessibility of the passenger to the two pools* on each side of the membrane (Vidaver, 1966). As expressed by Regen and Morgan (1964), "the combining site . . . can collide with sugar molecules from only one side of the membrane at a given instant." The shift need not involve actual relocation of the individual sites nor movement of a membrane-soluble complex (as will be seen in consideration of recent "noncarrier" models in Section II, D). Britton (1963) has discussed the appearance of counterflow behavior in terms of its relation to the Ussing flux-ratio criterion for active transport, and notes that in fact there is at hand *no true criterion of carrier mobility* in the rigorous sense. Wilbrandt (1972b) stresses that while counterflow behavior is a useful positive indication of carrier mediation, its absence cannot be taken as a negative criterion, since at low concentrations the phenomenon may escape detection, and at very high saturation it may require a very long time to develop.

It must be emphasized that the countertransport situation does *not* involve modification of unidirectional *flux* (cis → trans) by the composition of the trans-compartment. The increased *net* flow upon addition of the second substrate to the trans-side is strictly assignable to the resultant diminution in the flux, trans → cis, and is thus really only a special instance of hetero-cis-inhibition (Wyssbrod *et al.*, 1971). Superimposed upon this may be true trans-effects, as considered in the next section.

2. Flux-Coupling Phenomena

a. Trans-Effects. Significant acceleration of the *unidirectional* efflux of an isotopically labeled transport substrate from cells often occurs upon supplementation of a previously substrate-free medium with an appreciable concentration of the unlabeled substrate or of a related passenger species. This behavior superficially resembles the enhancement of outflow that might result from cis-inhibition of influx (the counterflow phenomenon just discussed above), but differs radically in that it entails a genuine increase in the *efflux* (not merely the *net exodus*); the inward movement of the externally supplied passenger evidently enhances the supply of occupiable sites at the inner interface. The reality of this distinction in the natural world (and the readiness with which it might be experimentally overlooked) is brought out by the fact that, for the two red blood cell species in which monosaccharide counterflow was originally observed, the complicating superimposition of a true trans-acceleration was subsequently established for the human cells (Mawe and Hempling, 1965; Levine *et al.*, 1965) but found to be completely absent in the rabbit cells (Regen and Morgan, 1964). Nevertheless, the two types of event have often been blurred together in the relevant literature.

Observations suggesting the true dependence of a flux on the passenger's trans-concentration were first presented by Heinz (1954) for amino acid transfer in Ehrlich ascites tumor cells, then by Britton (1956) for sugar transfer in human erythrocytes. Such phenomena have now been extensively confirmed in many systems (Section I, D) under such banners as trans-acceleration, exchange-flux enhancement, homo-(or hetero-)transstimulation, substrate-facilitated transport of the carrier, "antiport," and (somewhat improperly) as "accelerative exchange diffusion" or "positive exchange diffusion." (These misusages are not entirely unreasonable, in that "exchange diffusion" may indeed be pictured as the *most extreme* example of trans-acceleration (since here no transfer whatsoever occurs unless a trans-substrate is present). However, the original application of the term does not allow this extension, restricting its proper use to precisely this type of mechanism which is *incapable of effecting any net transfer of total substrate.*)

In terms of the classical mobile-carrier theory, a trans-acceleration might have its basis in an increased mobility of the carrier within the membrane following its complexing with substrate (Heinz and Durbin, 1957; Heinz and Walsh, 1958); i.e., in Fig. 1, $D_{\overline{CS}} > D_C$. This amounts to rescinding the second of Widdas's classical restrictions, as listed in the legend of Fig. 1. The carrier would then be delivered to the membrane's cis-interface more rapidly when trans-substrate is available than when it is not, thus providing a faster supply of ferries for the cis $\rightarrow$ trans flux. It is of course equally plausible that $D_{\overline{CS}} < D_C$, so as to give rise to trans-inhibition (or "negative exchange diffusion"), but this has been much less widely observed (Section I, D). Its most common form is the virtual immobilization of a reacting site on the exterior cell surface upon its reaction with a substrate analog which evidently cannot be brought into the membrane at all (e.g., inhibition by phlorhizin of glucose exit from various cells may be viewed in this way).

Both accelerative and decelerative trans-effects constitute a strong argument for the involvement in the translocation process of an intermolecular association, *separately at each interface, but interacting* through some shared reaction or exchanged component. However, in spite of many printed proclamations to the contrary, the observation of this behavior does not specifically argue for a true mobility in the carrier; comparable kinetics are predicted with various "noncarrier" models for membrane interaction, even including the quite general Patlak (1957) "gate"-type which requires no particular visualization of the critical steps in the process. In general, any hypothesis whereby *transmembranal exchange becomes a more likely event than a one-way migration* is likely to entail prediction of trans-stimulation. It is noted, however, by Hoare (1972a) that the assumption (in the general model of Fig. 1) of the necessary activation of carrier translocation upon complexing with the substrate will not result in the trans-stimulation behavior unless another of the classical assumptions (i.e., that the trans-membrane step is rate-limiting) is retained. It must also be borne in mind that an appreciable degree of trans-acceleration may be encountered (on a rather elusive physicochemical basis) in simple artificial membranes (Kitahara *et al.*, 1965; Van Breemen and Van Breemen, 1969), so that the observation of this behavior does not in itself establish that a specific biological mechanism is operative.

b. Cotransport. The appearance of *accelerative* rather than inhibitory interplay between passengers on the cis-side of the membrane requires even greater complication of the general carrier model than is demanded to account for trans-effects. Cooperative cis-interaction may take the form simply of a sigmoid dependence of transport rate on the input concentration of a given substrate (homo-cis-stimulation); e.g., the rate of loss from one pool of a given amount of a labeled form may be accelerated by inclusion of additional unlabeled substrate. Similar association of the flows of two or more chemical species (hetero-cis-stimulation, "symport," or in some cases "competitive stimulation") may be observed. The study by Jacquez (1963) of amino acid transport by Ehrlich ascites tumor cells probably provides the most thorough characterization of cotransport within a single group of many closely related substrates, while cooperative transport interaction between different chemical classes is especially prominent in the almost ubiquitous linkage of sodium ion transfer with the absorption of organic nutrilites (Schultz and Curran, 1970, 1974; and symposium edited by Heinz, 1972).

This behavior should be clearly distinguished from "competitive acceleration" as defined by Rosenberg and Wilbrandt (1963) or Wilbrandt (1972a); in the latter phenomenon, both passengers are present on each side of the membrane at such relative concentrations that the two cis-inhibitions of the opposed fluxes of one of the components summate so as to give an enhancement of its *net* transfer (as referred to that occurring in the absence of the second substrate). This is not a true cis-stimulation, in precisely the same way that a "counterflow" is not a true trans-stimulation; in fact, this variety of competitive acceleration differs from countertransport only trivially, in that the induced net flow happens to be downhill rather than uphill.

True cotransport phenomena (unlike the trans-effects and the cis-inhibition considered in preceding sections) cannot be accommodated by the model of Fig. 1 simply by juggling the relative magnitudes of the several rate constants (Jacquez, 1961). [However, Oxender and Christensen (1963) pointed out the possibility of a deceptive mimicry of a cotransport by what is really a case of trans-acceleration: a simple competitor might be rapidly carried to the trans-side by a second pathway not shared with the test substrate, and then serve to accelerate the return of the carrier. Jacquez (1967) in fact concludes that this is the basis for the stimulation of L-tryptophan uptake observed in Ehrlich ascites tumor cells when L-methionine is added.] Their readiest interpretation (Wong, 1965) is that the carriers are *polyvalent* in their interaction with the passengers, providing more than one site on each unit. However, Jacquez (1964) noted that cis-stimulation could occur even with a monovalent system if there were a *direct* exchange between free substrate and carrier-complexed substrate (Section II, C, 2) and if the adsorption–desorption process became rate-limiting for the overall transfer (in diametric contrast with the assumption of the classical model). These restrictions would severely limit applicability to actual transport systems, whereas the concept of the carrier's polyvalence can be at least qualitatively fitted to the whole gamut of observed phenomena. Heinz *et al.* (1972) have specifically developed the formal predictions from a number of imaginable types of coupling interactions between pairs of solutes, bringing out several possible mechanisms for activation of cotransport or trans-acceleration (which they regrettably refer to as "countertransport").

It is especially significant that such coupling could serve as a mechanism for energizing a "secondary active transport" in a carrier system which had no direct linkage of its own with a metabolic reaction. If some "primary" active-transport system established an asymmetric distribution of one class of substrate, and if this gradient were spontaneously dissipated by the mediation of a polyvalent carrier in another system, there could result an uphill translocation of a second class of substrate associated with the same carrier. The first specific recognition of this possibility appears to be in the suggestion by Christensen *et al.* (1952) that amino acid transport in Ehrlich ascites tumor cells might be driven by the energy inherent in the K^+ asymmetry across the cell membranes; and subsequently the same group (Riggs *et al.*, 1958) suggested the involvement in this system of a ternary complex (sodium–carrier–glycine) carrying both classes of substrate together. The widespread linkage of amino acid and cation transport and its mechanistic significance have been thoughtfully reviewed by Christensen (1970).

The simplest basis for producing a secondary uphill transfer by way of such a ternary complex would be the modification, upon proper occupation of one site on the carrier, of either the mobility of the complex or the affinities at the other site (or both) (Heinz *et al.*, 1972). Kinetics appropriate to each of these possibilities have been observed in intestinal absorption systems (Crane, 1965; Curran *et al.*, 1967; Schultz and Curran, 1969). In fact, such interaction, under the cognomen "gradient coupling" (Crane, 1967), is currently widely accepted as the basis for the marked $[Na^+]$ dependence of active intestinal transport of sugars (Crane, 1962, 1965) and amino acids (Curran, 1968; Schultz, 1969), though serious experimental objections to this interpretation have been raised in the last few years (Heinz, 1970; Kimmich, 1970; Schafer and Heinz, 1971; Lin and Johnstone, 1971; Johnstone, 1972; Geck *et al.*, 1972; Kimmich and Randles, 1973; Tucker and Kimmich, 1973). [Schultz (1969) noted that true cotransport (with genuinely coupled flows) is only one form of gradient coupling; for example, if Na^+ altered the accessibility of membrane sites to another solute, it might modify that solute's unidirectional fluxes without actual participation in the translocation, thus possibly effecting uphill transfer "energetically coupled to the Na-gradient." It is difficult to see, however, how the energy stored in the gradient could *power* the secondary transport without progressive dissipation of the gradient as the process continued.] Recently, Verhoff and Sundaresan (1972) have sought to match these phenomena with a model that assigns the critical properties to a region of "confined reaction" just inside a simple semipermeable membrane (instead of to membrane-bound carriers), but the generality of this schema has not yet been examined; moreover, it is not clear that the claimed conceptual distinction from a general carrier operation is semantically defensible.

3. SPECIFICITY

A certain degree of permeant specificity might be achieved in a biological translocation without involving any intermolecular association, by suitable evolution of the detailed morphology of any pores traversing the membranes. However, it is scarcely conceivable that such selective leaks would allow the ready passage of only one enantiomorph of each of a series of hexoses or amino acids while virtually excluding various smaller nonelectrolytes. The sterospecificity revealed in many studies of cell "permeability" is in fact reminiscent of enzymic substrate-specificities; but generally the structural minima for reaction with transport sites seem to be somewhat less restrictive.

The simple fact of this selectivity implies little more than that the permeants must interact with fairly complex membrane sites (through at least three points of contact) in order that translocation be achieved; beyond this, it provides no particular information as to the transport mechanism. However, in presenting a detailed rationale for continuing studies, "On the Meaning of Effects of Substrate Structure on Biological Transport," Christensen (1972) clearly illustrated how a variety of basic questions concerning the transport processes can sometimes be answered through this approach. In particular, he emphasized how the deliberate design of artificial substrate analogs can assist in further analysis of transport mechanisms (not merely to further delineate the specificities).

4. SUSCEPTIBILITY TO ENZYME INHIBITORS, PROTEIN REAGENTS, HORMONES

It would be surprising if any operation that depends on an association of a solute with specific membrane sites failed to show high sensitivity to at least some of the varieties of reactive agents identified with experimental modification of enzymes and other proteins; and the appearance of such specific sensitivities in biological permeation processes is often taken as evidence of involvement of a mediated-transport mechanism. The kinetics of inhibition with various classes of presumed interaction have been analyzed by Rosenberg and Wilbrandt (1962). However, such evidence is not at all compelling unless it deals with some highly passenger-specific response, since it is clear that chemical or physicochemical modification of the general membrane interfaces, or of the linings of pores, could alter the characteristics of "passive" membrane permeability as well. The principal potential usefulness of these chemical modifiers in transport physiology lies in the chemical identification of critical reacting groups, and perhaps in the "labeling" of such groups in membranes and membrane fractions or extracts, so as to assist in tracing their partition during attempts at isolation and physicochemical characterization (Section I, E).

A great deal of attention has been paid to the apparent role of the hormone *insulin* in the activation of cell membrane transport mechanisms, particularly for glucose and related molecules; and a substantial body of evidence suggests also the direct involvement in transport phenomena of aldosterone, pituitary growth hormone, and vasopressin (review by Levine, 1972). In the last few years, a number of striking permeability modifications have also been seen accompanying changes in the cells' general physiological state, such as with virus-induced malignant transformations (Hare, 1967; Foster and Pardee, 1969; Hatanaka *et al.*, 1969, 1970; Hatanaka and Hanafusa, 1970; Hatanaka and Gilden, 1970, 1971; Inbar *et al.*, 1971; Martin *et al.*, 1971; Isselbacher, 1972a,b; Kalckar *et al.*, 1973; Venuta and Rubin, 1973; Weber, 1973) and upon reaction with phytoagglutinins (Mendelsohn *et al.*, 1971; Peters and Hausen, 1971). Again, the mere observation of marked changes in cell permeation behavior in response to such agents does not in itself provide any argument that intermolecular association of the passengers with specific membrane components is a necessary step in the translocation event; but this may be strongly suggested if the response is found to be highly selective for particular substrates in a set of analogs, or if the response is readily analyzable kinetically in terms of a carrier operation. In short, any observations dealing with the sensitivity of transport to modifying agents can be accepted as indications of carrier involvement *only insofar as they are tied directly to the other classes of evidence* discussed above.

5. HIGH TEMPERATURE-DEPENDENCE

This consideration has no real place in this list as a criterion of carrier mediation, but is mentioned in order to bemoan its common inclusion in textbook discussions of such criteria. While it is true that a carrier operation should show a heat of activation well in excess of the 4500 cal per mole characterizing a free diffusion (Solomon, 1952), and while figures of 15,000–20,000 cal per mole (or Q_{10} of 2.5–3 or more) are commonly seen and taken as indicative of involvement of a chemical process, it must be stressed that such observations have no differentiative value in the identification of mediated permeation. High energy-barriers may have to be bridged also in passing by purely diffusive means from one phase to another, as in entering or leaving a lipid membrane in biological systems (Danielli, 1952). In at least one instance (that of glycerol transfer through erythrocyte membranes), where assorted species of mammals have been grouped into two sets on the basis of several linked characteristics in the transport behavior (Jacobs *et al.*, 1935, 1938), it is in fact the distinctly *lower* Q_{10}, rather than the higher, which is associated with the signs of a special mediation; moreover, this Q_{10} *rises* when the mediated permeation is suppressed by chemical agents to which the other group is insensitive.

D. The Documented Occurrence of Such Phenomena

From the considerations discussed in Section I, C, it appears that *none of the recognized "criteria" can be reliably taken as a definitive basis for concluding that a true carrier (ferryboat) system underlies the operation of a biological translocation*; and it is even possible (though less clearly so) to conceive in each case of some alternative that would avoid altogether any presumption of intermolecular association of the passenger with a membrane constituent. However, the conjoint appearance of a number of these somewhat equivocal signs, in a great variety of biological barriers and in connection with the transfer of many classes of passengers, has strongly argued for this unitary interpretation. Table I presents a substantial, but only representative, collection of such multiple-criterion citations; a few more limited observations are also listed because of their pioneer significance, or their concern with rather esoteric systems illustrating the very wide apparent occurrence of these phenomena.

E. Efforts to Isolate and Assay Carrier Entities

Clearly our understanding of the transport-mediating mechanisms would be dramatically advanced if the carrier entities (or at least the substrate-recognizing sites) could be sufficiently isolated as to allow their chemical identification or physicochemical characterization in more rigorous terms. Perhaps the most recurrent theme in discussions of the state of this field for the last 15 years has been the necessity for concentration on this type

TABLE I

REPORTED OCCURRENCE OF INDICATIONS OF CARRIER-MEDIATED TRANSPORT IN
BIOLOGICAL SYSTEMS

Organism and tissue	Permeant(s)	Type of evidence[a]	Citation[b]
Bacteria			
Escherichia coli	Na$^+$, K$^+$	Up, Sat, Sp	Schultz *et al.*, 1963. JGP 47:329
	β-Galactosides	Sat, Cf, MM, Cp, I, Up	Kepes and Monod, 1957. CrAS 244:809; Kepes, 1960. Ref.
		Sat, I, F	Koch, 1964. Ref.
		MM, I, Cf, tA, Cp, Sp, Up	Winkler and Wilson, 1966. Ref.; 1967. BBA 135:1030; Wong and Wilson, 1970. BBA 196:336; Wilson and Kusch, 1972. BBA 255:786
		MM, Sp (multiple)	Rotman *et al.*, 1968. JMol 36:247
		Cp, tA, tI	Robbie and Wilson, 1969. BBA 173:234
		MM, I	Batt and Schachter, 1971. BBA 233:189
	Galactose	Up, Cp, I	Horecker *et al.*, 1960. JBC 235:1586; 1961. MTM, p. 378; Vorisek and Kepes, 1972. EJB 28:364
	α-Methylgluco-side	MM, Cp, Cf	Gachelin, 1970. Ref.
	β-Glucuronides	Up, I, MM, Cp, tA	Stoeber, 1957. CrAS 244:1091
	Monosaccharides	MM, Cp, Up	Novotny and Englesberg, 1966. BBA 117:217
	Leucine, valine, isoleucine	Up, MM, Cp, Cf, I	Cohen and Rickenberg, 1955. CrAS 240:2086
	Glutamate	Cf	Halpern, 1967. BBA 148:718
	Amino acids (AA)	MM, I, tA, Cp, Sp(mult.)	Piperno and Oxender, 1968. JBC 243:5914
	Basic	MM, Cp, Sp(mult.)	Rosen, 1971. JBC 246:3653
	Cyclic	Up, Cp, Sp	Tager and Christensen, 1971. JBC 246:7572
	Oligopeptides	Sp, Cp	Payne, 1971. BcJ 123:245, 255
	Adenosine, inosine	Sat, I, Cp, Cf	Peterson and Koch, 1966. BBA 126:129

(Continued)

TABLE I (*Continued*)

Organism and tissue	Permeant(s)	Type of evidence[a]	Citation[b]
	Adenosine, uridine, cytidine	I, Up, Cp, MM, tA	Peterson *et al.*, 1967. BBA 135:771
	Dicarboxylic acids	MM, Up, Cp, tA	Kay and Kornberg, 1971. EJB 18:274
Streptococcus faecalis	Phosphate, arsenate	Cp, Up, MM	Harold and Baarda, 1966. JBt 91:2257
	Glutamate, Rb^+	I	Harold and Baarda, 1966. JBt 91:2257
	K^+, Rb^+	Up, Sp	Harold *et al.*, 1967. Bch 6:1777
	Disaccharides	I	Abrams, 1960. JBC 235:1281
	Amino acids	I, Up	Holden and Utech, 1967. BBA 135:351
	Citrulline	Sat	Bibb and Straughn, 1964. JBt 87:815
Bacillus subtilis	Di- and tricarboxylic acids	Up, Sp, Cp	Ghei and Kay, 1972. FEBSL 20:137; McKillen *et al.*, 1972. MBBT, p. 249
Staphylococcus aureus	Phosphate	Sat, Cp, I, F	Mitchell, 1954. SSEB 8:254; JGM 11:73
	Disaccharides, glycosides	Up, Cf	Hengstenberg *et al.*, 1968. JBC 243:1881
	Glutamate	Cp, Co-am.ac., Co-glucose	Gale and Van Halteren, 1951. BcJ 50:34
Salmonella typhimurium	SO_4^{2-}, $S_2O_3^{2-}$	Up, Sat	Dreyfuss, 1964. JBC 239:2292
	Monosaccharides	MM	Simoni *et al.*, 1970. PNAS 58:1963
	Histidine, aromatic amino acids	MM(multiple)	Ames, 1964. ABB 104:1
Aerobacter aerogenes	*myo*-Inositol	Up, I, MM, Cp	Deshusses and Reber, 1972. BBA 274:598
Bacterium lactis aerogenes	K^+, Rb^+	Cp	Carroll *et al.*, 1950. JCS, p. 946
Mycobacterium avium	Glutamate	Sp, Up, MM, I, Cp	Yabu, 1967. BBA 135:181
Mycoplasma	K^+	Sat	Cho and Morowitz, 1969. BBA 183:295
	Histidine, methionine	Sat, I, Sp, Up, Cp, tA	Razin *et al.*, 1968. JBt 95:1685
Assorted spp.	Amino acids	Up, Sat, I, Co-K^+	Gale, 1954. SSEB 8:242
	Oligopeptides	Sp, Cp, Co-am.ac.	Payne and Gilvarg, 1971. AdE 35:187

TABLE I (*Continued*)

Organism and tissue	Permeant(s)	Type of evidence[a]	Citation[b]
Yeasts			
Saccharomyces	Na+, K+, H+, Ca²+, Mg²+, Mn²+, SO₄²⁻, H₂PO₄⁻	Cp, Sp, I, MM	Rothstein, 1961. MTM, p. 270; 1964. CFMT, p. 23; Armstrong and Rothstein, 1964. JGP 48:61; 1967. JGP 50:967
	Monosaccharides	I, MM	Hurwitz and Rothstein, 1951. JCCP 38:437; Rothstein, 1954. SSEB 8:165
		I, Cp, Cf, Sp, tA	Cirillo, 1961. MTM, p. 343; 1962. JBt 84:485
		Cp, MM, I, tA	Kotyk and Kleinzeller, 1963. FMb 8:156; Kotyk, 1965. FMb 10:30; 1967. Ref.
		MM, Cp, Sp	Cirillo, 1968. JBt 95:603, Scharff and Kremer, 1962. ABB 97:192
	Hexoses, glucosides	MM, Cp, Cf, Sp, I, some Up	Van Steveninck, 1968. BBA 150:424; 163:386; 1970. BBA 203:376; 1972. BBA 274:575; Van Steveninck and Dawson, 1968. BBA 150:47
	Mono- and disaccharides	Cp, Sp	Burger et al., 1959. BcJ 71:233
	Disaccharides	I, Sat, Sp, Cp, F	Rothstein, 1954. SSEB 8:165; Sols and de la Fuente, 1961. MTM, p. 361; Halvorson et al., 1964. CFMT, p. 171
	Glucosamine	Sat, Cp	Burger and Hejmova, 1961. FMb 6:80
	L-Lysine	Cp, MM(double)	Grenson, 1960. BBA 127:339
	L-Arginine	Sp, Cp, MM	Grenson et al., 1966. BBA 127:325
	Amino acids	Sp, Cp	Gits and Grenson, 1967. BBA 135:507
		MM, I, Cp, Co-am.ac.	Magaña-Schwencke and Schwencke, 1969. BBA 173:313; Magaña-Schwencke et al., 1969. BBA 173:325
	Dicarboxylic amino acids	MM, Cp	Joiris and Grenson, 1969. AIPB 77:154

(*Continued*)

TABLE I (*Continued*)

Organism and tissue	Permeant(s)	Type of evidence[a]	Citation[b]
	α-Aminoiso-butyric acid	Up, MM, tI	Kotyk and Říhová, 1972. BBA 288:380
Candida	Monosaccharides	Sat, MM, Cp, Cf	Cirillo *et al.*, 1963. JBt 86:1259
Rhodotorula		Up, Cf, MM, tA	Höfer, 1970. JMB 3:73
"Sauternes"		Sat, Cp	Sols, 1956. BBA 20:62
Fungi			
Neurospora	K⁺	Sat, Sp, MM(double)	Slayman and Tatum, 1965. BBA 102:149; Slayman and Slayman, 1970. JGP 55:758
	L-Sorbose	MM, Cp, Sp, I, Cf	Crocken and Tatum, 1967. BBA 135:100
	Monosaccharides	Up, MM, Cf	Scarborough, 1970. JBC 245:1694,3985
		Sp, Cp, I, tA	Schneider and Wiley, 1971. JBt 106:479
	L-Tryptophan	MM, Sp, Cp	Wiley and Matchett, 1966. JBt 92:1698
	Arginine	MM, I, Cp	Roess and DeBusk, 1968. JGM 52:421
	L-Aspartate	Up(double)	Wolfinbarger *et al.*, 1971. BBA 249:63
	Amino acids	MM, Cp, Sp(triple)	Pall, 1968. BBA 173:113; 1970. BBA 211:513
Aspergillus	Hexoses	Sat, Up, Sp(double)	Mark and Romano, 1971. BBA 249:216
Lichen			
Peltigera	Hexoses	Sat, Cp, I	Harley and Smith, 1956. ABL 20:513
Algae			
Chlorella	SO₄²⁻	Sat, MM, Cp	Kotyk, 1959. FMb 4:363; Vallée, 1969. BBA 173:486
	Cl⁻	Up	Barber, 1969. ABB 130:389
	Na⁺	Up	Barber, 1968. BBA 150:618
	K⁺	MM, Sp	Barber, 1968. BBA 163:141
	Hexoses	MM, Cp, Up	Tanner, 1969. BBRC 36:278; Komor and Tanner, 1971. BBA 241:170
Nitellopsis	Na⁺	Up	MacRobbie and Dainty, 1958. JGP 42:335
Halicystis	Na⁺, Cl⁻	Up	Blount and Levedahl, 1960. APS 49:1
Chara	Na⁺	Up	Hope and Walker, 1960. AJBS 13:277

TABLE I (*Continued*)

Organism and tissue	Permeant(s)	Type of evidence[a]	Citation[b]
Valonia	Na+, K+	Up	Gutknecht, 1966. BiB 130:331
Chaetomorpha	Na+, K+	Up	Dodd *et al.*, 1966. AJBS 19:341
Griffithsia	Na+, K+, Cl−	Up	Findlay *et al.*, 1969. AJBS 22:1163
Scenedesmus	Sugars	Cp, Sp, I, MM	Taylor, 1960. PRSLB 151:400,483
Higher plants			
Sugar cane slices, immature	Sucrose	Sat, Up, I	Bielesky, 1960. AJBS 13:204, 221
Carrot disks, corn roots	Hexoses, pentoses	Up, Sp, MM, Cp	Grant and Beevers, 1964. PlP 38:78
Pea chloroplasts	Aldoses	Sp, Sat, Cp	Wang and Nobel, 1971. BBA 241:200
	Amino acids	Sat, Cp(double)	Nobel and Cheung, 1972. NatNB 237:207
Protozoa			
Tetrahymena	L-Arabinose	Cf(glucose)	Cirillo, 1962. JBt 84:754
	Phenylalanine	MM, I, Sp, Cp	Stephens and Kerr, 1962. Nat 194:1094
	Phenylalanine, methionine	Up, Sat, Cp, Sp, Cf	Hoffmann and Rasmussen, 1972. BBA 266:206
Euglena	Amino acids	Up, I	Lux and Müller, 1969. BBA 177:186
Plasmodium lophurae	Glutamate, lysine, arginine	Up	Sherman *et al.*, 1967. CBP 23:43
Marine pseudomonad	α-Aminosiobutyrate, D-fucose	Up, Co-Na+	Drapeau and MacLeod, 1963. BBRC 12:111
Coelenterate			
Fungia (coral)	Glucose	Sp, I	Stephens, 1962. BiB 123:648
Molluscs			
Squid giant axons	Na+, K+(coupled)	I, Sp, Up	Hodgkin and Keynes, 1955. JPh 128:28,61; Caldwell *et al.*, 1960. JPh 152:561, 591; Mullins and Brinley, 1967. JGP 50:2333; Brinley and Mullins, 1968. JGP 52:181
	Cl−	Up	Keynes, 1963. JPh 169:690
	Glucose	Sp	Hoskin and Rosenberg, 1965. JGP 49:47

(*Continued*)

TABLE I (*Continued*)

Organism and tissue	Permeant(s)	Type of evidence[a]	Citation[b]
	Choline	Sat, Co-Na$^+$	Hodgkin and Martin, 1965. JPh 179:26P
Squid Schwann cells	Glycerol	I	Villegas and Villegas, 1962. BBA 60:202
Chiton gut sacs	Hexoses	Up(mult.)	Lawrence and Lawrence, 1967. CBP 22:341
Mytilus gills	Phosphate	I	Ronkin, 1950. PSEBM 73:41
Anodonta gills	Phosphate	Up	Schoffeniels, 1951. AIP 59:245
Crustaceans			
Crab nerves	Glutamate	Co-Na$^+$	Baker and Potashner, 1971. BBA 249:616
	Dicarboxylic acids	MM, Co-Na$^+$, Sp	Evans, 1973. BBA 311:302
Crab gills	Na$^+$	Up, I	Koch, 1954. CoP 7:15
Worms			
Schistosoma	Serotonin	MM, I, Up	Bennett and Bueding, 1973. MoP 9:311
Calliobothrium	Glucose, galactose	Up, Cp, I, tA, Co-Na$^+$	Fisher and Read, 1971. BiB 140:46
	Amino acids	MM, Cp, tA	Read et al., 1960. BiB 119:120
Gyrocotyle	Glucose, galactose, α-methylglucose	Cp, I	Laurie, 1971. EPs 29:375
Insects			
Silkworm larval midgut	K$^+$, Rb$^+$, Cs$^+$	Cp, Up	Zerahn, 1971. PTRSLB 262:315
Malpighian tubules	K$^+$	Up	Ramsay, 1953. JEB 30:358
Fishes			
Gut: Assorted	Glucose	Up	Musacchia and Fisher, 1960. BiB 119:327
Catfish	Monosaccharides	Up, I, Sp	Musacchia et al., 1964. BiB 126:291
Scup, puffer	Glucose	Up, I	Musacchia et al., 1966. CBP 17:93
Dogfish		Sat, I	Carlisky and Huang, 1962. PSEBM 109:405
	Amino acids/ mono-saccharides	Mutual Cp	Read, 1967. BiB 133:630
Flounder	Tyrosine	Up, Sp	Rout et al., 1965. PSEBM 118:933
Assorted	Glucose	I, Co-Na$^+$	Smith, 1966. JPh 182:559

TABLE I (*Continued*)

Organism and tissue	Permeant(s)	Type of evidence[a]	Citation[b]
Killifish	Monosac-charides, amino acids	Up, I	Huang and Rout, 1972. AJP 212:799
Toadfish	Glucose	MM, Up, I	Farmanfarmaian *et al.*, 1972. BiB 142:427
Kidneys: Assorted	Urea	Up	Smith, 1931. AJP 98:279,296
Goosefish	Glucose	Up	Malvin *et al.*, 1965. JCCP 65:381
Liver: Toadfish	L-Leucine	MM, Cp	Haschemeyer and Hudson, 1973. BiB 145:439
Gills: Goldfish	Na$^+$, Cl$^-$	Up	Garcia-Romeu and Maetz, 1964. JGP 47:1195
Trout	Cl$^-$	Up, Sat	Kerstetter and Kirschner, 1972. JEB 56:263
Amphibians Skin, frog	Cl$^-$	I	Ussing and Zerahn, 1951. APS 23, Suppl. 80:110; Erlij, 1971. PTRSLB 262:153
		Up	Jørgensen *et al.*, 1954. APS 30:178; Martin and Curran, 1966. JCCP 67:367
	Na$^+$	I, Sat, H	Ussing, 1948. CSHSQB 13:193; 1949. APS 17:1
		MM	Cereijido *et al.*, 1964. JGP 47:879
Kidneys: *Necturus*	Glucose	Co-Na$^+$	Khuri *et al.*, 1966. FPr 25:899
Frog	Urea	I, Up	Forster, 1954. AJP 179:372
	Na$^+$, Ca^{++}	I, Co-K$^+$	Vogel *et al.*, 1966. PfA 288:359
	Glucose, *p*-amino-hippurate	I, Co-Na$^+$, MM	Vogel *et al.*, 1965. PfA 283:151; 1966. PfA 288:359
Bladder: Frog	Na$^+$, Li$^+$	Up, Cp, H	Natochin and Leont'ev, 1964. FPr 24:T403
Toad	Na$^+$	I	Koefoed-Johnsen, 1957. APS 42, Suppl. 145:87; Herrera, 1966. AJP 210:980
		Up, H	Leaf, 1961. MTM, p. 247; Crabbe, 1961. JCI 40:2103; End 69:673; Herrera, 1965. AJP 209:819

(*Continued*)

TABLE I (*Continued*)

Organism and tissue	Permeant(s)	Type of evidence[a]	Citation[b]
		Up, MM	Frazier *et al.*, 1962. JGP 45:529
	Cl⁻	Up, I	Finn *et al.*, 1967. AJP 213:179
	Acetamide, urea	Sat, Cp	Levine *et al.*, 1973. JCI 52:2083
	α-Aminoiso-butyrate	I, Co-Na⁺	Thier, 1968. BBA 150:253
Gut: Frog	Monosaccharides	Sp	Westenbrink and Gratama, 1937. ANP 22:326
		Sat	Csáky and Fernald, 1960. AJP 198:445
		Up, Co-Na⁺	Csáky and Thale, 1960. JPh 151:59; Csáky and Lassen, 1964. BBA 82:215; Csáky and Ho, 1966. LSc 5:1025; Lassen and Csáky, 1966. JGP 49:1029
		Up, Sp	Lawrence, 1963. CBP 9:69
	Amino acids	Co-Na⁺	Csáky, 1961. AJP 201:999
	Na⁺	Co-sugars, am.ac.	Quay and Armstrong, 1969. PSEBM 131:46
Stomach: Frog	Cl⁻	Up	Hogben, 1955. AJP 180:641
Muscle: Frog	Na⁺	I	Frazier and Keynes, 1959. JPh 148:362
		I, Up	Carey *et al.*, 1959. JPh 148:51; Conway and Mullaney, 1961. MTM, p. 117
	Na⁺, K⁺(coupled)	Up	Steinbach, 1951. AJP 167:284; 1952. PNAS 38:451
	K⁺	H	Manery *et al.*, 1956. CJBP 34:893; Smillie *et al.*, 1960. AJP 205:67
	Monosaccharides	I, MM, Cp, H	Narahara and Özand, 1963. JBC 288:40; Weiss and Narahara, 1969. JBC 244:3084
Nerve: Frog	Na⁺, K⁺(coupled)	I	Hurlbut, 1963. JGP 46:1191, 1223; 1965. AJP 209:1295
Cornea: Frog	Cl⁻	Sp, Cp, I	Zadunaisky *et al.*, 1971. AJP 221:1832
		Co-Na⁺	Zadunaisky, 1972. BBA 282:255

TABLE I (*Continued*)

Organism and tissue	Permeant(s)	Type of evidence[a]	Citation[b]
Blood/CSF barrier: Frog	Sugars	I, MM, Sp, Cp	Prather and Wright, 1970. JMB 2:150
Reptiles			
Gut: Turtle	Glucose, 3-*O*-methylglucose	Up, I	Fox, 1961. AJP 201:295; 1965. BiB 129:490
Bladder: Turtle	Cl⁻	Up	Gonzalez *et al.*, 1967. AJP 212:641
Birds			
Fibroblasts: Chick	Sugars	MM	Hatanaka and Hanafusa, 1970. Ref.; Weber, 1973. Ref.
		I, MM, Cp	Kletzien *et al.*, 1972. JBC 247:2964; Kletzien and Perdue, 1973. JBC 248:711
		Up, MM	Venuta and Rubin, 1973. Ref.
	Glucose, leucine, adenine, uridine	I, Sat	Scholtissek, 1968. BBA 158:435
Erythrocytes: Pigeon	Glycine	Cp, Sp	Vidaver *et al.*, 1964. ABB 107:82
		MM, I, Cf, Co-Na⁺	Vidaver, 1964. Bch 3:662; 1964a. Ref.
		MM	Wheeler *et al.*, 1965. BBA 109:620
		I, Cp, Co-Na⁺	Kittams and Vidaver, 1969. BBA 173:540
		tA, tI	Vidaver and Shepherd, 1968. JBC 243:6140
		Co-anions	Imler and Vidaver, 1972. BBA 288:153
	Amino acids	I, Cp, Sp, tA, partly Co-Na⁺, MM(mult.)	Eavenson and Christensen, 1967. JBC 242:5386; Wheeler and Christensen, 1967. JBC 242:3782; Koser and Christensen, 1971. BBA 241:9
	Na⁺	MM, Co-am.ac.	Koser and Christensen, 1971. BBA 241:9
Duck	Amino acids	Up, Co-Na⁺	Christensen *et al.*, 1952. JBC 194:41
Goose	Glucose, 3-*O*-methylglucose	MM, Cf	Wood and Morgan, 1969. JBC 244:1451
Kidney slices: Chicken	Urate	Up, I	Platts and Mudge, 1961. AJP 200:387

(*Continued*)

TABLE I (*Continued*)

Organism and tissue	Permeant(s)	Type of evidence[a]	Citation[b]
Gut; Pigeon	Monosaccharides	I, Sp	Westenbrink, 1937. Nat 138:203; Westenbrink and Gratama, 1937. ANP 21:433
Gut segments: Chicken	L-Methionine	Cp, Co-histidine	Lerner and Taylor, 1967. BBA 135:990
	D-Xylose	I, MM, tA, Co-Na$^+$	Alvarado, 1967. CBP 20:461
	Phenylglucosides	Up, MM, I, Cp, Cf, Co-Na$^+$	Alvarado and Monreal, 1967. CBP 20:471
Gut cells: Chicken	D-Galactose, L-valine	I, Up, MM, Co-Na$^+$	Kimmich, 1970. Ref.; Bch 9:3659
	Amino acids	Up, Co-Na$^+$, Cp, Cp-sugars	Tucker and Kimmich, 1973. Ref.
Brain: Chick	Amino acids	MM	Levi, 1970. ABB 138:347
Bones: Chick	Amino acids	Up, I, Cp(double)	Adamson and Ingbar, 1967. JBC 242:2646
Chorioallantoic membranes: Chick	Na$^+$, Cl$^-$	Up	Moriarty and Hogben, 1970. BBA 219:463
Mammals			
Erythrocytes: Human	K$^+$/Rb$^+$, Na$^+$/Li$^+$	Cp, Sat	Solomon, 1952. Ref.
	Na$^+$, K$^+$	Up, I	Schatzmann, 1953. HPA 11:346
	K$^+$	Up	Gárdos, 1954. Ref.
	Na$^+$, K$^+$(coupled)	Up, Sp, Cp, I, Sat	Hoffman, 1962. Circ 26:1201; JGP 45:837; Maizels, 1968. JPh 195:657; Garrahan and Glynn, 1967. JPh 192:159,175,189
	K$^+$	I, Sat (not MM)	Sachs and Welt, 1968. JCI 47:949
	Ca^{2+}	Up, I	Schatzmann and Vincenzi, 1969. JPh 201:369
		Sat, Cp	Porzig, 1970. JMB 2:324
	Urea	I	Hunter et al., 1965. JCCP 65:299
Assorted spp.	Urea, ethylene glycol, glycerol, erythritol	I	Jacobs and Corson, 1934. BiB 67:325; Jacobs and Parpart, 1937. BiB 73:380; Hunter et al., 1965. JCCP 65:299; Hunter, 1970. BBA 211:216

TABLE I (Continued)

Organism and tissue	Permeant(s)	Type of evidence[a]	Citation[b]
Human	Erythritol	I, Cp-monosacch.	Bowyer and Widdas, 1956. DFS 21:251; Wieth, 1971. JPh 213:435; LaCelle and Passow, 1971. JMB 4:270
	Glycerol	Sat, Cp, I	Bowyer, 1957. IRC 6:469
	Monosaccharides	Sat, Cp, I, F, Cf, Sp	LeFevre, 1961. PhaR 13:39; Wilbrandt, 1961. MTM, p. 388
		I, Sp, MM, ta	Lacko and Burger, 1961, 1963, 1964. Refs.; Lacko et al., 1961, 1972. Refs.
		Sat	Britton, 1964. Ref.
		MM, F	Harris, 1964. JPh 173:344
		Cf	Bican and Lacko, 1966. Trf 6:130; Levine and Levine, 1969. Ref.
		ta	Mawe and Hempling, 1965. Ref.; Levine et al., 1965. Ref.; Eilam and Stein, 1972. Ref.
		Sp, MM, I	Barnett et al., 1973. BcJ 131:211
	D-Glucose	MM, I	Sen and Widdas, 1960, 1962. Refs.; 1962. JPh 160:404
		I	Dawson and Widdas, 1963. JPh 168:644; Deuticke et al., 1964. PfA 280:275; Forsling et al., 1968. JPh 194:535; Baker and Rogers, 1972. BPh 21:1871
Rabbit	Hexoses	Sp	Hillman et al., 1959. AJP 196:1277
	D-Ribose	MM	Steinbrecht and Hofmann, 1964. ZPC 339:194
	Monosaccharides	Sp, Sat, Cf	Regen and Morgan, 1964. Ref.
Fetal, many spp.	Hexoses	Sat, Cp	Widdas, 1955. JPh 127:318
Human	Amino acids	Cp-hexoses	Rieser, 1961. ECR 24:165
	L-Valine	MM	Rieser, 1962. ECR 27:577
	Amino acids	Sat, Cp, Sp, Cf, F	Winter and Christensen, 1964. Ref.; 1965. JBC 240:3594
	Choline	Up, MM, ta, Co-Na+, I	Martin, 1968. JGP 51:497; 1971. Ref.

(Continued)

TABLE I (*Continued*)

Organism and tissue	Permeant(s)	Type of evidence[a]	Citation[b]
	Purines	Cp, MM	Lassen, 1961. BBA 53:557; Lassen and Overgaard-Hansen, 1962. BBA 57:111, 115; Lassen, 1967. BBA 135:146
	Nucleosides	I, Cp, MM, Cf	Oliver and Paterson, 1971. CJB 49:262; Paterson and Oliver, 1971. CJB 49:271; Packard and Paterson, 1972. Ref.
		Sp, tA, tI	Cass and Paterson, 1972. JBC 247:3314; 1973. BBA 291:734
Reticulocytes: Rabbit	Glycine	Up, I	Riggs *et al.*, 1952. JBC 194:53
	Amino acids	MM, Cp, F, Cf, Co-Na[+]	Winter and Christensen, 1965. JBC 240:3594
	Glycine, alanine	MM, Cf, Co-Na[+]	Wheeler and Christensen, 1967. JBC 242:1450
	Cationic am. ac.	MM, Cp	Christensen and Antonioli, 1969. JBC 244:1497
Human, rat	Amino acids	I, partly Co-Na[+], Up	Yunis and Arimura, 1965. JLCM 66:177
Leukocytes: Rabbit	K[+]	Up, I	Elsbach and Schwartz, 1959. JGP 42:883
Guinea pig	Monosaccharides	Sp	Luzzatto and Leoncini, 1961. IJB 10:249
	Glycine, α-amino-isobutyrate	Up, Co-Na[+]	Yunis *et al.*, 1962. JLCM 60:1028; 1963. JLCM 62:465
Mouse	Sugars, phosphate, amino acids, adenosine	Up, I	Kessel and Hall, 1970. BBA 211:88; Kessel and Dodd, 1972. BBA 288:190
Mouse, (leukemic)	Nucleosides	Sat	Kessel and Shurin, 1968. BBA 163:179
Rabbit		Cp, Sp	Taube and Berlin, 1972. BBA 255:6
	Purines	Up, Sat	Hawkins and Berlin, 1969. BBA 173:324
Mouse (leukemic)	Folic acid and analogs	MM, tA	Goldman, 1971. BBA 233:624
		Up, MM, Cp, Cf	Goldman *et al.*, 1968. JBC 243:5007

TABLE I (*Continued*)

Organism and tissue	Permeant(s)	Type of evidence[a]	Citation[b]
Lymphocytes: Guinea pig	Monosaccharides	Cp, I, tA	Helmreich and Eisen, 1959. JBC 234:1958
Human	α-Aminoiso-butyrate	I, MM	Mendelsohn *et al.*, 1971. Ref.
Bovine	Glucose, 3-*O*-methylglucose, leucine, choline, uridine	Up, I, MM, Cf	Peters and Hausen, 1971. Ref.; EJB 19:502
Fibroblasts: Mouse	Glucose, galactose	Up, Sat, I, Cp, Cf	Rickenberg and deMaio, 1961. MTM, p. 409
Human	L-Tryptophan	Cp	Platter and Martin, 1966. PSEBM 123:140
	α-Aminoiso-butyrate	MM, I	Mahoney and Rosenberg, 1970. BBA 219:500
Embryonic cells Hamster	L-Phenylalanine	Up, MM, Cp, tA, I	Hare, 1967. Ref.
	Glucose	MM	Hatanaka and Gilden, 1971. Ref.
Mouse	Monosaccharides	MM	Hatanaka *et al.*, 1969, 1970. Refs.
Ehrlich ascites tumor cells: Mouse	Amino acids	Sp, Up, I, Cp(multiple)	Christensen *et al.*, 1952. Ref.; Christensen and Riggs, 1952. Ref.; Riggs *et al.*, 1954. JBC 209:395
		Co-Na+, Cf-K+	Riggs *et al.*, 1958. JBC 233:1479; Eddy, 1968. BcJ 108:195; Kromphardt *et al.*, 1963. BBA 74:549
		MM, Cp, tA	Heinz, 1954. Ref.; Heinz and Walsh, 1958. Ref.
		tI	Paine and Heinz, 1960. JBC 235:1080
		Up, I, Cp, MM	Tenenhouse and Quastel, 1960. CJBP 38:1311; Christensen, 1964. JBC 239:3584
		MM, Cp, Sp, I, F, tI	Oxender and Christensen, 1963. Ref., Nat 197:765
		tA, Co-am. ac.	Jacquez, 1963, 1967. Refs.
		MM(double), Cp(mult.)	Christensen, 1964. PNAS 51:337

(*Continued*)

TABLE I (*Continued*)

Organism and tissue	Permeant(s)	Type of evidence[a]	Citation[b]
		Cp, Co-Na$^+$, MM	Inui and Christensen, 1966. JGP 50:203
		Cp, tI, Co-Na$^+$, Co-am. ac.	Schafer and Jacquez, 1967. BBA 135:741; Belkhode and Scholefield, 1969. BBA 173:290
	Natural amino acids	Cp, Co-Na$^+$	Christensen and Handlogten, 1969. FEBSL 3:14; Christensen et al., 1969. PNAS 63:948
	L-Tryptophan	Up, Sat, Sp, I, tA, Co-am. ac.	Jacquez, 1961. AJP 200:1063
	α-Aminoiso-butyrate	Up, Co-Na$^+$	Geck et al., 1972. Ref.
	Na$^+$	Co-glycine	Eddy, 1968. BcJ 108:195,489
	Hexoses	Sat, Cp, I, Sp	Crane et al., 1957. JBC 224:649
		Cp	Nirenberg and Hogg, 1958. JACS 80:4407
	L-Sorbose	Cf-glucose	Cirillo and Young, 1964. ABB 105:86
	Monosaccharides	Cp, MM, I, F, Sp	Kolber and LeFevre, 1967. JGP 50:1907
	myo-Inositol	Up, Co-Na$^+$	Johnstone and Sung, 1967. BBA 135:1052
	Nucleosides	Sat, I	Jacquez, 1962. BBA 61:265
Sarcoma 37 ascites tumor: Mouse	Amino acids	Cp, Up, I	Chirigos et al., 1962. CaR 22:1349
	Amino acids, aromatic	Cp, tA, Co-am.ac.	Guroff et al., 1964. Ref.
	Histidine	Sp, tI, tA, Co-Na$^+$, MM(double)	Matthews, 1972. BBA 282:374
Gardner lym-phosarcoma: Mouse	Galactose	Cp	Nirenberg and Hogg, 1958. JACS 80:4407
Hepatoma: Rat	Uridine, choline	Cp, MM, I	Plagemann and Roth, 1969. Bch 8:4782
	Hypoxanthine	I	Dybring, 1973. BPh 22:1981
	Glucose, analogs	MM, I, Cp	Renner et al., 1972. JBC 247:5765
		I	Mizel and Wilson, 1972. JBC 247:4102

TABLE I (*Continued*)

Organism and tissue	Permeant(s)	Type of evidence[a]	Citation[b]
	Glucose, glucosamine	MM, I	Estensen and Plagemann, 1972. PNAS 69:1430
Skeletal muscles			
Rat diaphragm	Glycine	Up, Cp, Sp, I	Christensen and Streicher, 1949. AB 23:96; Christensen *et al.*, 1949. AB 23:106
	α-Aminoiso-butyrate	H	Kostyo *et al.*, 1959. Sci 130:1653
		Up, Co-Na⁺	Parrish and Kipnis, 1964. JCI 43:1994
		Co-Na⁺, H	Bihler and Sawh, 1971. BBA 241:302
Rat levator ani		H	Metcalf and Gross, 1960. Sci 132:41
Rat diaphragm	L-Tyrosine	Sp	Guroff and Udenfriend, 1960. JBC 235:3518
	Amino acids	Up, H	Manchester, 1970. BcJ 117:457
Eviscerated dogs	Monosaccharides	H; Sp thereof	Goldstein *et al.*, 1953. AJP 173:207
Eviscerated rats			Park *et al.*, 1956. Ref.
Rat diaphragm		MM, Sp, Cp, I, H	Battaglia and Randle, 1959. Nat 184:1713; 1960. BcJ 75:408; Randle, 1961. MTM, p. 431
Rat soleus	Glucose	MM, H	Chaudry and Gould, 1969. BBA 177:527
Rat levator ani	D-Xylose	MM, I, Sp, H	Bergamini, 1969. BBA 193:193
Rat soleus	Glucose, 3-O-methylglucose	Cp, I, H	Kohn and Clausen, 1971. BBA 225:277
Assorted	Monosaccharides	Sat, Cp, I, Cf, H	Park *et al.*, 1961. RPHR 17:493; Morgan *et al.*, 1964. JBC 239:369; Henderson, 1964. CJB 42:933
Rat diaphragm	Monocarboxylic acid	Sat, Cp, I	Foulkes and Paine, 1961. JBC 236:1019
Smooth muscles			
Rat bladder	Monosaccharides	Sp, Sat, H	Bihler *et al.*, 1971. FPr 30:256
Rabbit uterus	α-Aminoiso-butyrate	H	Noall, 1960. BBA 40:180; Noall and Allen, 1961. JBC 236:2987

(*Continued*)

TABLE I (*Continued*)

Organism and tissue	Permeant(s)	Type of evidence[a]	Citation[b]
Rat uterus		Cp, H, I	Roskoski and Steiner, 1967. BBA 135:347,727
		Sat, Up, I, Cp, Co-Na+, H	Riggs *et al.*, 1968. BBA 150:92
	Hexoses	I, H	Spaziani and Gutman, 1965. End 76:470; Spaziani and Suddick, 1967. End 81:205
	3-*O*-Methylglucose	MM, Sp, Cp, Cf, H, I	Roskoski and Steiner, 1967. BBA 135:347,717
Cardiac muscle: Rat	Glucose	Sat, H	Bleehen and Fisher, 1954. JPh 123:260
		MM, H, Cp, Cf	Post *et al.*, 1961. JBC 236:269; Morgan *et al.*, 1961. MTM, p. 423, JBC 236:253,262
	Galactose	Sat, Cp, H	Fisher and Lindsay, 1956. JPh 131:526; Fisher, 1971. ABMG 26:975
	Pentoses	MM, I, HCp	Fisher and Zachariah, 1961. JPh 158:73; Bihler *et al.*, 1965. JPh 180:157,168; Fisher and Gilbert, 1970. JPh 210:287
	Monosaccharides	Sp, Cf	Morgan *et al.*, 1964. JBC 239:369
Cardiac cells: Rat	α-Aminoisobutyrate	Cp, I, Co-Na+	Dunand *et al.*, 1972. BBA 255:462
Cardiac muscle: Dog	Adenosine	Cp, I	Kübler and Bretschneider, 1964. PfA 280:141
Brain slices: Rabbit and guinea pig	K+	Up, Co-glutamate	Terner *et al.*, 1950. BcJ 47:139
Guinea-pig	Na+, K+	Up	Cummins and McIlwain, 1961. BcJ 79:330; Bachelard *et al.*, 1962. BcJ 84:225
Rat and mouse	Amino acids	MM, I, Up	Elliott and van Gelder, 1958. JNc 3:28; Neame, 1961. JNc 6:358; Smith, 1967. JNc 14:291
Rat, guinea-pig, rabbit, dog		I	Gonda and Quastel, 1962. BcJ 84:394; Nakamura and Nagayama, 1966. JNc 13:305

TABLE I (*Continued*)

Organism and tissue	Permeant(s)	Type of evidence[a]	Citation[b]
Rat		I, Sp, Cp, Up	Abadom and Scholefield, 1962. CJBP 40:1575,1591,1603; Neame, 1968. PBR 29:185;
Mouse			Battistin *et al.*, 1969. BrR 16:187
Guinea pig, rat		Co-Na⁺, I	Takagaki *et al.*, 1959. JNc 4:124; Yoshida *et al.*, 1963. Nat 198:191; Nakazawa and Quastel, 1968. CJB 46:355,363
Guinea pig, mouse		Sat, I, Cp, Sp, Co-Na⁺	Tsukada *et al.*, 1963. JNc 10:241; Lahiri and Lajtha, 1964. JNc 11:77
Rat		MM, I, Co-Na⁺	Iversen and Neal, 1968. JNc 15:1141
Mouse		Cp, Sp(multiple)	Blasberg and Lajtha, 1965. ABB 112:361
		Up, Co-Na⁺	Margolis and Lajtha, 1968. BBA 163:374
		I, Up, tA	Blasberg *et al.*, 1970. BBA 203:464
		Sat, Cp, Co-am.ac.	Levi *et al.*, 1966. ABB 114:339
Rat	L-Tyrosine	Sat, Cp	Chirigos *et al.*, 1960. JBC 235:2075
		Up, I, Cp, Co-hexoses	Guroff *et al.*, 1961. JBC 236:1773
	L-Histidine	I, Co-Na⁺	DeAlmeida *et al.*, 1965. BcJ 95:793
	L-Histidine, L-methionine	Up, Sp, I, Cp, tA, Cf-mutual	Nakamura, 1963. JBT 53:314
	L-Histidine, histamine	Cp, Sp, I	Neame, 1964. JNc 11:67,655
Guinea pig	L-Glutamate	Up, Sat, I, Cp, Co-K⁺	Tsukada *et al.*, 1963. JNc 10:241
Rat, guinea pig	γ-Aminobutyric acid	Up, Sat, I, Cp, MM, Co-Na⁺	Tsukada *et al.*, 1963. JNc 10:241; Iversen and Neal, 1968. JNc 5:1141; Iversen and Johnston, 1971. JNc 18:1939
Rat	L-Tryptophan	Up, Sp, Cp, MM, tA, Co-Na⁺	Barbosa *et al.*, 1970. CrSB 164:345

(*Continued*)

TABLE I (*Continued*)

Organism and tissue	Permeant(s)	Type of evidence[a]	Citation[b]
Rat and Dog	5-Hydroxy-tryptophan	Up, I, Cp	Schanberg and Giarman, 1960. BBA 41:556; Schanberg, 1963. JPET 139:191; Smith, 1963. BJPC 20:178
Cat and rat	Assorted biogenic amines	MM, I	Dengler *et al.*, 1961. Sci 133:1072; Abraham *et al.*, 1964. ABB 104:160; Creese and Taylor, 1967. JPET 157:406; Blackburn *et al.*, 1967. LSc 6:1653
Mouse	Cadaverine, DNP, creatine	Up, I, tA	Piccoli and Lajtha, 1971. BBA 225:356
	Choline	Up, MM, I, Co-Na$^+$	Schuberth *et al.*, 1966. JNc 13:347; 1967. LSc 6:293
	Acetylcholine	Up	Schuberth and Sundwall, 1967. JNc 14:807
	Thiamine	Sat, I, Co-Na$^+$	Sharma and Quastel, 1965. BcJ 94:790
Guinea pig	D-Xylose	Cp, MM, Up	Gilbert, 1965. Nat 205:87
	Monosaccharides	MM, Sp, Cp(*not* Up)	Joanny *et al.*, 1969. BcJ 112:367; Bachelard, 1971. JNc 18:213
Brain cortical cells: Rat	L-Glutamate, glycine	MM, tA	Laššánová and Brechtlova, 1971. PBo 20:235
Brain glial cells: Rabbit	γ-Aminobutyric ac., serotonin	Up, I, MM	Henn and Hamberger, 1971. PNAS 68:2686
Brain synapto-somes: Rat	Glycine, glutamate, aspartate	Up, Cp, partly Co-Na$^+$, MM(double)	Logan and Snyder, 1971. Nat 234:297; Bennett *et al.*, 1972. Sci 178:997
	L-Tryptophan	Up, Cp, I, tA, MM	Grahame-Smith and Parfitt, 1970. JNc 17:1339
	Amino acids	Up, Cp, MM, part Co-Na$^+$	Peterson and Raghupathy, 1972. JNc 19:1423
	γ-Aminobutyric acid	MM, Co-Na$^+$	Martin, 1973. JNc 21:345
Mouse		Up, I, Co-Na$^+$	Kuriyama *et al.*, 1969. BrR 16:479
Cat		Up, I, Co-Na$^+$, Cp	Snodgrass *et al.*, 1973. JNc 20:771
Rat	Norepinephrine, serotonin	Up, I, Co-Na$^+$	Bogdanski *et al.*, 1968. LSc 7,I:419
Blood/brain barrier: Rat	L-Tyrosine	Up, Sp, Cp	Chirigos *et al.*, 1960. JBC 235:2075; Guroff and

TABLE I (*Continued*)

Organism and tissue	Permeant(s)	Type of evidence[a]	Citation[b]
			Udenfriend, 1962. JBC 237:803
Mouse	L-Arginine	Cp, Sp thereof	Battistin and Lajtha, 1969. BSIBS 45:1532
Rat		Cp, Sat	Baños *et al.*, 1971. JPh 214:24P
Rat and mouse	Amino acids	Up, Sat, Cp, Sp(mult.)	Richter and Wainer, 1971. JNc 18:613; Battistin *et al.*, 1971. BrR 29:85; Oldendorf, 1972. AJP 221:1629; 1973. AJP 224:967
Rat	Amphetamine	MM, Cp	Pardridge and Connor, 1973. Exp 29:302
	Organic ac. (short)	Cp	Oldendorf, 1973. AJP 224:1450
Dog	Glucose	MM	Crone, 1960. APS 50, Suppl. 175:33; 1965. JPh 181:103
Rabbit		Sat, Cp	Bradbury and Davson, 1964. JPh 170:195
Cat		Sat, F, Cf	Cutler and Sipe, 1971. AJP 220:1182
Rat		MM	Bachelard *et al.*, 1973. PRSLB 183:71
Rat and mouse	Glucose, mannose	Sat, Cp	LeFevre and Peters, 1966. JNc 13:35
Rat	Hexoses	Sat, Cp, Cf	Bidder, 1967. JNc 15:867
	Glucose, 3-*O*-methylglucose	Sat, Cp, Cf	Buschiazzo *et al.*, 1970. AJP 219:1505
Blood/CSF barrier	Lysine, leucine	tA	Lajtha and Mela, 1961. JNc 7:210
Rat and mouse	Amino acids	Up	Lajtha and Toth, 1961. JNc 8:216
Rabbit and cat	SO_4^{2-}, $S_2O_3^{2-}$, I^-	Up, I, Sat, Cp	Robinson *et al.*, 1968. JNc 15:1169
Rabbit	Quaternary ammonium compounds	Up, I, Sat, Cp	Tochino and Schanker, 1965. AJP 208:666
Dog	Glucose	Sp, Sat, Cp, Cf	Fishman, 1964. AJP 206:836; Atkinson and Weiss, 1969. AJP 216:1120
Rabbit	Purines	Up, I, MM, Cp	Berlin, 1969. Sci 163:1194
Chorioid plexus: Dog	Hexoses	Up, I, Sat, Cp, Co-Na^+	Csáky and Rigor, 1964. LSc 3:431

(*Continued*)

TABLE I (*Continued*)

Organism and tissue	Permeant(s)	Type of evidence[a]	Citation[b]
CSF/blood barrier: Rat and Mouse	Amino acids	Up	Lajtha and Toth, 1961. JNc 8:216; 1962. JNc 9:199; 1963 JNc 10:909
Rabbit	I^-, SCN^-, p-amino-hippurate	Up, Sat, Cp	Pollay and Davson, 1963. Brn 86:137
Cat	SO_4^{2-}, I^-	Up, Sat, Cp	Cutler et al., 1968. AJP 214:448
	L-Lysine	MM, Co-am.ac.	Cutler, 1970. JNc 17:1017
	γ-Aminobutyric acid	MM, Up, Cp	Snodgrass and Lorenzo, 1973. JNc 20:761
CSF/brain barrier: Rabbit	Sugars	Cp	Bradbury and Davson, 1964. JPh 170:195
Cat	Cycloleucine	MM, Cp, Sp	Cutler and Lorenzo, 1968. Sci 161:1363
	Glucose, galactose	I, MM, Cp	Brøndsted, 1970. JPh 208:187, APS 79:523
Lenses: Rat	Glucose	I, MM, Cp	Patterson, 1965. IOp 4:667
	3-*O*-Methyl-glucose	I, MM, Cp	Elbrink and Bihler, 1972. BBA 282:337
Corneas: Rabbit	Na^+	Up	Donn et al., 1959. TNYAS II21:578
Kidneys, *in situ*: Dog	K^+	Up	Berliner and Kennedy, 1948. PSEBM 67:542; Giebisch et al., 1966. AJP 211:560
	Na^+	Up	Malnic et al., 1964. AJP 206:674; 1966. AJP 211:529,548
	Phosphate	Sat, Cp-glucose	Pitts and Alexander, 1944. AJP 142:648
	Phosphate, HCO_3^-	Cp	Malvin and Lotspeich, 1956. AJP 187:51
	Amino acids	Sp	Doty, 1941. PSEBM 46:129
		Sat, Cp(multiple)	Pitts, 1943. AJP 140:156,535; Kamen and Handler, 1951. AJP 164:654; Webber et al., 1961. AJP 200:380; Beyer et al., 1947. AJP 151:202
Rat		Cp, MM, Sp(multiple)	Wilson and Scriver, 1967. AJP 213:185

TABLE I *(Continued)*

Organism and tissue	Permeant(s)	Type of evidence[a]	Citation[b]
Human		Sat, Cp(multiple)	Scriver, 1968. JCI 47:823
	L-Lysine	MM(multiple)	Rosenberg et al., 1967. Sci 155:1426
Rabbit	Diodrast, p-amino-hippurate	Sat, Cp	Josephson et al., 1953. APS 30:11
Dog	Glucose	Sat	Shannon et al., 1941. AJP 133:752
Human	Ascorbic acid	Sat	Friedman et al., 1940. JCI 19:685
Dog		Cp	Selkurt, 1944. AJP 142:182
Kidneys, per-fused: Rat	Glucose	Up, Sat	Ruedas and Weiss, 1967. PfA 298:12
Renal tubules, perfused: Rat	Glucose	MM, I	Loeschke et al., 1969. PfA 305:118; Loeschke and Baumann, 1969. PfA 305:139
Rabbit		Sat, Up	Tune and Burg, 1971. AJP 221:580
Renal tubular fragments: Rabbit	Na⁺, K⁺	Up	Burg and Orloff, 1962. AJP 203:327; Burg et al., 1964. AJP 206:483
	Amino acids	Cp, I, Up, Co-Na⁺ MM(multiple)	Hillman et al., 1968. BBA 150:528; JBC 243:5566; JBC 244:4494
Kidney slices: Rabbit	Na⁺, K⁺	Up	Mudge, 1951. AJP 167:206
	p-Amino-hippurate	Up, I	Cross and Taggart, 1950. AJP 161:181
	Urate	Up, I	Platts and Mudge, 1961. AJP 200:387
		Up, Co-K⁺	Berndt and Beechwood, 1964. AJP 206:642
Rat	Amino acids	Cp, MM, I, Sp	Rosenberg et al., 1961. BBA 54:479; Scriver et al., 1964. JCI 43:374; Scriver and Wilson, 1964. Nat 202:92; Scriver and Goldman, 1966. JCI 45:1357
		Cp-sugars	Segal et al., 1962. BBA 65:567; Thier et al., 1964. BBA 93:106

(Continued)

TABLE I *(Continued)*

Organism and tissue	Permeant(s)	Type of evidence[a]	Citation[b]
	Dicarboxylic amino acids	Up, Cp, Sp, MM	Rosenberg et al., 1962. JBC 237:2265
	Imino acids, glycine	MM, I, Cp, Co-Na+, Co-imino acids	Mohyuddin and Scriver, 1968. BBRC 32:852; 1970. AJP 219:1
Rabbit	D-Galactose	Cp, I, Up, MM, Co-Na+	Krane and Crane, 1959. JBC 234:211; Kleinzeller and Kotyk, 1961. BBA 54:367
	Hexoses	Cp, MM, partly Co-Na+	Kolinska, 1970. BBA 219:200
Hamster		Up	Elsas and McDonell, 1972. BBA 255:948
Rat	myo-Inositol	Up, I, Co-Na+, Cp-glucose	Hauser, 1965. BBRC 19:696; 1969. BBA 173:267
Dog	Folic acid	Sat, Cp, F	Goresky et al., 1963. JCI 42:1841
Renal cortical cells: Rabbit	Hexoses	I, MM, Sp, Up, Co-Na+	Kotyk et al., 1965. BcZ 342:129; Kleinzeller et al., 1967. BcJ 104:843,852; 1970. BBA 211:293; Kleinzeller, 1970. BBA 211:264, 277
	α-Aminoisobutyrate	Cp, Co-Na+, MM(doub.)	Scriver and Mohyuddin, 1968. JBC 243:3207
Gut: Rat	Cl⁻	Up	Curran and Solomon, 1957. JGP 41:143; Curran and Schwartz, 1960. JGP 43:555; Edmonds and Marriott, 1967. JEn 39:517
Dog and human	HCO₃⁻	Up	Swallow and Code, 1967. AJP 212:717; Phillips and Summerskill, 1967. JLCM 70:686
Rabbit and rat	Na+	MM, Co-hexoses, Co-am.ac.	Esposito et al., 1964. Exp 20:122; Schultz and Zalusky, 1965. Nat 205:292; Barry et al., 1969. JPh 204:299; Curran et al., 1970. JGP 55:297
Rabbit	Na+, K+	Co-hexoses, am.ac.	Koopman and Schultz, 1969. BBA 173:338; Frizzell et al., 1973. JCI 52:215

TABLE I *(Continued)*

Organism and tissue	Permeant(s)	Type of evidence[a]	Citation[b]
Rat, rabbit, cat, hamster	Amino acids	Sp, Up, I	Gibson and Wiseman, 1951. BcJ 48:426; Agar et al., 1953. JPh 121:255; Matthews and Smith, 1954. JPh 126:96; Wiseman, 1955. JPh 127:414
Rat, rabbit, hamster		MM, Cp, Sp, I	Finch and Hird, 1960. BBA 43:278; Smyth, 1961. MTM, p. 488; Hagihira et al., 1962. AJP 203:637; Lassen et al., 1964. BBA 88:570; Matthews and Laster, 1965. AJP 208:601; Schaeffer et al., 1973. JGP 62:131
Rabbit		MM, Cp, I, Up, Co-Na$^+$	Schultz and Zalusky, 1963. BBA 71:503; 1965. Nat 205:292; Rosenberg et al., 1965. BBA 102:161; Schultz et al., 1967. JGP 50:1241; Curran et al., 1967. Ref.; Hajjar and Curran, 1970. JGP 56:673; Peterson et al., 1970. AJP 219:1027; Schultz et al., 1972. BBA 288:367
Rat		Cf	Munck, 1965. BBA 109:142
Rabbit		Co-am.ac.	Munck and Schultz, 1969. BBA 183:182
Rat, rabbit, hamster		MM, I, Cp-hexoses	Saunders and Isselbacher, 1965. BBA 102:397; Chez et al., 1966. Sci 153:1012; Alvarado, 1968. Nat 219:276
Rat, rabbit, hamster, mouse, guinea pig		MM, Cp-hexoses, Cf-hexoses	Alvarado, 1966. Sci 151:1010; Reiser and Christiansen, 1969. AJP 216:915; Alvarado et al., 1970. FEBSL 8:153; Robinson and Alvarado, 1971. PfA 326:48

(Continued)

TABLE I (*Continued*)

Organism and tissue	Permeant(s)	Type of evidence[a]	Citation[b]
Dog and rat		Cp-hexoses *and* Co-hexoses	Annegers, 1966. AJP 210:701; Bingham *et al.*, 1966. BBA 120:314; 130:281; Munck, 1968. BBA 150:82; 1972. BBA 266:639; Bihler and Sawh, 1973. CJPP 51:378
Rat		Sp, Cp(multiple)	Daniels *et al.*, 1969. BBA 173:575; 183:637
Hamster	L-Tryptophan	Up, Sp, Sat, Cp, Co-Na⁺	Cohen and Huang, 1964. AJP 206:647
Rat	Imino acids, glycine	Up, Cp, Sp	Munck, 1966. BBA 120:97
Monkey, hamster	N-substituted amino acids	Sp, Cp, MM	Hagihara *et al.*, 1962. AJP 203:637
Hamster	Cyclic amino acids	Sat, Cp, Sp	Tager and Christensen, 1971. JBC 246:7572
Rat	Glucose, galactose	Sat	Cori, 1925. PSEBM 22:497; Cori *et al.*, 1929. PSEBM 26:433; Verzár, 1935. BcZ 276:17
		Cp	Cori, 1926. PSEBM 23:290
		MM, Cp	Fisher and Parsons, 1953. JPh 119:210,224
Hamster			Crane, 1960. BBA 45:477; Jørgensen *et al.*, 1961. AJP 200:111
Human			Holdsworth and Dawson, 1964. ClS 27:371
Rat	Pentoses	Sp	Davidson and Garry, 1941. JPh 99:239
Hamster, rat	Monosaccharides	Sp, Up	Wilson and Crane, 1958. BBA 29:30; Wilson and Landau, 1960. AJP 198:99; Barnett *et al.*, 1968. BcJ 109:61; 1969. BcJ 114:569; 1970. BcJ 118:843
Assorted spp.		Sp, Sat, Cp, I, F	Crane, 1965. Ref.; 1960. PhyR 40:789
Rat, hamster		tA, Cf	Christensen and Gray, 1963. PSEBM 114:215; Crane, 1964. BBRC 17:481; Bihler *et al.*, 1969. CJPP 47:525

TABLE I (*Continued*)

Organism and tissue	Permeant(s)	Type of evidence[a]	Citation[b]
Assorted spp.		MM, I, Cp, Up, Co-Na+	Csáky and Zollicoffer, 1960. AJP 198:1056; Capraro *et al.*, 1963. Exp 19:347; Crane, 1965. Ref.; Bihler, 1969. BBA 183:169; Goldner *et al.*, 1969. JGP 53:362; Barnett *et al.*, 1970. BcJ 118:843
		Cp-am.ac., Cf-am.ac.	Hindmarsh *et al.*, 1966. JPh 186:166; Alvarado, 1966. Sci 151:1010; 1968. Nat 219:176; Robinson and Alvarado, 1971. PfA 326:48
Rat	D-Fructose	Cp, Up(slight)	Macrae and Neudoerffer, 1972. BBA 288:137
Hamster	L-Glucose	Up, I, Cp, Cf, Co-Na+	Caspary and Crane, 1968. BBA 163:395
	5-Thio-glucose	MM, Up, I, Cp, Cf	Critchley *et al.*, 1970. BBA 211:244
	D-Xylose, arbutin	MM, I, Cf, Co-Na+, Cp	Alvarado, 1966. BBA 112:292
	myo-Inositol	MM, I, Co-Na+	Caspary and Crane, 1970. BBA 203:308
Rat	Folic acid	Sat	Smith *et al.*, 1970. BBA 219:37
	Riboflavin	MM, Co-Na+	Rivier, 1973. Exp 29:756
Rodents, assorted	Biotin	Up, MM, I, Cp	Spencer and Brody, 1964. AJP 206:653
Hamster		Sat, Cp, Co-Na+	Berger *et al.*, 1972. BBA 255:873
Gut mucosal cells: Hamster	Galactose	Up	McDougal *et al.*, 1960. BBA 45:483
Rabbit	3-*O*-Methyl-glucose, L-tyrosine	Up, I, Cp	Huang, 1965. LSc 4:1201
Liver, *in situ*: Dog	D-Galactose	Sat, Sp	Goresky *et al.*, 1973. JCI 52:991
Liver, perfused: Rat	L-Glucose	Cp-D-glucose	Williams *et al.*, 1968. AJP 215:1200
Liver slices: Rat	Na+, K+(coupled)	Up, I	Flink *et al.*, 1950. AJP 163:598; Elshove and van Rossum, 1963. JPh 168:531

(*Continued*)

TABLE I (*Continued*)

Organism and tissue	Permeant(s)	Type of evidence[a]	Citation[b]
	Amino acids	H	Sanders and Riggs, 1963. FPr 22:417; 1964. FPr 23:535; Chambers *et al.*, 1965. MoP 1:66
	Amino acids, non-metabolizable	Up, I, Co-Na+	Tews and Harper, 1969. BBA 183:601
Gall bladder: Rabbit	Na+, Cl−	Up, Sat	Diamond, 1962. JPh 161:442, 474
Pancreas slices: Mouse	Amino acids	MM, Cp, Up, I, Co-Na+	Bégin and Scholefield, 1965. JBC 240:332; BBA 104:566
		tA	Clayman and Scholefield, 1969. BBA 173:277
Pancreas β-islet cells: Mouse	Glucose	Sp, MM, I	Hellman *et al.*, 1971. BBA 241:147
	L-leucine	Sp, Cp	Hellman *et al.*, 1972. BBA 266:436
Thyroid slices: Sheep	I−	MM, Cp, I, Co-Na+	Wolff, 1960. BBA 38:316; 1964. PhyR 44:45
	myo-Inositol	I, Co-Na+	Hauser, 1965. BBRC 19:696
Adrenal cortex	Ascorbic acid	I, Co-Na+	Sharma *et al.*, 1964. BcJ 92:564
Thymus nuclei: Calf	Amino acids	Sat, Cp, I, H, Co-Na+	Allfrey *et al.*, 1961. PNAS 47:907
	Thymidine, adenine, adenosine	Co-Na+	Allfrey *et al.*, 1961. PNAS 47:907
Fat cells: Rat	Glucose	H	Baker and Dutter, 1964. ABB 105:68
		Sat, H	Rodbell, 1964. JBC 239:375
		MM, Sp, Cp, Cf, I, H	Crofford and Renold, 1965. JBC 240:14,3237; Crofford, 1967. AJP 212:217
Mouse		H, Co-Na+	Letarte and Renold, 1969. BBA 183:366
Rat	3-*O*-Methyl-glucose	I, Cp	Blecher, 1966. BBRC 23:68
		H	Clausen, 1969. BBA 183:625
	Sorbitol	I, H	Crofford *et al.*, 1965. BBA 111:429

TABLE I (*Continued*)

Organism and tissue	Permeant(s)	Type of evidence[a]	Citation[b]
Mouse	α-Amino-isobutyrate	Sat, Up, H, Co-Na⁺	Touabi and Jeanrenaud, 1969. BBA 173:128
Bones: Rat	Amino acids	Up, Sat, Cp, Co-Na⁺	Finerman and Rosenberg, 1966. JBC 241:1487
	α-Amino-isobutyrate	Up, MM, Cp, part Co-Na⁺	Rosenbusch et al., 1967. BBA 135:732
Prostate and seminal vesicles: Rat	α-Amino-isobutyrate, 2-deoxyglucose	I, Cp, H, Co-Na⁺	Mills and Spaziani, 1968. BBA 150:435
Placenta: Sheep	Glucose	Sat	Widdas, 1952. Ref.
Human, macaque		MM	Chinard et al., 1956. JPh 132:289
Rabbit	L-Sorbose	Cp-glucose	Colbert et al., 1958. PSEBM 97:867
Guinea pig	Monosaccharides	Sp	Folkert et al., 1960. AJOG 80:221
Sheep			Karvonen and Leppänen, 1960. APF 6:53
Goat			Walker, 1960. BcJ 74:287
Guinea pig	Amino acids, non-metabolizable	Up	Feldman and Christensen, 1962. PSEBM 109:700
Yolk-sac: Rabbit	Amino acids	Up, Sat, Cp	Deren et al., 1966. DvB 13:370.

[a] Cf: Counterflow (inducer indicated if other than analog); Co: cotransport (co-agent indicated; am. ac. = amino acid); Cp: Competition (competitor indicated if other than analog; competitive basis for analog inhibition not established in every case); F: fluxes ≫ net transfer rates; H: hormonal stimulation (rarely, inhibition); I: inhibition by chemical inactivators; MM: Michaelis-Menten-like kinetics; Sat: saturation behavior not necessarily MM; Sp: specificity among allied substrates; often stereospecificity; tA: trans-acceleration, by whatever mechanism; tI: trans-inhibition; Up: uphill transport.

[b] Condensed literature abbreviation code is as follows: AB(B): *Arch. Biochem. (Biophys.);* ABL: *Ann. Bot. London, New Ser.;* ABMG: *Acta Biol. Med. Ger.;* AdE: *Advan. Enzymol.;* AePP: *Naunyn-Schmiedebergs Arch. exp. Pathol. Pharmakol.;* AIP(B): *Arch. Int. Physiol. (Biochim.);* AJBS: *Aust. J. Biol. Sci.;* AJOG: *Amer. J. Obstet. Gynecol.;* AJP: *Amer. J. Physiol.;* ANP: *Arch. Neer. Physiol.;* APF: *Ann. Paediat. Fenn.;* APS: *Acta Physiol. Scand.;* BBA: *Biochim. Biophys. Acta;* BBRC: *Biochem. Biophys. Res. Commun.;* Bch: *Biochemistry;* BcJ: *Biochem. J.;* BcZ: *Biochem. Z.;* BiB: *Biol. Bull.;* BJPC: *Brit. J. Pharmacol. Chemother.;* BPh: *Biochem. Pharmacol.;* Brn: *Brain;* BrR: *Brain Res.;* BSIBS: *Boll. Soc. Ital. Biol. Sper.;* CaR: *Cancer Res.;* CBP: *Comp. Biochem.*

(Continued)

of attack. Some success in this direction can be claimed for a number of the microbial systems; but with the vertebrate mechanisms, the lower affinities and lesser susceptibility to genetic or trophic manipulation have so limited technical progress that the several reported developments must still be classed as quite controversial. An obvious intrinsic difficulty in this approach is the destruction of the whole arena for transport phenomena in the course of dissolving the membranes, or otherwise separating the critical sites from the original structure. The best circumvention of this difficulty is probably the reconstitution of the functional system by replacement of the critical ingredient in artificial barriers, or in the membranes of cells that have lost the component or were never able to synthesize it properly by reason of a genetic defect. Lacking this, perhaps the best circumstantial evidence of involvement of the isolated materials in the carrier operation is demonstration of their selective binding of the passenger species *in vitro*, and the appropriate response in this process to transport-modifying factors. Such binding is most commonly indicated by way of equilibrium dialysis; a convenient summary of alternative procedures is given by Wyssbrod *et al.* (1971). Before considering this type of investigation, however, we shall first examine the evidence of survival

---

Footnote *b* (*Continued*):

Physiol.; CCh: *Clin. Chem.;* CFMT: "The Cellular Functions of Membrane Transport" (J. F. Hoffman, ed.), Prentice-Hall, Englewood Cliffs, New Jersey, 1964; Circ: *Circulation;* CJB(P): *Can. J. Biochem. (Physiol.);* CJPP: *Can. J. Physiol. Pharmacol.;* ClS: *Clin. Sci.;* CoP: *Colston Pap.;* CrAS: *Compt. rend. Acad. Sci.;* CrSB: *Compt. rend. Soc. Biol.;* CSHSQB: *Cold Spring Harbor Symp. Quant. Biol.;* DFS: *Discuss. Faraday Soc.;* DvB: *Develop. Biol.;* ECR: *Exp. Cell Res.;* EJB: *Eur. J. Biochem.;* End: *Endocrinology;* EPs: *Exp. Parasitol.;* Exp: *Experientia;* FEBSL: *FEBS Lett.;* FMb: *Folia Microbiol.;* FPr: *Fed. Proc.;* HPA: *Helv. Physiol. Acta;* IJB: *Ital. J. Biochem.;* IOp: *Invest. Ophthalmol.;* IRC: *Int. Rev. Cytol.;* JACS: *J. Amer. Chem. Soc.;* JBC: *J. Biol. Chem.;* JBt: *J. Bacteriol.;* JBT: *J. Biochem. (Tokyo);* JCCP: *J. Cell. Comp. Physiol.;* JCI: *J. Clin. Invest.;* JCS: *J. Chem. Soc.;* JEB: *J. Exp. Biol.;* JEn: *J. Endocrinol.;* JGM: *J. Gen. Microbiol.;* JGP: *J. Gen. Physiol.;* JLCM: *J. Lab. Clin. Med.;* JMB: *J. Membrane Biol.;* JMol: *J. Mol. Biol.;* JNc: *J. Neurochem.;* JPET: *J. Pharmacol. Exp. Ther.;* JPh: *J. Physiol. (London);* LSc: *Life Sci.;* MBBT: "The Molecular Basis of Biological Transport" (J. F. Woessner, Jr. and F. Huijing, eds.). Academic Press, New York, 1972; MoP: *Mol. Pharmacol.;* MTM: "Membrane Transport and Metabolism" (A. Kleinzeller and A. Kotyk, eds.). Academic Press, New York, 1961; Nat(NB): *Nature (New Biol.);* PBo: *Physiol. Bohemoslov.;* PBR: *Progr. Brain Res.;* PfA: *Pflügers Arch. ges. Physiol.;* PhaR: *Pharmacol. Rev.;* PhyR: *Physiol. Rev.;* PlP: *Plant Physiol.;* PNAS: *Proc. Nat. Acad. Sci. U.S.;* PRSLB: *Proc. Roy. Soc. London, Ser. B;* PSEBM: *Proc. Soc. Exp. Biol. Med.;* PTRSLB: *Phil. Trans. Roy. Soc. London, Ser. B;* Ref.: Full citation listed in *References* section; RPHR: *Recent Progr. Horm. Res.;* Sci: *Science;* SSEB: *Symp. Soc. Exp. Biol.;* TNYAS: *Trans. N. Y. Acad. Sci.;* Trf: *Transfusion;* ZPC: *Z. Physiol. Chem.*

of carrier phenomena in cell fractions retaining at least a partial structural barrier through which the substrate's movement can be detected.

1. Isolated Membranes, Membrane Particulates, and Vesicles

a. *Erythrocyte "Ghosts."* Lefevre (1961a) and Lacko (1966) reported that the human erythrocyte sugar-transport mechanism continued to operate at apparently full capacity in crude "ghosts," recovered after loss of the bulk of the cell contents by osmotic hemolysis. However, the extent of the transient binding entailed in this transport was so small that Lefevre (1961b) could not resolve it as a shift in the proportion of ^{14}C-labeled glucose held by the ghosts at low substrate levels. Residual carrier function even in virtually hemoglobin-free, extensively washed ghosts was observed by Jung (1971b) in the form of a selective permeability to D-glucose, and this further survived the removal of substantial protein by treatment with Pronase (Jung et al., 1973). In fact, Carter et al. (1973) have demonstrated retention, in vesicles prepared mechanically from trypsin-treated ghosts, of many aspects of the stereoselectivity and inhibitor-sensitivity of the intact cells' sugar-transport system.

By use of rather high concentrations of ammonium sulfate and rapid separation of ghosts on certain glass filters, Kahlenberg et al. (1971) were able to stabilize a stereoselective attachment of D-glucose to red cell membranes which retained no demonstrable transport function. In this case vesicular uptake did not seem to be involved, since the binding remained unaltered after sonication or repeated freeze-thawing of the preparations. In keeping with the intact-cell transport kinetics, the apparent affinity in this binding decreased at higher temperatures, without change in the total capacity (about 200,000 sugar molecules per original cell). Various inhibitors of the transport system also blocked the binding; and other monosaccharides inhibited in appropriate proportion to their ranking in the transport affinity sequence (Kahlenberg and Dolansky, 1972). Further study of this phenomenon has suggested participation of both phospholipid and protein components of the ghosts: Kahlenberg and Banjo (1972) report substantial reductions in the binding after treatment of the ghosts with phospholipases (particularly A_2), although similar treatment of the intact cells was essentially ineffective (Kahlenberg et al., 1972). Similarly, although proteases might remove as much as 27% of the membrane protein from intact cells without disturbing the sugar-transport function, a very modest digestion of the ghosts by trypsin or Pronase sufficed to augment markedly the $(NH_4)_2SO_4$-stabilized D-glucose binding, suggesting the exposure of latent adsorption sites. Binding was, however, depressed by more extreme proteolysis.

Vidaver and Shepherd (1968) found that the membranes of pigeon erythrocytes, restored after hemolysis, were still capable of even the active accumulation of glycine in response to applied sodium-ion gradients.

b. Adipose Cell Membranes. Rodbell (1967) reported that the glucose utilization in "ghosts" of rat adipose cells (retaining little cytoplasm and essentially no fat) continued to respond characteristically to insulin, lipolytic hormones and theophylline (metabolic responses generally accepted as involving modification of glucose transport); and Soderman *et al.* (1973) found that affinity binding by agarose-coupled insulin involved only the plasma membrane sacs, not the assorted cellular contaminants in crude fat-cell ghost preparations. Inhibition of transport of glucose or 2-deoxyglucose into these adipose cell ghosts by cytochalasin B has also recently been reported (Czech *et al.*, 1973).

Carter and Martin (1969) isolated "particles" from homogenates of these cells which were unable to oxidize or phosphorylate D-glucose, but which removed this sugar much more rapidly than L-glucose from an equimolar mixture of the enantiomers, and similarly favored the subsequent exodus of the D-form upon transfer to a sugar-free medium. This selectivity was abolished by phlorhizin; and a distinct homo-trans-acceleration was evident with D-glucose. Improved fractionation (Carter *et al.*, 1972) showed the plasma-membrane component to have the highest selectivity, which was further heightened and stabilized by encouraging vesicle formation through sonication. These vesicles exhibited competitive inhibition of glucose uptake, and uphill countertransport. Meanwhile, Illiano and Cuatrecasas (1971) had observed similar behavior in relatively intact ghosts, initial rates of D-glucose uptake (and the inhibition by 3-O-methylglucose) following Michaelis–Menten kinetics; responsiveness to insulin, however, was very irregular among several preparations.

c. Intestinal Brush Border. Although Csáky and Fernald (1961) demonstrated that the point of abrupt increase in sugar concentration in the process of absorption of 3-O-methylglucose-^{14}C by frog intestine was at the mucosal surface, they were unable to detect any binding of the sugar to sedimentable fractions containing 90% of the mucosal proteins. Nominally, the binding of glucose to rat-gut mucosal homogenates was examined by Fernie *et al.* (1967), but their data appear to deal only with the gross disappearance of the sugar from such mixtures. In the last few years, however, a number of investigations have indeed shown a preferential fixation of ^{14}C from labeled D-glucose at very low levels by various preparations of hamster-gut microvilli or brush-border fragments (recipes reviewed by Eichholz, 1969). However, it is not clear whether this "binding" is a true adsorption rather than a retained sugar metabolite; it is evident only at *micro*molar substrate levels, while the intestinal sugar transport itself saturates in the *milli*molar range. Moreover, Torres-Pinedo *et al.* (1972) have presented arguments for dismissing the "binding" as a sign of bacterial contamination. Faust *et al.* (1967, 1968) showed the process to be inhibited by phlorhizin, N-ethylmaleimide, Hg^{2+}, or p-hydroxy-mercuribenzoate, and (presumably competitively) by various monosaccharides or other glucose analogs, and thus suggested some involvement with the transport process. However, in extending the latter observations, Eichholz *et al.* (1969) noted that the apparent affinity sequence in this competition did not align at all with that of the absorptive system; and Parsons (1969) pointed out a parallel instead with the specificity of the intestinal hexokinase. Moreover, the behavior is less evident in relatively intact brush-border preparations than in those subjected to limited disruptive treatments; it appears to be localized in the "inner core" of the microvilli rather than at the membrane proper, and upon further fractionation in a Ficoll density gradient (Faust *et al.*, 1972), was found in a filamentous core fraction detached from the enzymic activities generally taken as membrane-bound. With the exception of this last type of preparation, the "binding" has also proved to be indifferent to Na^+ or ouabain and is decidedly more pronounced at the caudal end of the gut (Eichholz *et al.*, 1969), where sugar absorption is relatively poor. These several considerations led Eichholz to doubt the critical involvement of the phenomenon in the intestinal absorption process. In any case, it is hard to see how this very tight fixation of substrate within the cell would serve to generate the essential *uphill* transcellular delivery; but Faust *et al.* (1972) chose to view it as a "conduit" for access (mechanism unspecified) to some "energy-requiring process which would result in intracellular accumulation."

Burns and Faust (1969) and Faust *et al.* (1970) have claimed similar binding of the actively absorbed amino acids L-alanine, L-histidine, and L-proline (vs aspartate or glutamate as reference). The binding apparently had an absolute requirement for Na^+, and was maximal only if the $[Na^+]$ attained about 30 mM. The alanine fixation uniquely showed little evidence of saturation at 1 mM; and in mixtures of two of the substrates, there was a peculiar pattern of both inhibitory and cooperative interplay that is difficult to account for even if a multiplicity of types of sites be presumed.

More recently, Hopfer *et al.* (1973) have obtained a vesicular preparation from rat-gut brush-border fragments (separated from the core material) which, except for its specific Na^+ dependence, behaves essentially like Carter and Martin's (1969) preparation from adipose cells: it presents a phlorhizin-inhibitable, bidirectional facilitated-diffusion system for D-glucose, not shared with L-glucose, exhibiting saturation kinetics, competitive inhibition (by galactose in excess), and both homo- and hetero-trans-acceleration. The recovered permeant was identified as unaltered glucose, and its equilibrium distribution responded appropriately to osmotic manipulation of the medium by addition of (impermeant) celloboise, suggesting that any binding contributed only insignificantly to the measured uptake.

d. Renal Brush Border. A similar variety of preparations from the kidney tubular epithelial surface have also shown suggestive interactions with transport substrates. Binkley *et al.* (1968) separated "nephrosome" vesicles from rat renal homogenates (retaining many biochemical indices of membranal origin), which accumulated leucine even in the absence of soluble homogenate fractions. The process was inhibited in the presence of valine or isoleucine. Also, Hillman and Rosenberg (1970) described sodium-dependent binding of proline by brush-border preparations from rabbit proximal tubules, saturating at ca. 75 mM, and blocked by glycine or alanine.

Much attention has been accorded the *two-component* binding of the classical sugar-transport inhibitor, phlorhizin, by rat-kidney brush-border particles, as first described by Bode *et al.* (1970). The transport inhibition by this glucoside was associated with only about 1% of the total binding capacity, but the affinity of this small component was about 100 times that of the rest of the sites. Inactivation by N-ethylmaleimide was much more evident at the high-affinity sites, and p-chloromercuribenzoate interfered at both components. Frasch *et al.* (1970) showed a systematically competitive displacement by D-glucose at the high-affinity sites only (with an apparent K_i about 40 times higher than the transport K); but L-glucose appeared, surprisingly, to *enhance* the affinity for phlorhizin. Lowering of the [Na$^+$] reduced both the phlorhizin affinity and the apparent number of sites involved. Glossmann and Neville (1972) considerably extended the study of competitive displacement of phlorhizin in this system by various monosaccharides and derivatives, and (in spite of rather marked quantitative dissonance with the other group's binding figures) confirmed the basic phenomena and established a satisfactory parallel with the specificity of the renal reabsorptive transport. Bode *et al.* (1972) have also examined the structural correlations of the displacement of phlorhizin in this system by other glycosides. Thomas and Kinne (1972) describe further separation of the phlorhizin-binding sites from some of the brush-border enzymic activities, through the use of papain and deoxycholate. Thomas *et al.* (1972) have also shown that either phlorhizin or glucose offers partial protection of these sites from reaction with N-ethylmaleimide, and they suggested on the basis of the quantitation of this interaction that each phlorhizin molecule may tie up three potential glucose- or NEM-fixing sites. Similarly bi-affine binding of glucose itself was reported in rabbit renal brush-border fragments by Chesney (1971), and this was inhibited by phlorhizin. However, the glucose transport K falls squarely between (in terms of orders of magnitude) the two ranges of observed binding K; and the study of phlorhizin competition (Chesney et al., 1973) dealt with the sites of higher glucose affinity where the phlorhizin K_i vastly exceeds the glucose-binding K. Although there is no apparent requirement for Na+ in this binding, the [Ca²⁺] is fairly critical. Busse et al. (1972) interpreted similar observations on a bi-affine fixation of glucose by renal tubule membranes in terms of a *vesicular transport*, rather than a simple binding.

e. Other Membranes. Varon et al. (1964) and Weinstein et al. (1965) have interpreted their observations on the turnover of ¹⁴C-labeled γ-amino-butyric acid in suspensions of mouse-brain particulates as indicating a carrier-mediated transfer between medium and a rapidly exchanging pool, and in turn a slowly exchanging pool. Elevation of [Na+] enhanced these turnovers.

Kaback and Stadtman (1966) used osmotic shock and EDTA treatment of spheroplasts of *Escherichia coli* to prepare membranes which (so long as glucose was supplied) retained at least half of the proline-accumulating activity of the original spheroplasts. The accumulated amino acid was recoverable largely as unmodified proline. Trans-stimulation of its exit was induced by external proline or hydroxyproline, but not by ten other amino acids. Moreover, various metabolic inhibitors depressed this accumulation in the bacterial "ghosts." All these characteristics were essentially missing from similar ghosts prepared from a mutant deficient in respect to proline transport. Comparable findings have been reported for other amino acid transport systems in these cells (Kaback and Stadtman, 1968). Kaback (1972) provides an excellent general review of the study of the survival of this type of activity in bacterial membrane vesicles.

Ca²⁺-binding to vesicular fractions of sarcotubular membranes isolated from rat muscle was found to be rather selectively abolished by treatment with phospholipase C (Yu et al., 1968), though extensive digestion with phospholipases A or D did not significantly disturb the binding.

2. EXTRACTED PHOSPHOLIPIDS

The entrainment of sugars, amino acids, salts, and other distinctly hydrophilic materials into tissue lipid extracts by reason of association with amphiphilic phosphatides has been recognized for many years (Bing, 1899, 1901; Christensen, 1939; Folch and Van Slyke, 1939; Rouhi et al.,

1952, 1953, 1954; Silberman and Gaby, 1961; and many others). Moreover, it has often been demonstrated that the occurrence of this sort of association at a phase interface can effect the transfer of water-preferring solutes into various fat solvents, or *through* a fat-solvent phase into a second aqueous compartment (Freundlich and Gann, 1915; Cruickshank, 1920; Woolley, 1958; Hendler, 1959; Hoffman *et al.*, 1959; Woolley and Campbell, 1960, 1962; Vilkki, 1962; Ward and Fantl, 1963; Feinstein, 1964; Blaustein and Goldman, 1966). Since this grossly mimics hypothetical carrier behavior, there has been considerable discussion of a possible role for such complexes in the mediation of membrane transports, particularly of salts (Christensen and Hastings, 1940; Solomon *et al.*, 1956; Vogt, 1957; Kirschner, 1957, 1958; Vanatta, 1963, 1966; Schneider and Wolff, 1965); and Hokin and Hokin (1959, 1960a,b, 1961) present evidence that active Na^+ transport entails an actual cycle of synthesis and breakdown of phosphatidic acid serving as the Na^+ carrier. This proposition is seriously questioned, however, by the failure of label from ^{32}P-labeled ATP to appear appropriately in the lipid components during active transport of Na^+ and K^+ (Hokin and Hokin, 1964; Glynn *et al.*, 1965). The membrane Na^+, K^+-ATPase, now generally taken to be central to the "sodium pump" mechanism, is known to be a lipoprotein (Schatzmann, 1962) and is reversibly inactivated by removal and replacement of phospholipid components (Hegyvary and Post, 1969; Israel, 1969).

The extent to which such association of monosaccharides with phospholipids (mainly from human red blood cell membranes) might reflect the characteristics of the facilitated diffusion of sugars in these cells was studied quantitatively for some years in the author's laboratory (LeFevre *et al.*, 1964, 1968; LeFevre, 1967; Jung *et al.*, 1968). The complexing *in vitro* did show a number of suggestive parallels with the biological transport, and even served to mediate transfer of sugars through a chloroform layer (thus crudely mimicking a carrier operation); but the process failed altogether to duplicate the transport stereoselectivity reported for the intact cells (LeFevre and Marshall, 1958) or even their extensively washed ghosts (Jung *et al.*, 1971). Demonstrable differences in the rate of transfer of various sugars in the inanimate three-phase system were relatively small, and bore no relation to the sequence in the biological system (LeFevre *et al.*, 1968). A similar situation was seen in bimolecular "black membranes" prepared from erythrocyte total-lipid extracts or from their major phospholipid fractions (Jung, 1971a). Also, in total-lipid extracts of mouse intestine, Baker (1967) noted an utter lack of correlation between the sugarcomplexing selectivity and the pattern of the active intestinal sugar absorption. However, in similar complexing with rat-gut phospholipids, Reiser and Christiansen (1968) found distinct mutual inhibitions, of apparently competitive nature, between sugars and amino acids, somewhat reminiscent of the patterns in the gut transport.

Moore and Schlowsky (1969a) reported a peculiarity in the influence of red cell phospholipids on the migration of glucose and galactose from an aqueous phase into n-butanol, which they improperly interpreted as an explicit parallel with the carrier behavior: at low sugar levels, glucose traversed the phospholipid-laden interface faster than galactose, while the order was reversed when the sugars were at much higher concentrations. The superficially similar situation in the net movement of the two sugars in intact cells (Wilbrandt, 1956) is in precise accord with carrier theory (given the sugars' relative affinities), because of the differing saturations in the back-flux. However, the essential framework for expression of this kinetic peculiarity is missing in the two-phase situation studied by Moore and Schlowsky, as is evident in the very equation they apply in analyzing their data. In further study of the phenomenon with an assortment of individual phospholipids, Moore and Schlowsky (1969b) found that a major factor in the differential behavior of the two hexoses was the far higher butanol-to-water distribution ratio for galactose, when equilibrated in the presence of nearly all the tested lipids.

Cation transfer through lipid phases has also been accorded considerable attention. Woolley and Campbell (1960) found that the "serotonin receptor" which they extracted from hog duodenum, when supplied with serotonin, allowed the transfer of Ca^{2+} through a benzene–butanol phase. Papahadjopoulos (1971) has reported discrimination between Na^+ and K^+ in the permeability behavior of some highly purified artificial phospholipid bilayers. However, in general such membranes (whether examined as suspended "liposome" microvesicles, as macroscopic liquid-inflated bubbles, or as planar barriers across small apertures) offer much greater resistance to flow of water-soluble materials than do biological membranes, unless some protein or macrocyclic agent is added (review by Bangham, 1970). At the present time, there is little active attention to the phospholipids as possible mediators of carrier processes, and it appears unlikely that they will prove to be primary components of the essential reactions. On the other hand, it is evident that in at least some systems their *presence* is essential for normal function of the (presumably protein) critical components.

3. Solubilized Proteins

a. In Ionic Transfer. Soon after lipid or proteolipid bilayer membranes were first prepared, it was noted that their high electrical resistance could be brought much closer to that of a typical cell membrane by inclusion of small amounts of certain proteins; and that, in the presence of an ionic gradient, this might even give rise to a membrane potential which responded to applied current pulses in a manner reminiscent of some excitable cell membranes (Mueller et al., 1962; Mueller and Rudin, 1967b; Kushnir, 1968). A variety of proteins that associate readily with phospholipid vesicles by a combination of electrostatic and hydrophobic interactions profoundly reduced the vesicles' resistance to cation diffusion (Kimelberg and Papahadjopoulos, 1971a,b; Calissano and Bangham, 1971). Even nonelectrolytes like glucose were observed to leak from some types of liposomal vesicles much more rapidly when traces of particular proteins were added (Sweet and Zull, 1969, 1970; Gould and London, 1972). Although the specific protein used is critical in eliciting these phenomena in each case, it has not in general been circumstantially associated with any relevant transport system. However, cationophoric behavior in various preparations of $(Na^+ + K^+)$-activated membrane adenosinetriphosphatases has been of special concern, because of the striking parallels (brought out by Post et al., 1960; Post and Albright, 1961; Skou, 1961) between the behavior of this type of enzyme and that of the processes of Na^+ extrusion and K^+ accumulation that underlie the major cationic asymmetry across most animal cell membranes and appear to be critical to bioelectrical phenomena and to many specialized salt and nutrilite transports (reviews by Judah and Ahmed, 1964; Skou, 1965, 1969, 1973; Albers, 1967; Bonting, 1970; Hokin and Dahl, 1972). Preparations of this enzyme from rat brain were found to reduce appropriately the resistance of black membrane bilayers in the presence of adenosine triphosphate (Jain et al., 1969); however, the same can be said for a Na^+, K^+-independent ATPase (Redwood et al., 1969). Recently, two enzymically inactive polypeptides derived from the membrane $(Na^+ + K^+)$-ATPase solubilized from Electrophorus electric organs or from beef kidneys have been found to interact to increase the cationic conductance of bilayers prepared from oxidized cholesterol or 7-dehydrocholesterol (though not that of various bilayers of phospholipid) (Shamoo and Albers, 1973; Shamoo, 1974). In some circumstances, Na^+ is evidently an absolutely specific requirement for incorporation of this material into the membrane as an ionophore, but no distinct selectivity among cations is evident once the ionophore is established.

Vitamin D-induced Ca^{2+}-binding proteins found in association with calcium absorption and secretion processes have been isolated from gut tissue of many types of animals and from the kidneys and uterus (shell gland) of chickens (Wasserman et al., 1969; Ingersoll and Wasserman, 1971), rat-liver mitochondria (Lehninger, 1971), pig brains (Wolff and Siegel, 1972), and elsewhere. The review by Wasserman et al. (1969)

presents the arguments bearing on the question of the function of these binders as membrane carriers, intracellular carriers, or intracellular storage depots.

b. In Erythrocyte Sugar Transfer. A series of reports from Manchester (Bobinski and Stein, 1966; Bonsall and Hunt, 1966; Stein, 1967b, 1968; Levine and Stein, 1967) made it appear for a time that a protein obtained from human erythrocyte ghosts by NaI- or Triton X-100-extraction retained *in vitro* a capacity for preferential binding of D-glucose (from a mixture with the poorly transported isomers, L-glucose or L-sorbose). This was revealed by three techniques: retardation chromatography, equilibrium dialysis, and ultrafiltration; detectable excesses of D-glucose appeared to be retained with the protein material, and this retention was appropriately inhibited by transport-blocking agents. However, LeFevre and Masiak (1970) found that the retardation of D-glucose on chromatographic columns evidently reflected not binding per se, but a vastly larger entrapment into membrane-enclosed vesicles. When these residual transporting structures were destroyed, the (presumed) residual binding was below the limits of detection. Further, Masiak and LeFevre (1972) were altogether unable to corroborate the reported selective retention of D-glucose in equilibrium dialyses with proteins from erythrocyte ghosts. Møller (1971) also failed to detect any binding selectivity for D-glucose in this system, and concluded on this basis that some sort of hydrophilic pathway must be presumed in lieu of a mobile-carrier apparatus. He calculates from his negative data that the number of transport sites must be so small that their turnover rate would have to exceed considerably any reasonable maximum for the cycling of a carrier.

Several attempts have been made to tag the red-cell transport active centers selectively with irreversible inactivating agents. Pursuing circumstantial evidence that N-terminal histidine might be critical in the recognition site of the facilitated-diffusion mechanism for glycerol, Stein (1958) found that the competitive analog 1,3-dihydroxypropane prevented reaction of phenylisothiocyanate with stroma to produce the histidine derivative, but did not interfere with the production of other derivatives; moreover, after the glycol was washed off, a second treatment of the stroma yielded the N-terminal histidine product. A second approach also developed by Stein (1964) took advantage of the *second-order kinetics* of inactivation of the glucose-transport apparatus in these cells by fluorodinitrobenzene (FDNB); from a mixture of cells equivalently reacted with ³H-labeled FDNB and ¹⁴C-labeled FDNB at two widely varying concentrations, Stein examined a papain digest of the membranes for differential distribution of the two labels. The appropriate differential did appear, but unfortunately only diffusely throughout the residues of the undigested "core" material. More recently, Eady and Widdas (1973) have reported some indication of a differential labeling by FDNB of electrophoretically separated ghost proteins, in parallel with the substrate-induced acceleration of sugar-transport inhibition by this agent, but as yet the resolution of this phenomenon is unsatisfactory. Even more uncertain is the reality of the selectivity of the tagging of two red-cell membrane protein components with the potential affinity label, D-glucosyl isothiocyanate, recently described by Taverna and Langdon (1973).

   c. *In Bacterial Nutrilite Uptake.* A substantial number of proteins and lipoproteins derived from microbial membranes (or adjacent structures) have in recent years been strongly implicated in specific cellular transport mechanisms (reviews by Pardee, 1968; Kaback, 1970; Chavin, 1971; Oxender, 1972). The bulk of this work has rested on the relative ease with which appropriate bacterial mutant forms can be identified, isolated, and cultured, and specific protein syntheses often controlled by manipulation of culture conditions. These technical features have been invaluable in allowing experimental correlation of transport functions with the appearance of specific membrane protein components, leading in some instances to functional reconstitution in transport-defective mutants. The bacterial studies will be only briefly discussed here, since most have been detailed in a recent chapter in this series by Boos (1974).

   The isotopic double-labeling approach employed in the erythrocyte studies noted above has been useful also in the search for transport-related proteins in microbial systems. Batches of cells or cell membranes, differing appropriately in respect to their genetic makeup or inductive state, are treated with two isotopic forms of the agent, then mixed for extraction and fractionation of the labeled proteins. Kolber and Stein (1966) applied this technique to the permease expression of the *lac* operon in *Escherichia coli*: cells grown with phenylalanine-^{14}C in the presence of an inducer (β-thiomethylgalactoside) were mixed with cells grown without inducer in phenylalanine-^{3}H, and a soluble extract was prepared from the mixed spheroplast membranes. Upon fractionation of this extract with a linear saline gradient through a DEAE-cellulose column, relatively high ^{14}C:^{3}H ratios were then observed in three protein components. Two of these were identified with the two major enzymic activities controlled by the *lac* operon, leaving the third as presumably representing the permease; this band was appropriately missing from mutants lacking the permease function. Another component of this same system, designated the "M protein," was isolated by Fox and Kennedy (1965) by similar double-labeling (induced vs uninduced cells) with ^{14}C- and ^{3}H-labeled N-ethylmaleimide. The specificity of the labeling was enhanced by pretreatment with ordinary N-ethylmaleimide in the presence of the site-protecting agent, thiodigalactoside. The M-protein, still retaining a capacity for binding the transport substrates (β-galactosides), was solubilized from the membrane particulate fraction with Triton X-100 or by sonication, and physicochemically characterized by Jones and Kennedy (1969). It appears to make up about 0.35% of the total cell protein. The genetic control of its production, in correlation with that of the transport function, has been extensively examined by Fox et al. (1967). Haškovec and Kotyk (1969) have similarly used a double-labeling procedure to identify a carrier-associated component in the inducible galactose-transport system in baker's yeast.

Anraku (1968) has described an additional component isolated from bacterial membranes which, though not itself a substrate-binder, must be restored to the depleted cells in order for the binding protein itself to become fully functional in transport. Wong and MacLennan (1973) report the reactivation of an uphill lactose-transport system in a mutant form of *Escherichia coli* (which otherwise retained only the facilitated-diffusion system) by adding *fatty acids*, reducing the rate of lactose exit. The D-lactate-dependent transport of several amino acids and other substrates in *membrane vesicles* from *E. coli* mutants lacking the dehydrogenase has also recently been achieved (Reeves et al., 1973) by addition of the enzyme extracted from wild-type vesicles.

d. In Other Nutrilite Transfers. Thus far for the nonbacterial systems, the genetic element is missing from the evidence for involvement of membrane proteins in transport, and the arguments turn largely on the circumstances of appropriate location and binding characteristics (Pardee, 1970). In some instances, however, reconstitution in artificial membranes or in deficient natural membranes has been achieved. The papain solubilization of the sucrase–isomaltase complex from rabbit intestine (Cogoli et al., 1973) has allowed demonstration of specific reconstitution in artificial lipid bilayers of a translocation mechanism previously postulated for the intact gut: Storelli et al. (1972) and Vögeli et al. (1972) found that the transfer of the label from sucrose-^{14}C through a variety of lipid bilayers was increased over a thousandfold by incorporation of this enzyme complex (though not by its addition to preformed membranes). This arrangement delivered the enzymic reaction products glucose and fructose into the trans-compartment, without altering the leakage of these two hexoses as such, and without change in the membranes' permeabilities to water. Accordingly, it was suggested that the enzyme complex itself spans the membrane, contacting both aqueous interfaces, rather than opening up some sort of specific polar channel.

The glucose-inhibitable phlorhizin-binding sites in renal brush border, discussed in Section I, E, 1, d, have been harder to release from the microvillus matrix. However, by following papain digestion with deoxycholate treatment and gel electrophoresis, Thomas (1972) has separated the phlorhizin-binding component from the primary membrane enzymic activities. Thomas (1973) has also apparently obtained some degree of differential labeling of the glucose-binding sites by N-ethylmaleimide-^{14}C, by way of preliminary protection of these sites with phlorhizin, and has used this labeling to identify the sugar binding tentatively with a particular electrophoretic protein band. Similarly, a deoxycholate-soluble protein from rat skeletal muscle which also shows phlorhizin-inhibitable binding of D-glucose at micromolar levels has been presumed to be a transport protein (Brush and Krawczyk, 1969).

.Amino acid-binding proteins released from membranes by osmotic shock or solubilized by detergents have been reported in *Neurospora* conidia (Stuart and DeBusk, 1971; Wiley, 1970), baker's yeast (Voříšek, 1972, 1973), and rat brain mitochondria (Strasberg and Elliott, 1967). Horák and Kotyk (1973) have recently isolated from baker's yeast plasma membranes a lipoprotein which selectively binds the substrates of the intact yeast's monosaccharide-transport system, but these preparations are still somewhat crude and variable in performance.

4. MACROCYCLIC COMPOUNDS

Mueller and Rudin (1967a) and Andreoli *et al.* (1967) found that artificial lipid bilayers became highly cation-selective, and in particular K$^+$-selective (with a large increase in electrical conductance), upon addition of tiny quantities of such antibiotics as valinomycin, actins, enniatins, and gramicidins, consisting of ring formations of amino acids and carboxylic acids with inward-facing polar groups. In some instances, these additives appear to be acting as actual *carriers* in the strict sense, solubilizing the ions in the lipid phase by enclosing them (Tosteson, 1968; Eisenman *et al.*, 1968); while in other cases (Mueller and Rudin, 1968; Holz and Finkelstein, 1970) an indiscriminate opening up of microapertures in the bilayer has been indicated. Detailed consideration of the likely structure–functional relations are given in reviews by Pressman (1968), Pressman and Haynes (1969), and Shemyakin *et al.* (1969); a convenient condensed summary is that of Haydon (1970).

II. SPECIFIC PROBLEMS IN FORMAL APPLICATION OF CARRIER MODELS

A. Basis for Emergence of the "Classical" Model

To a very great extent, the history of carrier "model-building" has been dictated by the sequence of experimental revelations regarding the behavior of one particular system: monosaccharide transfer into and out of human red blood cells. This is largely because of technical considerations relating to the reproducibility of experimental manipulations, the discreteness and relative uniformity of the cells, and the relative ease of recording translocation events in this system. Thus the classical Widdas (1952, 1954) model for a mobile-carrier operation, though originally designed to fit observations on placental sugar transfer, became established primarily through its kinetic accord with quantitative data on the erythrocyte system. In particular, it was the demonstration of the induction of uphill countertransport of labeled glucose by unlabeled glucose or by mannose (Rosenberg and Wilbrandt, 1957) that made it necessary to reject simpler alternative models (LeFevre, 1948; LeFevre and LeFevre, 1952) which had made no provision for any intervening step separating the reacting sites on the two sides of the membrane.

As noted in the legend to Fig. 1, Widdas's schema allows the full description of the behavior of a given passenger species at any set of concentrations by stating (for a specified temperature and medium) only *two* constants (a K, or d/a as in Fig. 1, and a maximal rate determined by DC_T), rather analogous to the K_m and V_{max} of Michaelis–Menten enzyme kinetics. Although the model does not necessarily demand it, reassurance in applying this schema was lent by the finding (LeFevre, 1962) that the maximal rate does not differ very substantially among several sugars having affinities (Ks) spanning several orders of magnitude. However, the indications by Britton (1956) and by Lacko et al. (1961) that trans-stimulation effects might occur suggested that the classical model was still significantly oversimplified. Further systematic examination of these phenomena independently by Levine et al. (1965) and by Mawe and Hempling (1965) made it clear that the rate constant for the transmembranal migration of the passenger–carrier complex in this system ($D_{\overline{SC}}$) must be about four times that of the unoccupied carrier (D_C). A very similar situation has been described for transport of leucine in these same cells (Hoare, 1972a) and of a variety of monosaccharides in baker's yeast (Kotyk, 1967). It is thus necessary to replace the one V_{max} by two such terms in defining each substrate's behavior; generally, this is done by retaining the classical K and V_{max} and adding as a third constant the ratio r, which is $D_{\overline{SC}}/D_C$.

With this single complication, the Widdas model is found to accord very well with any *one* of the several classes of carrier behavior described in Section I, C, except for "cotransport," which does not seem to occur in this particular system. However, with increasing application of this accepted framework for quantification of the transport parameters for the substrate, D-glucose, in this system, it has become increasingly evident that the "constants" derived from one set of procedures often conflict seriously with those given by another, equally legitimate, approach. These conflicts will now be examined in some detail.

B. Inconsistencies in Application of Classical Model

1. THE ESTIMATION OF CONVENTIONAL TRANSPORT PARAMETERS

When permeation through cell membranes is studied in very dilute cell suspensions, the external concentration of the permeants remains virtually constant; thus, barring marked trans-effects, the unidirectional influx holds practically steady while the efflux either rises or falls toward the (equilibrium) influx value. Accordingly, in examining net transfer by carrier mediation in such suspensions, it is advantageous to record net *exodus* rather than cellular uptake, and to employ internal concentrations that nearly saturate the transfer mechanism. Under these circumstances, the efflux will also remain fairly constant over most of the equilibration period, so that the net exit will be essentially linear, greatly facilitating estimations of the rate. This is the basis of what has become the standard procedure (Sen and Widdas, 1960, 1962) for calculation of the transport parameters in the erythrocyte–monosaccharide system; the arrangement is diagrammed in the upper left panel of Fig. 2. Cells previously equilibrated with a high (presumably saturating) "infinite-cis" concentration of the sugar are transferred to a medium containing various lesser levels of the substrate, and the initial rates of net exit are measured. The systematic slowing of this exit as the trans-concentration is increased then provides a measure of the saturation characteristics of the *influx*. Specifically, in a plot of the reciprocal of net exit rate vs the external passenger concentration, the classical model would predict linearity such that the x intercept $= -K$, and the y intercept $= V_{max}^{-1}$. Sen and Widdas made such estimates, finding a marked diminution in both parameters (for D-glucose) as the temperature was reduced over the range 38°–ca. 3°C.

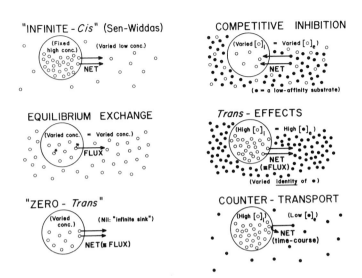

FIG. 2. Principal experimental arrangements revealing failings in "classical" model. Intracellular compartment (in human erythrocyte suspensions) is represented by a large circular enclosure in each panel; outside medium is generally so large that concentrations of (monosaccharide) permeants remain virtually invariant in a given run. The three left-hand panels deal with experiments on single chemical species (generally D-glucose), isotopically labeled where indicated in middle panel by asterisks. Arrows indicate the measured sugar translocation or flux in each case, with initial concentrations as diagrammed. The three right-hand panels deal with situations involving two passenger species, indicated by black balls and white balls; for two lower panels, however, these may simply be isotopically distinguishable forms of a given sugar. Subscripts i and e on the concentration terms connote intracellular and extracellular, respectively. The bent arrow in the lower-right panel indicates tracing of time course of *transient* (entry followed by exit) of the initially extracellular species. Further explanation accompanies the text discussion of the several types of experiment.

However, Levine and Stein (1966) were unable to bring these measurements into harmony with the (conceptually simpler) analysis of their findings on the rate of *equilibrium exchange* of the same sugar (middle-left panel of Fig. 2). When the rate constant for the exponential loss of glucose-^{14}C into a medium containing the same level of unlabeled glucose was determined at various concentrations, and the data analyzed in terms of equations derived from the conventional model, the indicated K was enormously higher than that given by the Sen–Widdas method. A smaller difference in the V_{max}s was interpretable in terms of the loading effect on carrier mobility (the r factor discussed in the preceding section). The conflict in the K estimations was further worsened by reducing the temperature, since this *in*creased the value given by the tracer-equilibration procedure, to the point that the discrepancy with the Sen–Widdas K was as much as 70-fold.

In a systematic direct comparison of the two procedures on identical suspensions in the same laboratory, Miller (1968a,b) showed that the anomaly could not be attributed to merely technical variables. Moreover, consideration of the r factor relieved the dissonance only slightly, allowing the theoretical ratio of the two apparent K's to rise only slightly above 2. Nor was there any improvement in the picture if it was assumed that the extramembranal diffusion of the glucose might become rate-limiting.

Levine and Stein (1966) also estimated the glucose K as its K_i in the competitive inhibition of the equilibrium exchange of labeled L-sorbose (a very low-affinity substrate in this system), and found fair agreement with the figure given by the glucose equilibrium-exchange experiments. [Again, however, the temperature characteristics differed (Levine *et al.*, 1971), the glucose K_i showing a minimal value at 25–30°C. These investigators verified that the K_i obtained was independent of which of several low-affinity sugars was used as the test passenger.] A slightly lower apparent K emerged from computer-fitting of data relating the half-times of *net uptake* to the applied glucose concentrations (in the manner of LeFevre, 1962), but this could be reconciled with the other figures by taking into account the loading effect (i.e., that r = ca. 4). Miller (1968a) also found a relatively high K_i for glucose in blocking the uptake of L-sorbose (as in the arrangement diagrammed in the upper-right panel of Fig. 2).

Karlish *et al.* (1972) analyzed the time course of glucose exit from preloaded cells into a very large volume of sugar-free medium (the "infinite-sink" or "zero-trans" situation illustrated in the lower left panel of Fig. 2), and thus calculated a value of K appreciably lower than the equilibrium-exchange figure, but still about ten times the Sen–Widdas K. Miller (1971) found a much smaller discrepancy between the zero-trans and infinite-cis exit Ks, but this does little to relieve the clash with the classical theory, since the model requires the former value to approach the latter from below, never to exceed it. In fact, Lieb and Stein (1971b) present a theoretical argument for the conclusion that *any* symmetrical carrier picture, formulated as in Fig. 1, must be rejected by reason of this finding, without regard to whether Widdas's particular restrictions are applied.

Thus, none of the other three independent measures of apparent affinity in this system appears to be reconcilable, by any variation of the schema of Fig. 1, with the K given by the (equally valid) Sen–Widdas procedure. A suggestion of similar problems in the nucleoside transport system in these same cells is noted by Packard and Paterson (1972).

2. Other Kinetic Problems

Also encountered by Miller (1965, 1968a,b, 1969) were inconsistencies in the quantitative prediction of the system's *countertransport* behavior. When, as diagrammed in the lower right panel of Fig. 2, cells were rather heavily preloaded with a sugar and then transferred to a low concentration of the labeled form of the same sugar or of a second type of substrate, the resultant transient accumulation of the label followed calculated expectations only with certain substrate combinations. With others, the accumulation peaked too early, and at too high or too low a level, to be fitted to the other measurements. Thus, self-induced counterflow adequately accorded with prediction for D-galactose or D-mannose, but the peak fell short with D-glucose; and though the induction of mannose accumulation by preloading with galactose was consistent with the respective patterns of homo-counterflow, the reciprocal combination (i.e., with mannose driving galactose) showed decidedly *too much* transient galactose accumulation. Levine and Levine (1969) have developed the general equations defining the occurrence of this type of induced uphill transport where r is not restricted to the classical value of unity (so that both simple counterflow and true trans-effects may contribute to the overall phenomenon). They noted that Miller's data for galactose and mannose accord with the assumption that $r = 1$ for all the combinations other than this troublesome one, which instead requires that $r = $ ca. 2. It is notable that no such anomalies were evident in the course of comparable studies with rabbit erythrocytes (Regen and Morgan, 1964).

Certain peculiarities in the characteristics of hetero-trans-stimulation noted by Miller (1968a,b, 1969) also appeared difficult to interpret by the conventional model. In the arrangement illustrated in the middle-right panel of Fig. 2, the acceleration of isotopic D-glucose exit by the presence of an equivalent external concentration (130 mM) of D-glucose or D-galactose distinctly exceeded the homo-trans-stimulation obtained with any of the three sugars separately. However, Levine et al. (1965) felt that their observations of somewhat similar phenomena could be attributed to predictable differences in the extent of cis-inhibition on the intracellular side which would develop during the period of measurement (because of the entry of some of the externally applied sugar). Moreover, on the basis of an exhaustive reexamination of the exchange behavior of these three sugars, Eilam and Stein (1972) challenged Miller's presumption of effective saturation of efflux at 130 mM, especially in the case of galactose. They concluded that the seemingly anomalous trans-stimulation interrelations follow reasonably from this consideration and from the measurable minor differences in apparent V_{max} for the three hexoses.

C. Attempts to Reconcile Mobile-Carrier Theory with Anomalies

1. Effects of Unstirred Layers

In the derivation of predictions from transport models, each of the two extramembranal compartments is commonly treated (either explicitly or tacitly) as a homogeneous pool within which mixing is instantaneous. In more realistic terms, this presumes that the transmembranal movement of the passenger species is sufficiently slow in comparison to its aqueous diffusion that no significant local gradients develop within the aqueous compartments. Thus, any "unstirred-layer effects" immediately adjacent to the membrane interfaces are ignored. The seriousness of this over-simplification as a source of error in the measurement of permeation rates should be at least crudely indicated by the sensitivity of these measurements to controlled variation of the stirring of one or both of the chambers. In line with the theoretical expectation (Levich, 1962), this factor has proved increasingly significant for solutes with higher mobilities, and is especially notable in the transfer of water itself, both in artificial membranes (Ginzburg and Katchalsky, 1964) and in various biological situations (Dainty, 1963; Dainty and House, 1966a,b; Sha'afi et al., 1967).

In connection with the several critical anomalies in red-cell sugar transport discussed in the preceding section, many discussants have informally called attention to the likelihood of temporary confinement of recently translocated sugar molecules within the immediate vicinity of the membranes (with a corresponding deficit on the other side). Hoos et al. (1972) noted particularly that this consideration could give rise to false apparent trans-effects even in a system intrinsically lacking this capacity. Naftalin (1971) reported that varying degrees of stirring of erythrocyte suspensions did indeed alter the apparent rates of glucose transfer, but not always in the obvious direction. Hankin and Stein (1972) were, however, quite unable to verify this, and are inclined to attribute Naftalin's findings to his failure to disperse the cells sufficiently rapidly at the nominal instant of mixing of his suspensions. Moreover, Lieb and Stein (1972a) have calculated, by considering the unstirred layer as a transport resistance in series with that of the membrane traversal proper, that the cell would have to immobilize about 350 times its own volume of external medium to give rise to a mere 1% increase in the overall resistance. Miller (1972) has also presented technical reasons for denying any quantitative significance of the unstirred-layer factor in generating the very low relative values for K obtained with the Sen–Widdas procedure. In addition, Lieb and Stein (1972b) argue that an unstirred layer on the trans-side would lead to an overestimation of the K by the "infinite-cis" method of Sen and Widdas

(1962) and an *under*estimate by the "zero-trans" procedure; consequently, *any allowance made for this factor would only worsen the conflict* in the experimental findings.

Recently, however, Regen and Tarpley (1974) have presented the case for a substantial diffusion resistance as a *critical* feature in an asymmetrical carrier mechanism, which might explain virtually all the kinetic data on this transport system. Here, the series resistance is pictured as residing not in an actual unstirred layer in the adjacent aqueous pool, but rather in superficial membrane structures overlying the carrier sites deeper within the membrane (and mainly on the intracellular side). This proposal will be taken up more fully in Section II, C, 3, a, since it involves abandonment of the simplifying assumptions of Widdas (1952) noted in the legend of Fig. 1.

2. SUBSTITUTIVE EXCHANGE IN THE ASSOCIATION–DISSOCIATION PROCESS

Another possible basis for trans-acceleration suggested by Lacko *et al.* (1961) and specifically developed by Jacquez (1964) centers on the partial coupling of the a and d processes of Fig. 1. Specifically, it is proposed that the detachment of a transport substrate from a complex at a membrane interface might less readily occur spontaneously than when an impact of another passenger molecule from the adjacent pool allows its immediate substitution in the complex for the detaching molecule. This amounts to postulating a conjoint a–d reaction, such that $C\dot{S} + \overset{\circ}{S} \to C\overset{\circ}{S} + \dot{S}$ (where $\dot{S}$ and $\overset{\circ}{S}$ are two species of substrates as in Fig. 1), without the appearance of free C as an intermediate (presumably by way of a transitory ternary complex, $\dot{S}C\overset{\circ}{S}$). It has even been suggested that differences in susceptibility to this type of exchange among various amino acids in some tumor cells (Guroff *et al.*, 1964; Jacquez and Sherman, 1965) provide a simpler explanation than the accepted multiple-carrier hypothesis (Oxender and Christensen, 1963) for the complex patterns of interplay between amino acid fluxes.

Lieb and Stein (1972a) pointed out that such replacement events would obviously not be reflected in the Sen–Widdas or other procedures dealing with *net* movements, but could enhance the fluxes in any *exchange* situation. Grossly, this would mimic the action of a parallel exchange diffusion; but here the greater rapidity of exchange is envisioned as occurring in the interfacial steps rather than in any transmembranal step. Lieb and Stein reported that their (unpublished) kinetic analysis of this proposition shows that the large ratio of the equilibrium-exchange K to the infinite-cis K could readily be accounted for on this basis; but the anomaly in respect to the high K with the zero-trans procedure would remain unrelieved.

3. Relaxation of Classical Restrictions on Relative Rate
Constants

a. Transmembranal Asymmetries. One of the earliest suggestions that
the classical model for red-cell sugar transport might be oversimplified was
the finding of Wilbrandt *et al.* (1956) that the rate constant for net glucose
entry was nearly twice that for exit. Bowyer and Widdas (1958) appeared
to have disposed of this problem on technical grounds (cumulative errors
arising from the approximations in the applied analytic equations). However, recently it has been reported independently from several laboratories
that this system shows distinctly different characteristics of influx and
efflux when set up in various ways with appropriately reciprocal sugar
gradients. Working with extremely brief incubations so as to ensure closer
approximation to *initial* entry rates, Lacko *et al.* (1972) found that the Ks
estimated from the concentration-dependence of zero-trans influx (or of
influx into fixed intracellular levels) are of the same low order as the
(otherwise unmatched) K given by the standard Sen–Widdas infinite-cis
net-exit procedure; but, in approximate confirmation of the findings of
Levine and Stein (1966) and Miller (1968a), the equilibrium-exchange K
was about 12 times larger, vastly exceeding the factor of 2 permitted by
the symmetrical model. Thus the exceptionally low apparent K was evidently associated not with the infinite-cis condition as such, but rather with
those two procedures which reflect essentially the saturation of *influx*;
accordingly, it was suggested that the glucose *affinity may be higher at the
outer surface* of the membrane than at the cytoplasmic surface, in spite of
the fact that the cells do not actively accumulate the sugar.

Dealing with these data of Lacko *et al.* prior to publication, Geck (1971)
examined the possibility of their theoretical accommodation by the *un-
restricted* form of the mobile-carrier model as given in Fig. 1. Here the
relative magnitude of the eight rate constants determining the transfer of
a single passenger species is restrained by only one equality (dictated by
the second law of thermodynamics, disallowing a spontaneous accumulation
against a gradient). Geck's analysis shows that the loosening of the conventional model's restrictions comes reasonably close to encompassing the
observed factor of 12 by which the equilibrium-exchange K of Lacko *et al.*
exceeds their Sen–Widdas K; it also is compatible with Miller's (1968a)
peculiar interrelations in the hetero-trans-stimulation behavior (Section
II, B, 2). Some improvement toward fitting the aberrant peaks in Miller's
counterflow experiments was also achieved; however, there remained no
accounting for the fact that the glucose K_i in the blocking of sorbose transfer is distinctly lower than its K in equilibrium exchange.

Paralleling the above indication that the entry K may be lower than the exit K is the finding of Baker and Widdas (1973) that, in order to exert an equivalent inhibitory effect on the glucose equilibrium exchange, the slow permeant, 4,6-O-ethylidene-α-D-glucopyranose, must attain a much higher concentration inside than outside the cells. Baker and Widdas follow Geck in postulating an asymmetric nonaccumulative model wherein the inequality of Ks is balanced by a corresponding inequality in the Vs; but unlike Geck, they treat this in terms of *two distinct membrane components* operating for influx and efflux, and they superimpose an *additional exchange process* having its own independent rate constant. Thus in addition to the familiar K and V_{max}, the Baker–Widdas schema entails two more fixed parameters for each substrate: the separate V for the exchange process, and a ratio expressing the basic carrier asymmetry. In new glucose-counterflow experiments similar to those of Miller (1968a), they found an asymmetry factor of about 10 to provide excellent computer fits to the observed movements of both the driving sugar and the driven sugar. The distinct superiority of Baker and Widdas's model in matching data of this type lies in the fact that the initial rapid uptake of radioactivity is here governed mainly by the exchange parameter, whereas a quite independent parameter becomes critical after the peak accumulation (because relatively little of the second sugar then remains in the cells, so that the tracer exit approaches the total exodus). This model would also account for the relatively large K_i, if a high rate constant for sorbose-glucose exchange were assumed.

Batt and Schachter (1973) agreed with the above investigators in finding experimentally an apparent flux asymmetry in the system, but came to quite conflicting conclusions regarding how this is arranged. Lineweaver–Burk analysis of their measurements of glucose movements inward and outward under the respective zero-trans conditions confirmed that the V_{max} for influx is only about one-third that for efflux, but showed the apparent influx K to be distinctly *larger* than the efflux K. (Both remained much smaller than the apparent equilibrium-exchange K, however.) Batt and Schachter's chief emphasis is that this asymmetry correlates with the presence or absence of appreciable *intracellular sugar*, rather than with the type of kinetic data involved: with whatever test procedure, the ratio V_{max}/K is about 5–6 times higher when the cells contain glucose than when they are nearly sugar-free. Accordingly, they suggested that the carrier exists in two states, the equilibrium between which is governed by the intracellular glucose level. Edwards (1973a) came to a similar conclusion on the basis of asymmetry in the inactivation of the system by fluoro-dinitrobenzene. Batt and Schachter further reported that the exit process is far more susceptible than the entry to inhibition by several classes of chemical agents; while Barnett *et al.* (1973) found that transport in either direction shows unequal sensitivity at the inner and outer surfaces to competitive inhibition by various alkyl-substituted sugars. However, Beneš *et al.* (1972) demonstrated, by incorporation of the inhibitor phloretin within red-cell ghosts at the time of lysis and resealing, that the asymmetry in inhibition by this agent is *reversed* when it is thus applied mainly to the inner surface of the membranes. They therefore assigned the usual pattern of greater sensitivity in the efflux to an apparent impermeability of the intact cells to phloretin, and discredit any inference of an intrinsic asymmetry in the system itself.

Schultz (1971, 1972) has provided a theoretical analysis of how such asymmetry might arise simply by confinement of a carrier-mediated transport to a layer making up less than the full thickness of the barrier traversed. If the permeant must move through a region in which the transfer is not a linear function of the concentration difference, and this be in series with a region of diffusive transfer (which may be simply an unstirred extramembranal layer), Schultz finds it likely that asymmetrical behavior will turn up in zero-trans or infinite-cis studies, but not at all in tracer-exchange or competitive-inhibition tests. In any case, the several indications above of an asymmetrical sugar transport apparatus in human red cells are not overly surprising in view of the evidence in the same cells of asymmetry in the transport of certain amino acids (Winter and Christensen, 1964; Christensen, 1972; Hoare, 1972b) and of choline (Martin, 1971; Edwards, 1973b). However, Batt and Schachter (1973) could not duplicate in rabbit erythrocytes (reticulocyte-enriched) the phenomena they observed in the human cells, confirming Regen and Morgan's (1964) conclusion of apparently complete symmetry in the rabbit system. Moreover, Hankin *et al.* (1972) have argued that *any* form of asymmetry in the parameters of Fig. 1 must be discarded as the basis for incompatibilities in the glucose-transport Ks and Vs given by the several procedures with human red cells. They arrive at this conclusion by noting quantitative inconsistencies in the interrelations of the several operational "constants" when considered in terms of a set of equations derived from the *completely general* schema given by Britton (1965). Hankin *et al.* made the necessary additional measurement of the K apparent in infinite-cis *entry*, and found that it differed only slightly from that widely observed in the reciprocal (Sen–Widdas) situation. However, when this new measure of K was related to those derived from equilibrium-exchange or zero-trans exit experiments so as to allow estimation of an apparent asymmetry ratio, the appropriate equations gave a figure of 6.5 by one procedure and 17.6 by another. On this basis, Lieb and Stein (1972a) reject any further consideration of the relaxation of the classical model's symmetry constraint as a possible means of retaining a mobile-carrier picture for human erythrocyte sugar transfer. Thus the very system which for years has served as the prime biological illustration of the classical model now ironically appears to have provided the most distinctly dissonant behavior.

However, recently a number of technical uncertainties regarding the applicability of these criteria for rejection of an asymmetric model were brought out in the analysis by Regen and Tarpley (1974) mentioned in Section II, C, 1. Possible reconciliation of the basic schema of Fig. 1 with the kinetic observations may be attainable if an asymmetrical carrier apparatus is presumed to lie sufficiently deep within the membrane that the substrate encounters a significant diffusive resistance before penetrating to this point. Regen and Tarpley's calculations indicate that the critical aspect of the asymmetry in this system is the relative difficulty of *movement of the empty carrier outward* from the inner side of the membrane. (Thus, if compartment 2 in Fig. 1 represents the cell interior, $D_{C_2 \to 1}$ is less than $\frac{1}{10}$ of the other D's.) Accordingly, in the absence of sugar most of the carrier would reside at the inner interface. This may represent an inherently greater stability in this orientation, preference for the inner surface extending also to the more mobile glucose-loaded carrier. However, Regen and Tarpley bring out several lines of argument for the interesting possibility that the carrier in itself is symmetrical, its entrapment at the inner side arising from the presence there of a membrane-bound nontransportable competitive inhibitor. For full delineation of a given substrate's behavior, the Regen–Tarpley analysis requires specification of the inner and outer diffusive resistances and five transport constants, each of which is a complex function of several of the carrier-cycle rate constants (essentially analogous to those in Fig. 1). One of the five transport constants is an "activity" parameter expressing a basic rate; the other four (any three of which define the fourth) are in the nature of *pseudo*-dissociation constants, and are expressed as concentrations. Two of these should be identical for all substrates of a given system, while the other two characterize the individual substrate's interactions. Regen and Tarpley supply a procedural recipe for measuring the parameters for a given substrate from a set of five experimental runs.

Thus at present it is difficult to judge whether the complication of the classical carrier picture by allowing asymmetric mobilities (and affinities) will provide the most acceptable framework for interpretation of the diverse kinetic findings. Although the major anomalies may certainly be accounted for in terms of this type of schema, the several specific models that have been advanced entail rather divergent postulates and secondary conclusions. It consequently appears that as yet no distinctive operational characteristic can be unambiguously cited for further evaluation of the general proposition.

b. *Rate Limitation by Association-Dissociation Processes.* In contrast to the above revisions of the conventional model, an amendment offered by Wilbrandt (1972b) retains transmembranal symmetry in affinities and translocation activations, but dismisses the classical restriction assigning the overall rate limitation to the transmembranal step. Wilbrandt's schema thus allows the rates of the adsorption–desorption to assume finite values and to differ at the two interfaces, as in the pre-Widdas schema of LeFevre and LeFevre (1952). This yields a general transfer equation containing four transport constants: the single K, a reciprocal-rate parameter for the transmembranal step, and two complex reciprocal-rate parameters associated with the interfacial events at the two sides. This provides good accord with many observations that are not accommodated by the classical schema. Analysis of D-glucose exit experiments indicates that the a–d processes at the inner surface are much slower than at the exterior, and have a substantially higher activation energy. The numerical parameters thus derived reconcile precisely the apparent Sen–Widdas K (which here approximates the "true" K) with the much larger K apparent in zero-trans experiments. Wilbrandt also reported a variability of the apparent glucose K_i according to which sugar is used as the test penetrant; his schema accounts for this phenomenon if it be presumed (as with the later Regen–Tarpley model discussed above) that a substantial resistance to sugar flow is interposed between the aqueous pool and the site of carrier-interaction, especially on the inner side. The calculated magnitude of this additional resistance is much too great to be assigned to an unstirred layer in the usual sense, and was taken as an indication that the carrier sites lie rather deep within the membrane structure (a conclusion reached independently by Regen and Tarpley, 1974).

D. Alternative ("Noncarrier") Models for Specific Membrane Transport Sites

In view of the many shortcomings in efforts to adapt conventional mobile-carrier theory to the various peculiarities in experimental behavior considered above, a number of recent alternative suggestions concern possible rearrangements whereby the individual selective sites might interact with the transported species without accompanying them across the barrier, effecting the translocation through interchanges between neighboring sites.

1. Channel-Lining "Bucket-Brigade" Sites

The possibility of mimicking much of the behavior of a true carrier system by confining the transmembranal movement of a solute to a hopping or *"creep" between adjacent fixed sites along a polar pathway* was brought out by Stein and Danielli (1956) shortly after the Widdas model was proposed. Detailed analysis of the theoretical behavior of two-site-long pores of this type, restricting the solutes to a "single-file" passage, was provided by Heckmann (1956a,b, 1968). Uphill countertransport may be expected here in spite of the "no-passing" proviso, but there are some systematic differences (Heckmann and Passow, 1967) from a true carrier model in respect to the circumstances that may elicit discrimination between two passengers species. Heckmann (1972) also defines conditions under which an apparent cotransport might seem to be occurring with this type of model.

The single-file feature is specifically eliminated from the proposal by Naftalin (1970) that such a "one-dimensional diffusion" could serve as the basis for the red cell's sugar-transfer behavior. In fact, in order to match the biological phenomena, Naftalin's model provides that exchange in position of two adjacent passengers in a pore occur *more* readily than a simple migration into an adjacent vacancy. (This is somewhat analogous to the interfacial replacement effect discussed in Section II, C, 2.) In terms of the diagrammatic view of Naftalin's schema in Fig. 3, this means that (given the appropriate occupancies and impacts) an event of type x

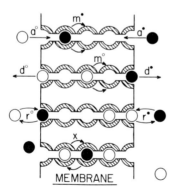

Fig. 3. Naftalin's lattice-membrane model. Chains of fixed binding sites line water-filled channels spanning the membrane. Symbols are essentially as in Fig. 1, and processes a and d have similar significance. Process $r°$ entails linked $a°$ and $d·$ events, while $r·$ involves the reverse exchange at an interface. Process m covers the migration of a passenger from one site in a chain to an adjacent vacancy, while x involves linkage of two such migrations of different passenger species in opposite directions between adjacent sites. Further discussion appears in Section II, D, 1.

is more likely to occur than m, and perhaps that r is more likely than d. To arrive at theoretical predictions for the several critical experimental situations, Naftalin set up a computer simulation with a 4×10 array of such sites, separating an infinite exterior chamber from a finite interior. Assigning various combinations of probabilities to the several steps in the translocation process, he generated *pseudo*-random numbers to decide whether each applicable event was to occur at each point in the model matrix, and executed each event accordingly. Within a few thousand iterations of this point-by-point transformation, a steady state was attained in each case, wherein the simulated fluxes (or net transfers) could be compared with the corresponding experimental findings.

This system readily mimicked the Sen–Widdas behavior, competitive inhibition, and the phenomenon of uphill counterflow; however, in line with the general theoretical derivations of Jacquez (1961) for this class of model, trans-effects were limited to small inhibitions (never stimulations), unless one presumed a significant unstirred-layer effect at the exterior membrane interfaces. Accepting this complication makes it possible, however, to account not only for trans-acceleration in general, but also for its greater prominence (Miller, 1968a) when the inducer is a somewhat lower-affinity sugar. The quantitative patterns of the glucose counterflow experiments are also more satisfactorily accommodated (Naftalin, 1972). Moreover, at least the general sequence of the differing magnitudes of the Ks derived from different experimental approaches is reproduced. [According to Naftalin's (1972) analysis, even the lowest Ks given experimentally are *gross* overestimates of the true carrier-passenger dissociation constant.] However, the dubious applicability of the unstirred-layer factor in the resolution of these anomalies has already been noted above (Section II, C, 1).

2. Oligomeric Bi-affine "Internal-Transfer" Sites

A rather involved model, but one in keeping with current notions of cell membrane structure and molecular enzymology, was developed and analyzed in some detail by Lieb and Stein (1970) as a conceivable basis for the kinetic complexities of erythrocyte sugar transfer. This model pictures the sugar-binding components as protein *tetramers* embedded in (and spanning) the membrane, in such a way that two binding sites of rather *differing affinities* are presented both at the inner and at the outer membrane interface, or alternatively are held facing an *internal cavity* enclosed by the tetramer. Within this cavity, migration of an adsorbed sugar molecule, or exchange of two sugars, may occur between apposed pairs of such sites (i.e., one of low affinity and one of high affinity). The confor-

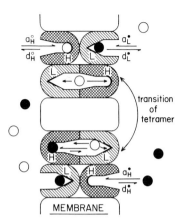

FIG. 4. Lieb and Stein's internal-transfer oligomeric model. Binding sites of lower affinity (L, singly hatched) and of higher affinity (H, cross-hatched) are coupled as indicated, in tetramers which undergo transition between two diagrammed conformations. Processes a and d are as in previous figures, differing characteristically according to identity of site and substrate, as symbolized. Distribution of substrates within internal cavity joining internalized apposed sites is determined by statistical likelihood of alternatives. Transition of tetramer is governed by state of occupancy by substrates. Further discussion appears in Section II, D, 2.

mational transition can occur only when the activation-energy barrier is reduced by attachment of at least one sugar to the tetramer; and the transition frequency is taken to be proportional to the number of occupied sites. The transition involves the entire tetramer as a unit, and in the original proposal effected the simultaneous internalization of all four sites; the arrangement diagrammed in Fig. 4 follows a later revision (Lieb and Stein, 1971a), preferred by reason of the identical energy state of the two conformations (this does not entail any modification in the predicted kinetic behavior).

When, with such a system, a single substrate molecule is carried to the internal cavity, it will generally leave in association with the high-affinity site. Accordingly, in a zero-trans situation, the flux will become appreciable only when cis-concentrations are sufficiently high to effect significant loading of the *low*-affinity sites; the measured K will then reflect essentially only this lower affinity. When two sugars are present in the cavity, their statistical distribution between the apposed sites will be dictated by the proportions among the four affinity constants; if one species is simply a labeled form of the other, the next transition will have an equal chance of placing the label in either pool. Thus, the infinite-cis K is set by the high-affinity binding, while the equilibrium exchange should mirror the low affinity (higher K) of the zero-trans situation, and should have *twice* as high a V_{max} as that apparent in the net transfers. Similar considerations predict an intermediate apparent K_i conforming to the observations with glucose inhibition of sorbose transfer; and a consistent assignment of constants was in fact shown to be quantitatively compatible with all of these classes of data in the red-cell sugar-transport system. The failure of some other systems to restrict the extent of their homo-trans-acceleration to an approximate doubling of the V_{max} can also readily be accommodated by appropriate modification of the postulate regarding the dependency of the transition frequency on the number of sites occupied.

The internal-transfer model rather grossly fails to accord with Miller's (1968a) observations on countertransport, but Lieb and Stein (1970) presented an analysis of these data which suggests a serious internal inconsistency. The failure to fit with Miller's conclusions from his hetero-trans-stimulation experiments was dismissed on the basis of the analysis by Eilam and Stein (1972) discussed in Section II, B, 2. Stein *et al.* (1973) have also developed a modification of the tetrameric internal-transfer model allowing its extension to active transport of Na^+ and K^+; this entails a *transposition* of the high- and low-affinity sites, in company with the transition between the two conformations of the tetramers.

3. "Introverting" Hemiport Sites

In seeking to prepare a simplified computer simulation of Naftalin's (1970) fixed-lattice model, LeFevre (1973) found that the critical kinetic properties might have their basis simply in a lack of randomness in the decision as to the *direction* of the one-dimensional movement from a given site in the lattice. Specifically, at the *boundary* sites the marked contrast in the physicochemical prospect facing inward into the lattice vs that facing outward into the aqueous pool suggests the likelihood of marked asymmetry in the ease of moving an adsorbed substrate in the two directions; moreover, this asymmetry might well be modified by the identity of the occupant substrate species. Thus, the relative probabilities that these interfacial sites will participate in a pool-oriented event as compared to an intramembranal event might depend not only on the relative magnitude of the intrinsic rate constants concerned (as in Fig. 3), but also on the relative disposition of the sites to favor one or the other orientation or conformation. LeFevre concluded that this consideration could be so critical in defining the overall patterns of translocation that it was in fact unnecessary to presume any other distinctive parameter among the several sugars sharing the human erythrocyte transport mechanism. He accord-

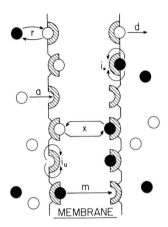

Fig. 5. LeFevre's introverting-hemiport model. Symbols are essentially as in prior figures; processes a, d, r, m, and x are as with Naftalin's schema in Fig. 3, except that here they are identical for all passenger species. Transition of individual sites is between "extroverted" conformation allowing a and d (or r) interchanges with pool, and "introverted" conformation allowing transmembranal events m or x if partner site is also introverted. Introversion equilibrium is expressed by constant i, which is determined by site's state of occupancy (indicated by subscript). Further explanation accompanies text discussion in Section II, D, 3.

ingly proposed a model in which the bucket-brigade channels were reduced to the interfacial sites only; this amounts to a *bilayer* of sites comparable to Stein's (1969) *hemiports*. Each substrate is characterized by a numerical *introversiveness* index (between 0 and 1), expressing the fraction of time that the sites spend in an "introverted" or internalized conformation from which interaction with an apposed introverted site on the other face of the membrane may occur; it is presumed that for the remainder of the time they are available for interaction with the adjacent pool.

A picture of this model is given in Fig. 5. To accord with the behavior observed in the biological system, it appears that the unoccupied sites must have an introversiveness (i_u in the figure) of about 0.3, while the substrate-induced i's cover a wide range (up to perhaps 0.99 or higher for the sugars classically characterized as having the highest affinities). It is clear that although this schema evolved from that of Naftalin (Fig. 3), it shares perhaps a greater number of conceptual attributes with the Lieb–Stein model of Fig. 4—notably the conformational transition of sites, its governance by reaction with substrate, and the transmembranal interaction between apposed monomeric elements at the two interfaces. The most critical distinctions are (a) that the transition of apposed sites in the introversion model is completely unlinked,[2] while in the oligomeric model all four sites in a unit, thus at *both* interfaces, are necessarily "flipped" at once; (b) that the introversion–extroversion is viewed as a rapid spontaneous oscillation, while the oligomeric transition occurs only rarely and only when at least one site is occupied; (c) that there is only a single class of binding sites in the introversion model; and (d) that these sites interact equally readily with all species of substrates, holding all with equal avidity (constants *a*, *d*, *r*, *x*, and *m* in Fig. 5 all being *substrate-independent*).

It was in fact this substantial reduction in necessary postulates that led the author to suggest this model in spite of serious reservations regarding its realism. Even when the number of independent parameters is further reduced by equating the probabilities of exchange processes and the corresponding one-way events (i.e., in Fig. 5, $r = d$, and $x = m$), only a single set of constants (LeFevre, 1973) is needed to provide adequate quantitative accord with all of Miller's (1968a) anomalies discussed in Section II, B. A relatively high apparent K in the zero-trans situation is also anticipated. A major objection, however, is that this fit with the biological system appears to require that the membrane sites constitute a substantial reservoir for bound substrate, significant in relation to the total substrate content of the intracellular pool at the experimental concentrations. This unrealistic aspect of the model's original computer simulation came about as an incidental by-product of the need to achieve appreciable "substrate transfer" within a reasonable number of computer iterations. It has not as yet been possible to adjust the introversion model so that a formal *steady-state* analysis yields appropriate predictions—a deficiency that would seem to demand some critical modifying postulate if the model is to become conceptually fully acceptable.

Thus, none of these alternative models, which substitute a transmembranal interaction between linked sites for the classical migration of a true carrier, appears to be entirely satisfactory in accommodating all types of kinetic data. Nevertheless, this type of functional organization of the presumed transport sites would not only quite readily account for many of the complicating observations, but would also fit harmoniously (Stein, 1969; Lieb and Stein, 1972a) with some of the concepts that are emerging from modern physicochemical studies of cell-membrane ultrastructure.

---

[2] In fact, it is not necessary to specify in this schema whether the interacting apposed hemiports remain paired as a structural fixture, or merely happen randomly to become momentarily juxtaposed by way of two-dimensional diffusion (Stein, 1969) within each liquid monolayer.

III. ALTERNATIVE FRAMEWORKS FOR INTERPRETATION OF BIOLOGICAL SOLUTE DISTRIBUTION AND TRANSLOCATION

From the days of the earliest systematic studies of cell permeability to the present era of intensive investigation of biological transport phenomena, there has persisted a significant and outspoken minority opposition to the general notion that a differentiated structure at the cell surface is critical in determining the access of solutes into the protoplasm or the distribution of materials between cells and their surroundings. The very existence of the plasma membrane as a distinct organelle was repeatedly questioned during the first half of this century. The alternative suggested was that the naked protoplasm constitutes a discrete semisolid phase, capable (in the manner of a colloidal gel) of imbibing or extruding water and of adsorbing or partially excluding specific solutes. Today, the cell membrane's reality as at least a mechanical enclosure enveloping the protoplasm is no longer denied, but its involvement in establishing or regulating the cell's composition is still strongly challenged.

The two major publications in English detailing the bases for these "sorption" hypotheses for control of biological permeation and distribution are Ling's (1962) text, "A Physical Theory of the Living State: the Association–Induction Hypothesis," [for a less demanding, exceptionally clear and systematic summary of his theory and the supportive evidence, Ling's (1969) presentation is recommended] and Troshin's (1966) supplemented translation, "Problems of Cell Permeability." (Both of these are broad treatises posing an unusual view not only of translocation phenomena, but of the fundamental functional organization of living material.) A more recent overview of the major lines of evidence currently under development in these schools is available as Vol. 204 (1973) of the *Annals of the New York Academy of Sciences*, covering a 3-day symposium in 1972 on "Physicochemical State of Ions and Water in Living Tissues and Model Systems." There are a number of distinctions in detailed mechanism that may be drawn between the views advanced by Ling and Troshin and other spokesmen for abandonment of "membrane theory," but all seem to share two primary theses: (1) that the macromolecules in the bulk phase of protoplasm are rich in selective adsorption sites (particularly for inorganic cations) and that this is critical in determining the cellular retention of specific components; and (2) that most, perhaps all, of the protoplasmic water is rather tightly structured, differing radically from ordinary water in its solvency and other physicochemical properties. We will first consider the latter proposition.

A. Semicrystalline or Solid-State Protoplasmic Structure

There is extensive evidence of imperfection in the osmometric behavior of many tissues, especially skeletal muscles, and this has been perhaps the main basis for the recurrent discussions of "bound water" in the biological literature. Cope (1967a) concluded that a wide variety of classical data of this type accords well with the presumption that the distribution of water is governed by its adsorption on cell proteins (following a Bradley isotherm) rather than by osmotic factors. Klotz (1958) has summarized a number of arguments for the icelike character of the water in the hydration sheath of proteins. This partial fixation of the solvent would presumably in turn modify significantly the cellular content of other materials. Ling (1965a) emphasized the formation of "polarized multilayers" of structured water around the macromolecules, and that such water may exclude not only ionic solutes but even nonelectrolytes. Moreover, there may result surprising retention of some solutes only poorly soluble in normal water, such as esters, fats, and various hydrocarbons (Fischer and Suer, 1938). On the basis of the exchange and distribution characteristics of Na^+, K^+, and Cl^- and the membrane-potential behavior in giant barnacle muscle fibers, Hinke (1970) and Gayton and Hinke (1971) estimated that only a little more than half of the fiber water acts as a solvent for these ions. However, Hill (1930) and Dydyńska and Wilkie (1963) reported close adherence to simple osmotic theory in the shifts of muscle water induced even by decidedly hypertonic solutions.

The appearance of similar solvency changes in water held by *ion-exchange resins* has been noted by Ling (1952), Ling *et al.* (1973), Damedian (1973), and Czeisler and Swift (1973); and Troshin (1966) emphasized the similarity of solute tissue:medium equilibria, in many instances, to those seen with *coacervates*: retention of an excess of sugar or other solute at low concentration ranges (reflecting association or binding), and partial exclusion at higher concentrations (expressing lowered solubility in the structured water). Unfortunately, this very adaptability of the adsorptive-iceberg hypothesis, in allowing essentially any qualitative pattern in the equilibrium distributions, makes it difficult to cite the appearance of any particular pattern as evidence for the applicability of the concepts.

The notion of substantial restraint on the motion of protoplasmic water molecules has been borne out in many studies of biological preparations by *nuclear magnetic resonance* (NMR) methods. Thus Bratton *et al.* (1965), using the spin-echo technique, concluded from the behavior of the proton signal from frog muscle that a significant fraction of the myoplasmic water is in a phase where random thermal rotational motion is restricted. Perhaps 20% of this "bound" water was evidently released during tetanic isometric contraction. Chapman and McLauchlan (1967) reported that the proton resonance in rabbit nerve sections (reflecting dipolar interactions between protons in the water molecules) depended on the orientation of the nerve in the magnetic field. This suggested that most of the axoplasmic water is partially oriented with respect to the nerve axis, presumably by reason of electrostatic interactions near the fiber surfaces or structural-protein molecular surfaces. In a high-resolution NMR study of frog nerves, Fritz and Swift (1967) also found a major deficit in the water signal, which was decidedly augmented upon depolarization of the tissue by electrical or chemical means. The restricted water in rat and mouse muscles was further resolved by Hazlewood et al. (1969) into two components, the smaller of which so broadened the signal that it had failed to appear at all in high-resolution scanning. By prefeeding rats with 50% D_2O, Cope (1969) was able to eliminate a number of technical alternatives in the interpretation of these spectra, greatly reinforcing the certainty of the semicrystallinity of much of the water in muscle and brain tissue, and confirming that a significant fraction of this is even more highly structured than the major component. Confirmation of the motional restriction of most of the water in frog muscle has also come from the study of ^{17}O resonance (Swift and Barr, 1973). There persists, however, some argument (Cooke and Wien, 1973) as to whether the NMR behavior reflects the *extremely* tight binding of a *small* water component in rapid exchange with the bulk of (free) cell water, or the more modest restriction of the *main body* of the water (the very tightly bound component then exchanging too slowly to make any contribution at all to the NMR behavior).

Cope (1963) has suggested a general view of enzyme function whereby the macromolecule is considered as a conductive solid bearing multiple active sites. Intersite charge migration within the solid is governed by the laws of solid-state physics, while electron exchanges at the surface behave as at liquid/solid electrode junctions. In extending these concepts to biological ion-transport phenomena, Cope (1965a, 1970) pictures a cell as a spongelike skeleton of such conductive solids, and the membrane as an interface or junction between two dissimilar materials, acting like a semiconductor with charge-transfer properties differing from those of either of the adjacent phases. In support of the proposition that ion transfer through cell membranes resembles electron transfer across solid/liquid interfaces, Cope replots a variety of *selected* biological-transfer data from the literature, which accord well with the "Elovich rate equation" (whereby the *logarithm* of the transfer rate should be linear with the cis-concentration of the transferred component).

Hazlewood et al. (1971) correlated the gradual development of the ordering of water in rat muscles (as reflected in the NMR behavior) during the first 30–40 days of the animal's life with the progressive loss of muscle Na⁺ content. They concluded that the process involved interactions both of the nuclear spins of the water molecules with the macromolecular surfaces and between water molecules, resulting in multilayers of bound water in which the Na⁺ is not readily accommodated. In this framework, the almost universal predominance of K⁺ in the cellular cation makeup might be assigned simply to the ease of its fitting into such structured water because of the relative weakness of this ion's hydration. Thus, in picturing the protoplasm as a three-dimensional network of fixed charges (mainly on terminal amino acid groups), Ling (1952) had initially suggested that the energy of the electrostatic bonding between cations and fixed anionic charges would be determined by the mean dielectric constant of the hydrate shell and the "distance of least separation" dictated by the radius of the hydrated ion. However, Eisenman (1961) and others noted the modification of the cationic bonding sequence with greater strength of the anion field about the fixed charge. Accordingly, Ling (1960, 1962) refined his theory by taking into account statistical variation in the ions' effective diameters, and long-range "inductive" changes in the sites' "effective charges" by reason of conformational rearrangements in neighboring sites. The possible role of such adjustable adsorptions in phenomena generally attributed to membrane carrier function must now be considered.

B. Specific Adsorption, Induction, and Cooperativity

1. The Apparent Compartmentalization of Cellular Inorganic Ions

Much of the argument for the sorptionists' views concerns the characteristics of distribution and exchange of Na⁺, K⁺ and the less biologically prominent alkali-metal ions in animal tissues, most notably in skeletal muscle (Simon et al., 1959; Troshin, 1961; Ling, 1966b). Although some prominent reports to the contrary may be cited (Hodgkin and Horowicz, 1959; Sjödin and Henderson, 1964; Allen and Hinke, 1971), there is substantial kinetic evidence of inhomogeneity in the cellular cations (e.g., Bozler et al., 1958; Hashish, 1958; Ernst and Hajnal, 1959; Troshin, 1961; Allen and Hinke, 1970); and interaction of both competitive (Hechter and Lester, 1960; Ling and Ochsenfeld, 1966) and cooperative (Jones and Karreman, 1969; Ling and Bohr, 1970; Gulati, 1973; Karreman, 1973) nature has been observed and interpreted in terms of the fitting of the ions into a semicrystalline lattice and the modification of the free energy of adsorption at individual sites by occupation of neighboring sites in the lattice.

The *form of the time course of ionic transfers* has been taken to indicate that the site of the principal resistance to diffusion is the bulk phase of the cytoplasm, rather than the membranes. Thus Ling (1966b) and Ling *et al.* (1967) reported that an "influx profile analysis" of the uptake of Na^+, Ca^{2+}, Mg^{2+}, glucose and other materials into frog muscle and tendon (and even of labeled water into frog eggs) failed to show the inflection, in a plot of the cell content vs time$^{1/2}$, that would be consistent with a rate limitation at a superficial barrier. In frog muscle, only K^+ uptake appeared to be surface-limited rather than bulk phase-limited (Ling, 1966a). Fenichel and Horowitz (1963) find that the washout of assorted nonelectrolytes from frog muscles also suggests the primacy of cytoplasmic diffusion in the process.

Cope's (1965b) NMR study of the state of the muscle Na^+ indicates that even the limited amount that does manage to gain access into the myoplasm is only about 28% in the form of free ions, the remainder being complexed with macromolecules. Rotunno *et al.* (1967) similarly concluded from nuclear spin-resonance characteristics of frog skin that the free Na^+ here is only about 40% of the total tissue content (in keeping with the exchange behavior of labeled Na^+ in this tissue), and they noted that this offers an explanation for the asymmetric distribution of the ion across the outer surface of the skin, where there is little sign of Na^+-"pump" activity. Homogenization of frog muscle had little effect on the size of the "NMR-invisible Na" component (Cope, 1967c), and even in preparations of actinomyosin from rabbit psoas muscle the Na^+ appeared to be nearly half in a bound state. However, Berendsen and Edzes (1973) have raised some question as to whether the characteristics of tissue sodium magnetic resonance necessarily show that the ion is in the two forms, bound and free.

Perfusion of frog muscle with isotonic sucrose (Cope, 1967b) brought out that the most readily leaking fraction of Na^+ was the *complexed* component, the apparently "free" ions emerging only very slowly. Cope therefore concluded that, as postulated by Simon (1959), the noncomplexed fraction is confined within the reticular or vacuolar compartments in the muscle. [Ashley (1970) concluded that the effluxes of isotopic Na^+, K^+, and Cl^- from skeletal muscles are interpretable rather simply in terms of such morphological compartmentalization, though Ca^{2+} appears to be genuinely subject to a rapid intracellular binding.] Czeisler *et al.* (1970) verified the greater lability of the bound Na^+ upon soaking of muscles in K^+ media, and established by technical improvements that the "missing" Na^+ can be rendered visible by NMR methods (thus eliminating some challenges to the earlier interpretations). Ling and Cope (1969) found similarly that the appearance of a very large *excess* of muscle Na^+, in apparent exchange for K^+ lost upon soaking of the tissues for several days in K+-deficient media, involves no change in the NMR-visible "free" Na+; it is thus concluded that the K+ that was replaced had similarly been in an adsorbed state. Cope and Damadian (1970) were able to reinforce this proposition by the use of very high-gauss NMR methods (to detect the weak ^{39}K signal) in certain bacteria containing extremely high levels of potassium. Damadian (1969) also demonstrated by equilibrium dialysis an extensive binding of K+ by both intact and fragmented cells of *Escherichia coli*, the selectivity for K+ over Na+ being less pronounced in the fragments. The specific binding activities of a variety of protein fractions turned out not to differ substantially; about one-fourth of the cells' normal complement of K+ could be assigned to this association with proteins.

2. Ionic Activities and Bioelectric Potentials

In view of the intimate relationship of Na+ and K+ distribution across cell membranes to bioelectric phenomena as conventionally interpreted, serious problems are raised in this field also if substantial fractions of these ions are not "free" within excitable cells. Troshin (1966) concluded (largely by insisting on only the most simplistic presentation of "membrane theory") that the details of observed behavior of cell potentials "compel" a search for a radical alternative to the accepted physicochemical bases for these phenomena. In general, the sorption theories hold that the distributions of the major ions between cells and extracellular fluids or between protoplasmic compartments do not entail true *activity* differences. Troshin (1966) suggested instead that resting potentials result from the operation of *the measurement itself*, the insertion of intracellular electrodes (or, in an earlier era, the establishment of a local zone of injury) disturbing the protein-cation binding so as to give rise to interfacial "salt potentials" and other diffusion-linked potentials not originally present. However, Kurella (1961) found this type of argument to be superfluous, deeming it quite reasonable that K+ held in a polyelectrolyte gel structure by ionic bonds should generate a sustained potential difference from a surrounding medium. He in fact demonstrated similar phenomena in suspensions of cation-exchanger granules, and assigned them to Donnan-like potentials at the resin boundaries (the ionic activitiee in the polyelectrolyte phase being quite high even though the ions are associated with fixed charges).

Hodgkin and Keynes (1953) concluded from the movement of ^{42}K ions inside short lengths of *Sepia* giant axons that their diffusivity and electrokinetic mobility in the protoplasm were essentially like those in a half-molar KCl solution, and that at least 90% of the cell K+ was free to exchange. Harris (1954) reported similar findings in frog muscle fibers. Kushmerick and Podolsky (1969) did find about a 50% lowering of the diffusivity of tagged ions after injection into muscle fibers, but since the same was true for nonelectrolytes, this was assigned to a viscosity consideration rather than to any specific binding. Similarly, Carpenter *et al.* (1973) suggested that the quite low conductivities which they have observed in *Aplysia* axoplasm (in spite of apparently high activity coefficients for Na^+ and K^+) must derive from the structuring of the water. However, Caillé and Hinke (1972), in studying the diffusion of labeled solutes through giant barnacle muscle fibers, noted a deviation from the theoretical profile in the concentration-distance curve for Na^+ which did not occur with sorbitol-^{14}C and was suggestive of an excess local accumulation of the Na^+ along the diffusion pathway.

Varying estimates of intracellular Na^+ and K^+ activities have been made with specific ion-selective glass microelectrodes in squid giant axons (Hinke, 1961), frog muscle fibers (Lev, 1964), and giant barnacle muscle fibers (McLaughlin and Hinke, 1966; Hinke and McLaughlin, 1967); all agree that the major part of the cell K^+, but substantially less of the cell Na^+, appears to be free. Hinke established that the transmembranal difference in Na^+ potential properly exceeded to a small degree the height of the squid axon's action potential. When the barnacle muscles were irreversibly shortened by elevated temperatures, the Na^+ activity more than doubled (Hinke and McLaughlin, 1967), while the K^+ activity scarcely changed.

3. PROBLEMS RELATING TO ACTIVE-TRANSPORT PHENOMENA

A very notable advantage of the sorption theories of solute distribution lies in their elimination of the need for investing a considerable fraction of the tissue's metabolic energy output (perhaps even more than is available: Minkoff and Damadian, 1973) in the maintenance of normal transmembranal gradients.[3] Ling (1956b) particularly criticized the undeservedly special consideration accorded to Na^+ by the membranologists, in designing the "sodium pump" to account for the failure of this ion to attain a Donnan distribution in the face of its fairly ready permeation of most membranes. Ling noted that, with application of the same criteria to other components (notably Ca^{2+} and Mg^{2+}), one quickly arrives at ridiculously high figures for the calculated energy demand (further discussed by Ling *et al.*, 1973).

The sorptionists seldom discuss the many situations in which biological translocations effect an increase in (electro)chemical activity of the transferred species in the passage between two simple aqueous compart-

[3] However, Kornacker (1972) argued on the basis of statistical mechanics that conventional calculations very substantially overstate the efficiency required in biological active transports.

ments, as in most transepithelial active transports, but do not deny the necessity for intermolecular association with a vectorial component in these situations. Thus Troshin (1966) acknowledged that metabolic energy must be expended in linkage to a "pump" in such epithelia, but concluded that "the morphology of these cells shows clearly that such a 'pump' mechanism is to be found inside, and not on the surface of, the cells." He presented for this type of pump a model that is indistinguishable from an activated version of the mobile-carrier model of Fig. 1, except that the whole cell, rather than its membrane, is taken to be the barrier. That the fundamental process involved here is directly identifiable with that underlying cellular accumulation is, however, strongly indicated by the experiments of Oxender and Christensen (1959). In essence, they converted packed layers of Ehrlich ascites tumor cells or of HeLa cells into temporary amino acid-secreting *pseudo*-epithelia by addition to the *cis*-compartment of pyridoxal, 4-nitrosalicylaldehyde, or extra K^+ (agents known to enhance cellular accumulation of the same amino acids which respond in the artificial-layer arrangement). Moreover, Christensen and Riggs (1952) established that osmotic swelling of these cells accompanied intracellular accumulation of the amino acids, in appropriate proportion; also, the cell lysates failed to concentrate amino acids appreciably in a dialysis equilibrium, and in some instances the total cellular accumulation exceeded any reasonable stoichiometric limits for binding (Christensen, 1955). Detailed studies of the simple exponential form of equilibration (from either direction) of glycine distribution in these cells led Heinz (1957) also to the conclusion that at least 90% of the intracellular glycine must be free and in a single compartment. Also in spheroplasts from *Escherichia coli*, Sistrom (1958) has correlated osmotic swelling quantitatively with a wide variety of factors governing the operation of the β-galactoside permease system, such as to suggest that only a quite trivial fraction of the intracellular galactosides could be bound (if any). Schultz *et al.* (1966) and Csáky and Esposito (1969) similarly observed a proper degree of swelling of rabbit and frog gut mucosal epithelium upon accumulation of amino acids or sugars, suggestive of full osmotic activity of the intracellular nutrilites. Neville (1973), however, disputed the conclusiveness of any of these arguments for ruling out specific amino acid adsorption as the basis for the apparently concentrative uptake, and even chose to interpret the observed cotransport and preloading phenomena on the basis of inductive interplay between adsorptive sites.

A related type of argument for the membranal localization of the critical event in cellular accumulations is the alleged ability of red blood cell ghosts, properly fortified but lacking the bulk of the normal cell contents, to continue active cation transport. Gárdos (1954) showed accumulation of K^+ by such preparations, but there is uncertainty as to whether contamination by unlysed cells may be responsible for this (Freedman, 1973); Hoffman (1962) found Na^+ extrusion by a strophanthidin-sensitive mechanism in rather pure ghost preparations, but this involved no definite demonstration of transfer against an activity gradient. Ellory and Tucker's (1970) finding that a ouabain-sensitive uptake of K^+ in low-K sheep erythrocytes could be activated by application of anti-L antibodies also suggests a surface localization of the critical apparatus. Ghosts have also been found to accumulate glycine in the direction of an applied Na^+ gradient (Vidaver, 1964a) or to extrude it in indirect response to a change in the Donnan potential (Vidaver, 1964b); Crane (1965) has similarly induced reversal of the direction of uphill intestinal glucose transfer *in vitro* by reversal of the Na^+ gradient.

IV. EVALUATION OF PRESENT POSITION AND FUTURE PROSPECTS

In summary of the several considerations dealt with in this chapter, the author would emphasize the following points of intermixed fact and personal opinion.

1. An enormous body of evidence strongly suggests that *nearly all* functionally significant biological permeations involve some sort of transient *intermolecular association* between the permeant species and rather specific components of the traversed membranes.

2. *No completely reliable criterion* is at hand for establishing in a given case whether these membrane components are "true carriers" forming a membrane-soluble *mobile* complex with the permeant, and thus providing a ferryboat for its traversal of the membrane; but the bulk of the experimental data are *compatible* with this interpretation.

3. Although many permeant-binding constituents of *microbial* membranes have been identified and strongly implicated in transport processes, there is still substantial uncertainty regarding the functional role of comparable ingredients thus far reported in *animal* cell membranes. It is, however, anticipated that major developments in this type of biochemical attack on the issue will be soon forthcoming, and will probably establish in due course a *multiplicity* of molecular organizations underlying different transport mechanisms. It is likely that both protein and phospholipid components will prove critical, and that the gross physicochemical assembly of the membrane structure will prove essential to the endowment of the specific components with their permeant-selective characteristics.

4. A significant fraction of protoplasmic water in many types of cell is at least to some degree structured, such that its solvency properties may differ appreciably from those of simple water. Also, substantial fractions of some cellular ions appear to be compartmentalized within intracellular barriers and partially restricted by association with macromolecules. These considerations will almost certainly receive increasing attention, as factors greatly complicating the analysis of solute translocation mechanisms and masking the interfacial processes. In some instances, such studies may well entail a critical reevaluation of the reality of membrane transport mechanisms previously presumed on the basis of a more simplistic view of the cytoplasm.

5. Certain quantitative data on the behavior of nonaccumulative transport systems cannot be accommodated by the long-accepted symmetrical mobile-carrier models; these inconsistencies can be largely resolved by presuming various *asymmetries* in the carrier operation in the two directions, but several recent proposals along these lines suggest quite conflicting conclusions regarding the basic mechanism.

6. The anomalous observations are also encompassed to varying degrees by several alternative models, which share the feature that the critical membrane sites *individually interact with the permeant only on one side* of the membrane; the transmembranal interplay then arises only by way of the characteristics of *exchanges between linked sites* on the two sides rather than by their actual traversal of the barrier. Although no such model as yet appears wholly adequate to cover existing observations, the prospects in this direction are quite promising, and are bolstered by the ready compatibility of such models with currently developing hypotheses regarding membrane morphology at the molecular level.

ACKNOWLEDGMENT

The major part of the literature search in the preparation of this chapter was carried out during the summer of 1973 at the Marine Biological Laboratory, Woods Hole, Massachusetts.

REFERENCES

Albers, R.W. (1967). Biochemical aspects of active transport. *Annu. Rev. Biochem.* **36**, 727-756.

Allen, R.D., and Hinke, J.A.M. (1970). Sodium compartmentalization in single muscle fibers of the giant barnacle. *Can. J. Physiol. Pharmacol.* **48**, 139–146.

Allen, R.D., and Hinke, J.A.M. (1971). Single influx rate for ^{42}K in the single fiber of the barnacle. *Can. J. Physiol. Pharmacol.* **49**, 624-626.

Andreoli, T.E., Tieffenberg, M., and Tosteson, D.C. (1967). The influence of valinomycin on the ionic permeability of thin lipid membranes. *J. Gen. Physiol.* **50**, 2527-2545.

Anraku, Y. (1968). Transport of sugars and amino acids in bacteria. III. Studies on the restoration of active transport. *J. Biol. Chem.* **243**, 3128-3135.

Ashley, C.C. (1970). Ion movements in skeletal muscle. *In* "Membranes and Ion Transport" (E.E. Bittar, ed.), Vol. 2, pp. 1-31. Wiley (Interscience), New York.

Baker, G.F., and Widdas, W.F. (1973). The asymmetry of the facilitated transfer system for hexoses in human red cells and the simple kinetics of a two component model. *J. Physiol. (London)* **231**, 143-165.

Baker, R.D. (1967). Solubilization of sugars in non-polar solvent by intestinal lipid: Possible relation to membrane transport. *Tex. Rep. Biol. Med.* **25**, 223-239.

Bangham, A.D. (1970). The liposome as a membrane model. *In* "Permeability and Function of Biological Membranes" (L. Bolis *et al.*, eds.), pp. 195-206. North-Holland Publ., Amsterdam.

Barnett, J.E.G., Holman, G.D., and Munday, K.A. (1973). An explanation of the asymmetric binding of sugars to the human erythrocyte sugar-transport systems. *Biochem. J.* **135**, 539-541.

Batt, E.R., and Schachter, D. (1973). Transport of monosaccharides. I. Asymmetry in the human erythrocyte mechanism. *J. Clin. Invest.* **52**, 1686-1697.

Beneš, I., Kolínská, J., and Kotyk, A. (1972). Effect of phloretin on monosaccharide transport in erythrocyte ghosts. *J. Membrane Biol.* **8**, 303-309.

Berendsen, H.J.C., and Edzes, H.T. (1973). The observation and general interpretation of sodium magnetic resonance in biological material. *Ann. N. Y. Acad. Sci.* **204**, 459-485.

Bing, H.J. (1899). Untersuchungen über die reducirenden Substanzen im Blute. *Skand. Arch. Physiol.* **9**, 336-411.

Bing, H.J. (1901). Ueber Lecithinverbindungen. *Skand. Arch. Physiol.* **11**, 166-175.

Binkley, F., King, N., Milckin, E., Wright, R.K., O'Neal, C.H., and Wundram, I.J. (1968). Brush border particulates of renal tissue. *Science* **162**, 1009-1011.

Blaustein, M.P., and Goldman, D.E. (1966). Action of anionic and cationic nerve-blocking agents: Experiment and interpretation. *Science* **153**, 429-432.

Bobinski, H., and Stein, W.D. (1966). Isolation of a glucose-binding component from human erythrocyte membranes. *Nature (London)* **211**, 1366-1368.

Bode, F., Baumann, K., Frasch, W., and Kinne, R. (1970). Die Bindung von Phlorrhizin an die Bürstensaumfraktion der Rattenniere. *Pfluegers Arch.* **315**, 53-65.

Bode, F., Baumann, K., and Diedrich, D.E. (1972). Inhibition of [³H] phlorizin binding to isolated kidney brush porder membranes by phlorizin-like compounds. *Biochim. Biophys. Acta* **290**, 134-149.

Bonsall, R.W., and Hunt, S. (1966). Solubilization of a glucose-binding component of the red cell membrane. *Nature (London)* **211**, 1368-1370.

Bonting, S.L. (1970). Sodium-potassium activated adenosinetriphosphatase and cation transport. *In* "Membranes and Ion Transport" (E.E. Bittar, ed.), Vol. 1, pp. 257-363. Wiley (Interscience), New York.

Boos, W. (1974). Pro and contra carrier proteins; sugar transport via the periplasmic galactose-binding protein. *Curr. Top. Membranes Transp.* **5**, 51-136.

Bowyer, F., and Widdas, W.F. (1958). The action of inhibitors on the facilitated hexose transfer system in erythrocytes. *J. Physiol. (London)* **141**, 219-232.

Bozler, E., Calvin, M.E., and Watson, D.W. (1958). Exchange of electrolytes in smooth muscle. *Amer. J. Physiol.* **195**, 38-44.

Bratton, C.B., Hopkins, A.L., and Weinberg, J.W. (1965). Nuclear magnetic resonance studies of living muscle. *Science* **147**, 738-739.

Britton, H.G. (1956). The permeability of the human red cell to labelled glucose. *J. Physiol. (London)* **135**, 61P-62P.

Britton, H.G. (1963). Induced uphill and downhill transport: Relationship to the Ussing criterion. *Nature (London)* **198**, 190-191.

Britton, H.G. (1964). Permeability of the human red cell to labelled glucose. *J. Physiol. (London)* **170**, 1-20.

Britton, H.G. (1965). Fluxes in passive, monovalent and polyvalent carrier systems. *J. Theor. Biol.* **10**, 28-52.

Brush, J.S., and Krawczyk, M.E. (1969). Identification and partial purification of a probable glucose transport protein from rat muscle. *Fed. Proc., Fed. Amer. Soc. Exp. Biol.* **28**, 463.

Burns, M.J., and Faust, R.G. (1969). Preferential binding of amino acids to isolated mucosal brush borders from hamster jejunum. *Biochim. Biophys. Acta* **183**, 642-645.

Busse, D., Elsas, L.J., and Rosenberg, L.E. (1972). Uptake of D-glucose by renal tubule membranes. I. Evidence for two transport systems. *J. Biol. Chem.* **247**, 1188-1193.

Caillé, J.P., and Hinke, J.A.M. (1972). Evidence for Na sequestration in muscle from Na diffusion measurements. *Can. J. Physiol. Pharmacol.* **50**, 228-237.

Calissano, P., and Bangham, A.D. (1971). Effect of two brain specific proteins (S100 and 14.3.2) on cation diffusion across artificial lipid membranes. *Biochem. Biophys. Res. Commun.* **43**, 504-509.

Carpenter, D.O., Hovey, M.M., and Bak, A.F. (1973). Measurements of intracellular conductivity in *Aplysia* neurons: Evidence for organization of water and ions. *Ann. N. Y. Acad. Sci.* **204**, 502-533.

Carter, J.R., and Martin, D.B. (1969). Glucose uptake by isolated particles from rat epididymal adipose tissue cells. *Proc. Nat. Acad. Sci. U.S.* **64**, 1343-1348.

Carter, J.R., Jr., Avruch, J., and Martin, D.B. (1972). Glucose transport in plasma membrane vesicles from rat adipose tissue. *J. Biol. Chem.* **247**, 2682-2688.

Carter, J.R., Jr., Avruch, J., and Martin, D.B. (1973). Glucose transport by trypsin-treated red blood cell ghosts. *Biochim. Biophys. Acta* **291**, 506-518.

Chapman, G., and McLauchlan, K.A. (1967). Oriented water in the sciatic nerve of rabbit. *Nature (London)* **215**, 391-392.

Chavin, S.J. (1971). Isolation and study of functional membrane proteins. Present status and future prospects. *FEBS (Fed. Eur. Biochem. Soc.) Lett.* **14**, 269-282.

Chesney, R.W. (1971). Binding of D-glucose to the isolated rabbit renal brush border. *Fed. Proc., Fed. Amer. Soc. Exp. Biol.* **30**, 1116 Abs.

Chesney, R.W., Sacktor, B., and Rowen, R. (1973). The binding of D-glucose to the isolated luminal membrane of the renal proximal tubule. *J. Biol. Chem.* **248**, 2182-2191.

Christensen, H.N. (1939). The contaminants of blood phospholipids. *J. Biol. Chem.* **129**, 531-538.

Christensen, H.N. (1955). Mode of transport of amino acids into cells. *In* "Amino Acid Metabolism" (W.D. McElroy and B. Glass, eds.), pp. 63-106. Johns Hopkins Press, Baltimore, Maryland.

Christensen, H.N. (1960). Reactive sites and biological transport. *Advan. Protein Chem.* **15**, 239-314.

Christensen, H.N. (1970). Linked ion and amino acid transport. *In* "Membranes and Ion Transport" (E.E. Bittar, ed.), Vol. 1, pp. 365-394. Wiley (Interscience), New York.

Christensen, H.N. (1972). On the meaning of effects of substrate structure on biological transport *J. Bioenerg.* **4**, 31-61.

Christensen, H.N., and Hastings, A.B. (1940). Phosphatides and inorganic salts. *J. Biol. Chem.* **136**, 387-398.

Christensen, H.N., and Riggs, T.R. (1952). Concentrative uptake of amino acids by the Ehrlich mouse ascites carcinoma cell. *J. Biol. Chem.* **194**, 57-68.

Christensen, H.N., Riggs, T.R., Fischer, H., and Palatine, I.M. (1952). Amino acid concentration by a free cell neoplasm: Relations among amino acids. *J. Biol. Chem.* **198**, 1-15.

Cogoli, A., Mosiman, H., Vock, C., von Balthazar, A-K., and Semenza, G. (1973). A simplified procedure for the isolation of the sucrase-isomaltase complex from rabbit intestine. *Eur. J. Biochem.* **30**, 7-14.

Cohen, G.N., and Monod, J. (1957). Bacterial permeases. *Bacteriol. Rev.* **21**, 169-194.

Cooke, R., and Wien, R. (1973). Nuclear magnetic resonance studies of intracellular water protons. *Ann. N. Y. Acad. Sci.* **204**, 197-209.

Cope, F.W. (1963). A theory of enzyme kinetics based on electron conduction through the enzymatic particles, with applications to cytochrome oxidases and to free radical decay in melanin. *Arch. Biochem. Biophys.* **103**, 352-365.

Cope, F.W. (1965a). A theory of ion transport across surfaces by a process analogous to electron transport across liquid-solid interfaces. *Bull. Math. Biophys.* **27**, 99-109.

Cope, F.W. (1965b). Nuclear magnetic resonance evidence for complexing of sodium ions in muscle. *Proc. Nat. Acad. Sci. U.S.* **54**, 225-227.

Cope, F.W. (1967a). A theory of cell hydration governed by adsorption of water on cell proteins rather than by osmotic pressure. *Bull. Math. Biophys.* **29**, 583-596.

Cope, F.W. (1967b). A non-equilibrium thermodynamic theory of leakage of complexed Na^+ from muscle, with NMR evidence that the non-complexed fraction of muscle Na^+ is intra-vacuolar rather than extra-cellular. *Bull. Math. Biophys.* **29**, 691-704.

Cope, F.W. (1967c). NMR evidence for complexing of Na^+ in muscle, kidney, and brain, and by actinomyosin. The relation of cellular complexing of Na^+ to water structure and to transport kinetics. *J. Gen. Physiol.* **50**, 1353-1375.

Cope, F.W. (1969). Nuclear magnetic resonance evidence using D_2O for structured water in muscle and brain. *Biophys. J.* **9**, 303-319.

Cope, F.W. (1970). The solid-state physics of electron and ion transport in biology. *Advan. Biol. Med. Phys.* **13**, 1-42.

Cope, F.W., and Damadian, R. (1970). Cell potassium by ^{39}K spin echo nuclear magnetic resonance. *Nature (London)* **228**, 76-77.

Crane, R.K. (1962). Hypothesis for mechanism of intestinal active transport of sugars. *Fed. Proc., Fed. Amer. Soc. Exp. Biol.* **21**, 891-895.

Crane, R.K. (1965). Na^+-dependent transport in the intestine and other animal tissues. *Fed. Proc., Fed. Amer. Soc. Exp. Viol.* **24**, 1000-1006.

Crane, R.K. (1966). Structural and functional organization of an epithelial cell brush border. *Symp. Int. Soc. Cell. Biol.* **5**, 71-102.

Crane, R.K. (1967). Gradient coupling and the membrane transport of water-soluble compounds: A general biological mechanism? *Protides Biol. Fluids, Proc. Colloq.* **15**, 227-235.

Crane, R.K. (1968). Digestive-absorptive surface of the small bowel mucosa. *Annu. Rev. Med.* **19**, 57-68.

Cruickshank, J. (1920). The adsorption of dyes and inorganic salts by solutions of lecithin. *J. Pathol. Bacteriol.* **23**, 230-232.

Csáky, T.Z., and Esposito, G. (1969). Osmotic swelling of intestinal epithelial cells during active sugar transport. *Amer. J. Physiol.* **217**, 753-755.

Csáky, T.Z., and Fernald, G.W. (1961). Localization of the "sugar pump" in the intestinal epithelium. *Nature (London)* **191**, 709-711.

Curran, P.F. (1968). Coupling between transport processes in intestine. *Physiologist* **11**, 3-23.

Curran, P.F., Schultz, S.G., Chez, R.A., and Fuisz, R.E. (1967). Kinetic relations of the Na-amino acid interaction at the mucosal border of intestine. *J. Gen. Physiol.* **50**, 1261-1286.

Czech, M.P., Lynn, D.G., and Lynn, W.S. (1973). Cytochalasin B-sensitive 2-deoxy-D-glucose transport in adipose cell ghosts. *J. Biol. Chem.* **248**, 3636-3641.

Czeisler, J.L., and Swift, T.J. (1973). A comparative study of sodium ion in muscle tissue and ion exchange resins through the application of nuclear magnetic resonance. *Ann. N. Y. Acad. Sci.* **204**, 261-273.

Czeisler, J.L., Fritz, O.G., and Swift, T.J. (1970). Direct evidence from nuclear magnetic resonance studies for bound sodium in frog skeletal muscle. *Biophys. J.* **10**, 260-268.

Dainty, J. (1963). Water relations of plant cells. *Advan. Bot. Res.* **1**, 270-326.

Dainty, J., and House, C.R. (1966a). "Unstirred layers" in frog skin. *J. Physiol. (London)* **182**, 66-78.

Dainty, J., and House, C.R. (1966b). An examination of the evidence for membrane pores in frog skin. *J. Physiol. (London)* **185**, 172-184.

Damadian, R. (1969). Ion exchange in Escherichia coli: Potassium-binding proteins. *Science* **165**, 79-81.

Damadian, R. (1973). Biological ion exchanger resins. *Ann. N. Y. Acad. Sci.* **204**, 211-248.

Danielli, J.F. (1952). Structural factors in cell permeability and secretion. *Symp. Soc. Exp. Biol.* **6**, 1-15.

Danielli, J.F. (1954). The present position in the field of facilitated diffusion and selective active transport. *Proc. Symp. Colston Res. Soc.* **7**, 1-14.

Dydyńska, M., and Wilkie, D.R. (1963). The osmotic properties of striated muscle fibres in hypertonic solutions. *J. Physiol. (London)* **169**, 312-329.

Eady, R.P., and Widdas, W.F. (1973). The use of sugars and fluorodinitrobenzene (*FDNB*) to differentially label red cell membrane components involved in hexose transfers. *Quart. J. Exp. Physiol. Cog. Med. Sci.* **58**, 59-66.

Edwards, P.A.W. (1973a). The inactivation by fluorodinitrobenzene of glucose transport across the human erythrocyte membrane. The effect of glucose inside or outside the cell. *Biochim. Biophys. Acta* **307**, 415-418.

Edwards, P.A.W. (1973b). Evidence for the carrier model of transport from the inhibition by N-ethylmaleimide of choline transport across the human red cell membrane. *Biochim. Biophys. Acta* **311**, 123-140.

Eichholz, A. (1969). Fractions of the brush border. *Fed. Proc., Fed. Amer. Soc. Exp. Biol.* **28**, 30-34.

Eichholz, A., Howell, K.E., and Crane, R.K. (1969). Studies on the organization of the brush border in intestinal epithelial cells. VI. Glucose binding to isolated intestinal brush borders and their subfractions. *Biochim. Biophys. Acta* **193**, 179-192.

Eilam, Y., and Stein, W.D. (1972). A simple resolution of the kinetic anomaly in the exchange of different sugars across the membrane of the human red blood cell. *Biochim. Biophys. Acta* **266**, 161-173.

Eisenman. G. (1961). On the elementary atomic origin of equilibrium ionic specificity. *In* "Membrane Transport and Metabolism" (A. Kleinzeller and A. Kotyk, eds.), pp. 163-179. Academic Press, New York.

Eisenman, G., Ciani, S.M., and Szabo, G. (1968). Some theoretically expected and experimentally observed properties of lipid bilayer membranes containing neutral molecular carriers of ions. *Fed. Proc., Fed. Amer. Soc. Exp. Biol.* **27**, 1289-1304.

Ellory, J.C., and Tucker, E.M. (1970). A specific antigen-antibody reaction affecting ion transport in sheep LK erythrocytes. *In* "Permeability and Function of Biological Membranes" (L. Bolis *et al.*, eds.), pp. 120-127. North-Holland Publ., Amsterdam.

Ernst, E., and Hajnal, M. (1959). Distribution and condition of potassium in muscle. (Translated for *Chem. Abstr.*) *Acta Physiol.* **16**, 77-86.

Faust, R.G., Wu, S.M.L., and Faggard, M.L. (1967). D-glucose: Preferential binding to brush borders disrupted with tris(hydroxymethyl)aminomethane. *Science* **155**, 1261-1263.

Faust, R.G., Leadbetter, M.G., Plenge, R.K., and McCaslin, A.J. (1968). Active sugar transport by the small intestidine. The effects of sugars, amino acids, hexosamines, sulfhydryl-reacting compounds, and cations on the preferential binding of D-glucose to Tris-disrupted brush borders. *J. Gen. Physiol.* **52**, 482-494.

Faust, R.G., Burns, M.J., and Misch, D.W. (1970). Sodium-dependent binding of L-histidine to a fraction of mucosal brush borders from hamster jejunum. *Biochim. Biophys. Acta* **219**, 507-511.

Faust, R.G., Shearin, S.J., and Misch, D.W. (1972). Sodium-dependent binding of D-glucose to a filamentous fraction of Tris-disrupted brush borders from hamster jejunum. *Biochim. Biophys. Acta* **255**, 685-690.

Feinstein, M.B. (1964). Reaction of local anesthetics with phospholipids. A possible chemical basis for anesthesia. *J. Gen. Physiol.* **48**, 357-374.

Fenichel, I.R., and Horowitz, S.B. (1963). The transport of nonelectrolytes in muscle as a diffusional process in cytoplasm. *Acta Physiol. Scand.* **60**, Suppl. 221, 1-63.

Fernie, S.M., Hart, S.L., and Nissim, J.A. (1967). Glucose binding by homogenates of intestinal mucosa. *Nature (London)* **213**, 985-987.

Fischer, M.H., and Suer, W.J. (1938). Physicochemical state of protoplasm. *Arch. Pathol.* **25**, 51-69.

Folch, J., and Van Slyke, D.D. (1939). Nitrogenous contaminants in petroleum ether extracts of plasma lipids. *J. Biol. Chem.* **129**, 539-546.

Foster, D.O., and Pardee, A.B. (1969). Transport of amino acids by confluent and nonconfluent 3T3 and polyoma virus-transformed 3T3 cells growing on glass cover slips. *J. Biol. Chem.* **244**, 2675-2681.

Fox, C.F., and Kennedy, E.P. (1965). Specific labeling and partial purification of the M protein, a component of the β-galactoside transport system of Escherichia coli. *Proc. Nat. Acad. Sci. U.S.* **54**, 891-899.

Fox, C.F., Carter, J.R., and Kennedy, E.P. (1967). Genetic control of the membrane protein component of the lactose transport system of Escherichia coli. *Proc. Nat. Acad. Sci. U.S.* **57**, 698-705.

Frasch, W., Frohnert, P.P., Bode, F., Baumann, K., and Kinne, R. (1970). Competitive inhibition of phlorizin binding by D-glucose and the influence of sodium: A study on isolated brush border membrane of rat kidney. *Pfluegers Arch.* **320**, 265-284.

Freedman, J.C. (1973). Do red cell ghosts pump sodium or potassium? *Ann. N. Y. Acad. Sci.* **204**, 609-615.

Freundlich, H., and Gann, J.A. (1915). Über kolloide Lösungen in Chloroform. *Int. Z. Phys.-Chem. Biol.* **2**, 1-18.

Fritz, O.G., and Swift, T.J. (1967). The state of water in polarized and depolarized frog nerves. A proton magnetic resonance study. *Biophys. J.* **7**, 675-687.

Gachelin, G. (1970). Studies on the α-methylglucoside permease of Escherichia coli. A two-step mechanism for the accumulation of α-methylglucoside-6-phosphate. *Eur. J. Biochem.* **16**, 342-357.

Gárdos, G. (1954). Akkumulation der Kaliumionen durch menschliche Blutkörperchen. *Acta Physiol.* **6**, 191-199.

Gayton, D.C., and Hinke, J.A.M. (1971). Evidence for the heterogeneous distribution of chloride in the barnacle muscle. *Can. J. Physiol. Pharmacol.* **49**, 323-330.

Geck, P. (1971). Eigenschaften eines asymmetrischen Carrier-Modells für den Zuckertransport am menschlichen Erythrozyten. *Biochim. Biophys. Acta* **241**, 462-472.

Geck, P., Heinz, E., and Pfeiffer, B. (1972). The degree and the efficiency of coupling between the influxes of Na$^+$ and α-aminioisobutyrate in Ehrlich cells. *Biochim. Biophys. Acta* **288**, 486-491.

Ginzburg, B.Z., and Katchalsky, A. (1964). The frictional coefficients of the flows of non-electrolytes through artificial membranes. *J. Gen. Physiol.* **47**, 403-418.

Glossmann, H., and Neville, M., Jr. (1972). Phlorizin receptors in kidney brush border membranes. *Hoppe-Seyler's Z. Physiol. Chem.* **355**, 708-709.

Glynn, I.M., Slayman, C.W., Eichberg, J., and Dawson, R.M.C. (1965). The adenosine-triphosphatase system responsible for cation transport in electric organ: Exclusion of phospholipids as intermediates. *Biochem. J.* **94**, 692-699.

Gould, R.M., and London, Y. (1972). Specific interaction of central nervous system myelin basic protein with lipids. Effects of basic protein on glucose leakage from liposomes. *Biochim. Biophys. Acta* **290**, 200-218.

Gulati, J. (1973). Cooperative interaction of external calcium, sodium, and ouabain with the cellular potassium in smooth muscle. *Ann. N. Y. Acad. Sci.* **204**, 337-357.

Guroff, G., Fanning, G.R., and Chirigos, M.A. (1964). Stimulation of aromatic amino acid transport by *p*-fluorophenylalanine in the Sarcoma 37 cell. *J. Cell. Comp. Physiol.* **63**, 323-331.

Hankin, B.L., and Stein, W.D. (1972). On the temperature dependence of initial velocities of glucose transport in the human red blood cell. *Biochim. Biophys. Acta* **288**, 127-136.

Hankin, B.L., Lieb, W.R., and Stein, W.D. (1972). Rejection criteria for the asymmetric carrier and their application to glucose transport in the human red blood cell. *Biochim. Biophys. Acta* **288**, 114-126.

Hare, J.D. (1967). Location and characteristics of the phenylalanine transport mechanism in normal and polyoma-transformed hamster cells. *Cancer Res.* **27**, 2357-2363.

Harris, E.J. (1954). Ionophoresis along frog muscle. *J. Physiol. (London)* **124**, 248-253.

Hashish, S.E.E. (1958). Effects of low temperatures and heparin on potassium exchangeability in rat diaphragm. *Acta Physiol. Scand.* **43**, 189-199.

Haškovec, C., and Kotyk, A. (1969). Attempts at purifying the galactose carrier from galactose-induced baker's yeast. *Eur. J. Biochem.* **9**, 343-347.

Hatanaka, M., and Gilden, R.V. (1970). Virus-specified changes in the sugar-transport kinetics of rat embryo cells infected with murine sarcoma virus. *J. Nat. Cancer Inst.* **45**, 87-89.

Hatanaka, M., and Gilden, R.V. (1971). Induction of sugar uptake by a hamster pseudotype sarcoma virus. *Virology* **43**, 734-736.

Hatanaka, M., and Hanafusa, H. (1970). Analysis of a functional change in membrane in the process of cell transformation by Rous sarcoma virus; alteration in the characteristics of sugar transport. *Virology* **41**, 647-652.

Hatanaka, M., Huebner, R.J., and Gilden, R.V. (1969). Alterations in the characteristics of sugar uptake by mouse cells transformed by murine sarcoma viruses. *J. Nat. Cancer Inst.* **43**, 1091-1096.

Hatanaka, M., Augl, C., and Gilden, R.V. (1970). Evidence for a functional change in the plasma membrane of murine sarcoma virus-infected mouse embryo cells: Transport and transport-associated phosphorylation of ^{14}C-2-deoxy-D-glucose. *J. Biol. Chem.* **245**, 714-717.

Haydon, D.A. (1970). The organization and permeability of artificial lipid membranes. *In* "Membranes and Ion Transport" (E.E. Bittar, ed), Vol. 1, pp. 64-92. Wiley (Interscience), New York.

Hazlewood, C.F., Nichols, B.L., and Chamberlain, N.F. (1969). Evidence for the existence of a minimum of two phases of ordered water in skeletal muscle. *Nature (London)* **222**, 747-750.

Hazlewood, C.F., Nichols, B.L., Chang, D.C., and Brown, B. (1971). On the state of water in developing muscle: A study of the major phase of ordered water in skeletal muscle and its relationship to sodium concentration. *Johns Hopkins Med. J.* **128**, 117-131.

Hechter, O., and Lester, G. (1960). Cell permeability and hormone action. *Recent Progr. Horm. Res.* **16**, 139-186.

Heckmann, K. (1965a). Zur Theorie der "Single File"-Diffusion I. *Z. Phys. Chem. (Frankfurt am Main)* [N.S.] **44**, 184-203.

Heckmann, K. (1965b). Zur Theorie der "Single File"-Diffusion II. *Z. Phys. Chem. (Frankfurt am Main)* [N.S.] **46**, 1-25.

Heckmann, K. (1968). Zur Theorie der "Single File"-Diffusion III. Sigmoide Konzentrationsabhängigkeit unidirektionaler Flüsse bei "Single File"-Diffusion. *Z. Phys. Chem. (Frankfurt am Main)* [N.S.] **58**, 206-219.

Heckmann, K. (1972). Single file diffusion. *In* "Biomembranes" (F. Kreuzer and J.F.G. Slegers, eds.), Vol. 3, pp. 127-153. Plenum, New York.

Heckmann, K., and Passow, H. (1967). Asymmetrische Diskriminierung bei Carrier- und Single-file-Diffusion. *Ber. Bunsenges. Phys. Chem.* **71**, 839-843.

Hegyvary, C., and Post, R.L. (1969). Reversible inactivation of (Na$^+$-K$^+$)-ATPase by removing and restoring phospholipids. *In* "The Molecular Basis of Membrane Function" (D.C. Tosteson, ed), pp. 519-528. Prentice-Hall, Englewood Cliffs, New Jersey.

Heinz, E. (1954). Kinetic studies on the "influx" of glycine-1-C^{14} into the Ehrlich mouse ascites carcinoma cell. *J. Biol. Chem.* **211**, 781-790.

Heinz, E. (1957). The exchangeability of glycine accumulated by carcinoma cells. *J. Biol. Chem.* **225**, 305-315.

Heinz, E. (1970). On the function of Na-ions in the transport of amino acids in Ehrlich carcinoma cells. *In* "Permeability and Function of Biological Membranes" (L. Bolis *et al.*, eds.), pp. 326-332. North-Holland Publ., Amsterdam.

Heinz, E., ed. (1972). "Na-linked Transport of Organic Solutes." Springer-Verlag, Berlin and New York.

Heinz, E., and Durbin, R.P. (1957). Studies of the chloride transport in the gastric mucosa of the frog. *J. Gen. Physiol.* **41**, 101-117.

Heinz, E., and Walsh, P.M. (1958). Exchange diffusion, transport, and intracellular level of amino acids in Ehrlich carcinoma cells. *J. Biol. Chem.* **233**, 1488-1493.

Heinz, E., Geck, P., and Wilbrandt, W. (1972). Coupling in secondary active transport. Activation of transport by co-transport and/or counter-transport with the fluxes of other solutes. *Biochim. Biophys. Acta* **255**, 442-461.

Hendler, R.W. (1959). Passage of radioactive amino acids through "nonprotein" fractions of hen oviduct during incorporation into protein. *J. Biol. Chem.* **234**, 1466-1473.

Hill, A.V. (1930). The state of water in muscle and blood and the osmotic behaviour of muscle. *Proc. Roy. Soc., Ser. B* **106**, 477-505.

Hillman, R.E., and Rosenberg, L.E. (1970). Amino acid transport by isolated mammalian renal tubules. III. Binding of L-proline by proximal tubule membranes. *Biochim. Biophys. Acta* **211**, 318-326.

Hinke, J.A.M. (1961). The measurement of sodium and potassium activities in the squid axon by means of cation-selective glass micro-electrodes. *J. Physiol. (London)* **156**, 314-335.

Hinke, J.A.M. (1970). Solvent water for electrolytes in the muscle fiber of the giant barnacle. *J. Gen. Physiol.* **56**, 521-541.

Hinke, J.A.M., and McLaughlin, S.G.A. (1967). Release of bound sodium in single muscle fibers. *Can. J. Physiol. Pharmacol.* **45**, 655-667.

Hoare, D.G. (1972a). A new kinetic analysis of the leucine transport carrier in erythrocytes. In "Biomembranes" (F. Kreuzer and J.F.G. Slegers, eds.), Vol. 3, pp. 107-116. Plenum, New York.

Hoare, D.G. (1972b). The temperature dependence of the transport of L-leucine in human erythrocytes. *J. Physiol. (London)* **221**, 331-348.

Höber, R. (1899). Über Resorption im Dünndarm. *Arch. Gesamte Physiol. Menschen Tiere* **74**, 246-271.

Hodgkin, A.L., and Horowicz, P. (1959). Movements of sodium and potassium in single muscle fibers. *J. Physiol. (London)* **145**, 405-432.

Hodgkin, A.L., and Keynes, R.D. (1953). The mobility and diffusion coefficient of potassium in gaint axons from *Sepia. J. Physiol. (London)* **119**, 513-528.

Hoffman, J.F. (1962). The active transport of sodium by ghosts of human red blood cells. *J. Gen. Physiol.* **45**, 837-859.

Hoffman, J.F., Schulman, J.H., and Eden, M. (1959). Specific Na⁺ carriage by cephalin in a model system. *Fed. Proc., Fed. Amer. Soc. Exp. Biol.* **18**, 70.

Hokin, L.E., and Dahl, J.L. (1972). The sodium-potassium adenosinetriphosphatase. In "Metabolic Pathways" (L.E. Hokin, ed.), 3rd ed., Vol. 6, pp. 269-315. Academic Press, New York.

Hokin, L.E., and Hokin, M.R. (1959). Evidence for phosphatidic acid as the sodium carrier. *Nature (London)* **184**, 1068-1069.

Hokin, L.E., and Hokin, M.R. (1960a). Studies on the carrier function of phosphatidic acid in sodium transport. I. The turnover of phosphatidic acid and phosphoinositide in the avian salt gland on stimulation of secretion. *J. Gen. Physiol.* **44**, 61-85.

Hokin, L.E., and Hokin, M.R. (1960b). The role of phosphatidic acid and phosphoinositide in transmembrane transport elicited by acetylcholine and other humoral agents. *Int. Rev. Neurobiol.* **2**, 99-136.

Hokin, L.E., and Hokin, M.R. (1961). Studies on the enzymic mechanism of the sodium pump. In "Membrane Transport and Metabolism" (A. Kleinzeller and Z. Kotyk, eds.), pp. 204-218. Academic Press, New York.

Hokin, M.R., and Hokin, L.E. (1964). The synthesis of phosphatidic acid and protein-bound phosphorylserine in salt gland homogenates. *J. Biol. Chem.* **239**, 2116-2122.

Holz, R., and Finkelstein, A. (1970). The water and nonelectrolyte permeability induced in thin lipid membranes by the polyene antibiotics nystatin and amphotericin B. *J. Gen. Physiol.* **56**, 125-145.

Hoos, R.T., Tarpley, H.L., and Regen, D.M. (1972). Sugar transport in beef erythrocytes. *Biochim. Biophys. Acta* **266**, 174-181.

Hopfer, U., Nelson, K., Perrotto, J., and Isselbacher, K.J. (1973). Glucose transport in isolated brush border membrane from rat small intestine. *J. Biol. Chem.* **248**, 25-32.

Horák, J., and Kotyk, A. (1973). Isolation of a glucose-binding lipoprotein from yeast plasma membrane. *Eur. J. Biochem.* **32**, 37-41.

Illiano, G., and Cuatrecasas, P. (1901). Glucose transport in fat cell membranes. *J. Biol. Chem.* **246**, 2472-2479.

Inbar, M., Ben-Bassat, H., and Sachs, L. (1971). Location of amino acid and carbohydrate transport sites in the surface membrane of normal and transformed mammalian cells. *J. Membrane Biol.* **6**, 195-209.

Ingersoll, R.J., and Wasserman, R.H. (1971). Vitamin D_3-induced calcium-binding protein. *J. Biol. Chem.* **246**, 2808-2814.

Israel, Y. (1969). Phospholipid activation of $(Na^+ + K^+)$-ATPase. *In* "The Molecular Basis of Membrane Function" (D.C. Tosteson, ed.), pp. 529-537. Prentice-Hall, Englewood Cliffs, New Jersey.

Isselbacher, K.J. (1972a). Sugar and amino acid transport by cells in culture—differences between normal and malignant cells. *N. Engl. J. Med.* **286**, 929-933.

Isselbacher, K.J. (1972b). Increased uptake of amino acids and 2-deoxy-D-glucose by virus-transformed cells in culture. *Proc. Nat. Acad. Sci. U.S.* **69**, 585-589.

Jacobs, M.H., Glassman, H.N., and Parpart, A.K. (1935). Osmotic properties of the erythrocyte. VII. The temperature coefficients of certain hemolytic processes. *J. Cell. Comp. Physiol.* **7**, 197-225.

Jacobs, M.H., Glassman, H.N., and Parpart, A.K. (1938). Osmotic properties of the erythrocyte. XI. Differences in the permeability of the erythrocytes of two closely related species. *J. Cell. Comp. Physiol.* **11**, 479-494.

Jacquez, J.A. (1961). The kinetics of carrier-mediated active transport of amino acids. *Proc. Nat. Acad. Sci. U.S.* **47**, 153-163.

Jacquez, J.A. (1963). Carrier-amino acid stoichiometry in amino acid transport in Ehrlich ascites cells. *Biochim. Biophys. Acta* **71**, 15-33.

Jacquez, J.A. (1964). The kinetics of carrier-mediated transport: Stationary-state approximations. *Biochim. Biophys. Acta* **79**, 318-328.

Jacquez, J.A. (1967). Competitive stimulation: Further evidence for two carriers in the transport of neutral amino acids. *Biochim. Biophys. Acta* **135**, 751-755.

Jacquez, J.A., and Sherman, J.H. (1965). The effect of metabolic inhibitors on transport and exchange of amino acids in Ehrlich ascites cells. *Biochim. Biophys. Acta* **109**, 128-141.

Jain, M.K., Strickholm, A., and Cordes, E.H. (1969). Reconstitution of an ATP-mediated active transport system across black lipid membranes. *Nature (London)* **222**, 871-872.

Jardetzky, O., and Snell, F.M. (1960). Theoretical analysis of transport processes in living systems. *Proc. Nat. Acad. Sci. U.S.* **46**, 616-622.

Johnstone, R.M. (1972). Glycine accumulation in absence of Na^+ and K^+ gradients in Ehrlich ascites cells: Shortfall of the potential energy from the ion gradients for glycine accumulation. *Biochim. Biophys. Acta* **282**, 366-373.

Jones, A.W., and Karreman, G. (1969). Potassium accumulation and permeation in the canine carotid artery. *Biophys. J.* **9**, 910-924.

Jones, T.H.D., and Kennedy, E.P. (1969). Characterization of the membrane protein component of the lactose transport system of *Escherichia coli*. *J. Biol. Chem.* **244**, 5981-5987.

Judah, J.D., and Ahmed, K. (1964). The biochemistry of sodium transport. *Biol. Rev. Cambridge Phil. Soc.* **39**, 160-193.

Jung, C.Y. (1971a). Permeability of bimolecular membranes made from lipid extracts of human red cell ghosts to sugars. *J. Membrane Biol.* **5**, 200-214.

Jung, C.Y. (1971b). Evidence of high stability of the glucose transport carrier function in human red cell ghosts extensively washed in various media. *Arch. Biochem. Biophys.* **146,** 215-226.

Jung, C.Y., Chaney, J.E., and LeFevre, P.G. (1968). Enhanced migration of glucose from water into chloroform in presence of phospholipids. *Arch. Biochem. Biophys.* **126,** 664-676.

Jung, C.Y., Carlson, L.M., and Whaley, D.A. (1971). Glucose transport carrier activities in extensively washed human red cell ghosts. *Biochim. Biophys. Acta* **241,** 613-627.

Jung, C.Y., Carlson, L.M., and Balzer, C.J. (1973). Effect of proteolytic digestion on glucose transport carrier of human erythrocyte ghosts. *Biochim. Biophys. Acta* **298,** 108-114.

Kaback, H.R. (1970). Transport. *Annu. Rev. Biochem.* **39,** 561-598.

Kaback, H.R. (1972). Transport across isolated bacterial cytoplasmic membranes. *Biochim. Biophys. Acta* **265,** 367-416.

Kaback, H.R., and Stadtman, E.R. (1966). Proline uptake by an isolated cytoplasmic membrane preparation of *Escherichia coli. Proc. Nat. Acad. Sci. U.S.* **55,** 920-927.

Kaback, H.R., and Stadtman, E.R. (1968). Glycine uptake in *Escherichia coli.* II. Glycine uptake, exchange, and metabolism by an isolated membrane preparation. *J. Biol. Chem.* **243,** 1390-1400.

Kahlenberg, A., and Banjo, B. (1972). Involvement of phospholipids in the D-glucose uptake activity of isolated human erythrocyte membranes. *J. Biol. Chem.* **247,** 1156-1160.

Kahlenberg, A., and Dolansky, D. (1972). Structural requirements of D-glucose for its binding to isolated human erythrocyte membranes. *Can. J. Biochem.* **50,** 638-643.

Kahlenberg, A., Urman, B., and Dolansky, D. (1971). Preferential uptake of D-glucose by isolated human erythrocyte membranes. *Biochemistry* **10,** 3154-3162.

Kahlenberg, A., Dolansky, D., and Rohrlick, R. (1972). D-Glucose uptake by isolated human erythrocyte membranes versus D-glucose transport by human erythrocytes. Comparison of the effects of proteolytic and phospholipase A$_2$ digestion. *J. Biol. Chem.* **247,** 4572-4576.

Kalckar, H.M., Ullrey, D., Kijomoto, S., and Hakomori, S. (1973). Carbohydrate catabolism and the enhancement of uptake of galactose in hamster cells transformed by polyoma virus. *Proc. Nat. Acad. Sci. U.S.* **70,** 839-843.

Karlish, S.J.D., Lieb, W.R., Ram, D., and Stein, W.D. (1972). Kinetic parameters of glucose efflux from human red blood cells under zero-*trans* conditions. *Biochim. Biophys. Acta* **255,** 126-132.

Karreman, G. (1973). Cooperative specific adsorption. *Ann. N. Y. Acad. Sci.* **204,** 393-409.

Kedem, O. (1961). Criteria of active transport. *In* "Membrane Transport and Metabolism" (A. Kleinzeller and A. Kotyk, eds.), pp. 87-93. Academic Press, New York.

Kepes, A. (1960). Etudes cinétiques sur la galactoside-perméase d'*Escherichia coli. Biochim. Biophys. Acta* **40,** 70-84.

Kepes, A. (1969). Carrier properties of β-galactoside permease: The role of permease in the leak of β-galactosides from E. coli. *In* "The Molecular Basis of Membrane Function" (D.C. Tosteson, ed), pp. 353-389. Prentice-Hall, Englewood Cliffs, New Jersey.

Kimelberg, H.K., and Papahadjopoulos, D. (1971a). Phospholipid-protein interactions: Membrane permeability correlated with monolayer "penetration." *Biochim. Biophys. Acta* **233,** 805-809.

Kimelberg, H.K., and Papahadjopoulos, D. (1971b). Interactions of basic proteins with phospholipid membranes. *J. Biol. Chem.* **246**, 1142-1148.

Kimmich, G.A. (1970). Active sugar accumulation by isolated intestinal epithelial cells. A new model for sodium-dependent metabolic transport. *Biochemistry* **9**, 3669-3677.

Kimmich, G.A., and Randles, J. (1973). Interaction between Na^+-dependent transport systems for sugars and amino acids. Evidence against a role for the sodium gradient. *J. Membrane Biol.* **12**, 47-68.

Kirschner, L.B. (1957). Phosphatidylserine as a possible participant in active sodium transport in erythrocytes. *Arch. Biochem. Biophys.* **68**, 499-500.

Kirschner, L.B. (1958). The cation content of phospholipids from swine erythrocytes. *J. Gen. Physiol.* **42**, 231-241.

Kitahara, S., Heinz, E., and Stahlmann, C. (1965). Effect of *trans*-solutes on the fluxes of chloride ions across artificial membranes. *Nature (London)* **208**, 187-189.

Klotz, I.M. (1958). Protein hydration and behavior. *Science* **128**, 815-822.

Koch, A. (1964). The role of permease in transport. *Biochim. Biophys. Acta* **79**, 177-200.

Kolber, A.R., and Stein, W.D. (1966). Identification of a component of a transport "carrier" system: Isolation of the permease expression of the *lac* operon of *Escherichia coli. Nature (London)* **209**, 691-694.

Kornacker, K. (1972). Living aggregates of nonliving parts: A generalized statistical mechanical theory. *Prog. Theor. Biol.* **2**, 1-22.

Kotyk, A. (1967). Mobility of the free and of the loaded monosaccharide carrier in *Saccharomyces cerevisiae. Biochim. Biophys. Acta* **135**, 112-119.

Kotyk, A. (1973). Mechanisms of nonelectrolyte transport. *Biochim. Biophys. Acta* **300**, 183-210.

Kotyk, A., and Janáček, K. (1970). "Cell Membrane Transport. Principles and Techniques." Plenum, New York.

Kurella, G. (1961). Polyelectrolyte properties of protoplasm and the character of resting potentials. *In* "Membrane Transport and Metabolism" (A. Kleinzeller and A. Kotyk, eds.), pp. 54-68. Academic Press, New York.

Kushmerick, M.J., and Podolsky, R.J. (1969). Ionic mobility in muscle cells. *Science* **166**, 1297-1298.

Kushnir, L.D. (1968). Studies on a material which induces electrical excitability in bimolecular lipid membranes. I. Production, isolation, gross identification and assay. *Biochim. Biophys. Acta* **150**, 285-299.

Lacko, L. (1966). Transport of sugars into human red blood cell ghosts. *J. Cell. Physiol.* **67**, 501-506.

Lacko, L., and Burger, M. (1961). Common carrier system for sugar transport in human red cells. *Nature (London)* **191**, 881-882.

Lacko, L., and Burger, M. (1963). Kinetic comparison of exchange transport of sugars with nonexchange transport in human erythrocytes. *J. Biol. Chem.* **238**, 3478-3481.

Lacko, L., and Burger, M. (1964). Effect of phlorizin on exchange and non-exchange transport of sugars in human erythrocytes. *Biochim. Biophys. Acta* **79**, 563-567.

Lacko, L., Burger, M., Hejmová, L., and Rejnková, J. (1961). Exchange transfer of sugars in human erythrocytes. *In* "Membrane Transport and Metabolism" (A. Kleinzeller and A. Kotyk, eds.), pp. 399-408. Academic Press, New York.

Lacko, L., Wittke, B., and Kromphardt, H. (1972). Zur Kinetik der Glucose-Aufnahme in Erythrocyten. Effekt der Trans-Konzentration. *Eur. J. Biochem.* **25**, 447-454.

Läuger, P. (1973). Ion transport through pores: A rate-theory analysis. *Biochim. Biophys. Acta* **311**, 423-441.

LeFevre, P.G. (1948). Evidence of active transfer of certain non-electrolytes across the human red cell membrane. *J. Gen. Physiol.* **31**, 505-527.

LeFevre, P.G. (1961a). Persistence in erythrocyte ghosts of mediated sugar transport. *Nature (London)* **191**, 970-972.

LeFevre, P.G. (1961b). Upper limit for number of sugar transport sites in red cell surface. *Fed. Proc., Fed. Amer. Soc. Exp. Biol.* **20**, 139.

LeFevre, P.G. (1962). Rate and affinity in human red blood cell sugar transport. *Amer. J. Physiol.* **203**, 286-290.

LeFevre, P.G. (1963). Absence of rapid exchange component in a low-affinity carrier transport. *J. Gen. Physiol.* **46**, 721-731.

LeFevre, P.G. (1967). The behavior of phospholipid-glucose complexes at hexane/aqueous interfaces. *Curr. Mod. Biol.* **1**, 29-38.

LeFevre, P.G. (1973). A model for erythrocyte sugar transport based on substrate-conditioned "introversion" of binding sites. *J. Membrane Biol.* **11**, 1-19.

LeFevre, P.G., and Davies, R.I. (1951). Active transport into the human erythrocyte: Evidence from comparative kinetics and competition among monosaccharides. *J. Gen. Physiol.* **34**, 515-524.

LeFevre, P.G., and LeFevre, M.E. (1952). The mechanism of glucose transfer into and out of the human red cell. *J. Gen. Physiol.* **35**, 891-906.

LeFevre, P.G., and McGinniss, G.F. (1960). Tracer exchange *vs.* net uptake of glucose through human red cell surface: New evidence for carrier-mediated diffusion. *J. Gen. Physiol.* **44**, 87-103.

LeFevre, P.G., and Marshall, J.K. (1958). Conformational specificity in a biological sugar transport system. *Amer. J. Physiol.* **194**, 333-337.

LeFevre, P.G., and Masiak, S.J. (1970). Reevaluation of use of retardation chromatography to demonstrate selective monosaccharide "binding" by erythrocyte membranes. *J. Membrane Biol.* **3**, 387-399.

LeFevre, P.G., Habich, K.I., Hess, H.S., and Hudson, M.R. (1964). Phospholipid-sugar complexes in relation to cell membrane monosaccharide transport. *Science* **143**, 955-957.

LeFevre, P.G., Jung, C.Y., and Chaney, J.E. (1968). Glucose transfer by red cell membrane phospholipids in $H_2O/CHCl_3/H_2O$ three-layer systems. *Arch. Biochem. Biophys.* **126**, 677-691.

Lehninger, A.L. (1971). A soluble, heat-labile, high-affinity Ca^{2+}-binding factor extracted from rat liver mitochondria. *Biochem. Biophys. Res. Commun.* **42**, 312-318.

Lev, A.A. (1964). Determination of activity and activity coefficients of potassium and sodium ions in frog muscle fibres. *Nature (London)* **201**, 1132-1134.

Levi, H., and Ussing, H.H. (1948). The exchange of sodium and chloride ions across the fibre membrane of the isolated frog sartorius. *Acta Physiol. Scand.* **16**, 232-249.

Levich, V.G. (1962). "Physiochemical Hydrodynamics." Prentice-Hall, Englewood Cliffs, New Jersey.

Levine, M., and Levine, S. (1969). Kinetics of induced uphill transport of sugars in human erythrocytes. *J. Theor. Biol.* **24**, 85-107.

Levine, M., and Stein, W.D. (1966). The kinetic parameters of the monosaccharide transfer system of the human erythrocyte. *Biochim. Biophys. Acta* **127**, 179-193.

Levine, M., and Stein, W.D. (1967). Techniques for analysis of glucose binding by human erythrocyte membranes. *Biochim. Biophys. Acta* **135**, 710-716.

Levine, M., Oxender, D.L., and Stein, W.D. (1965). The substrate-facilitated transport of the glucose carrier across the human erythrocyte membrane. *Biochim. Biophys. Acta* **109**, 151-163.

Levine, M., Levine, S., and Jones, M.N. (1971). The effect of temperature on the competitive inhibition of sorbose transfer in human erythrocytes by glucose. *Biochim. Biophys. Acta* **225**, 291-300.

Levine, R. (1972). Some mechanisms for hormonal effects on substrate transport. *In* "Metabolic Pathways" (L.E. Hokin, ed.), 3rd ed., Vol. 6, pp. 627-641. Academic Press, New York.

Lieb, W.R., and Stein, W.D. (1970). Quantitative predictions of a noncarrier model for glucose transport across the human red cell membrane. *Biophys. J.* **10**, 585-609.

Lieb, W.R., and Stein, W.D. (1971a). New theory for glucose transport across membranes. *Nature (London) New Biol.* **230**, 108-109.

Lieb, W.R., and Stein, W.D. (1971b). Rejection criterion for some forms of the conventional carrier. *J. Theor. Biol.* **30**, 219-222.

Lieb, W.R., and Stein, W.D. (1972a). Carrier and non-carrier models for sugar transport in the human red blood cell. *Biochim. Biophys. Acta* **265**, 187-207.

Lieb, W.R., and Stein, W.D. (1972b). The influence of unstirred layers on the kinetics of carrier-mediated transport. *J. Theor. Biol.* **36**, 641-645.

Lin, K.T., and Johnstone, R.M. (1971). Active transport of glycine by mouse pancreas. Evidence against the Na^+ gradient hypothesis. *Biochim. Biophys. Acta* **249**, 144-158.

Ling, G. (1952). The role of phosphate in the maintenance of the resting potential and selective ionic accumulation in frog muscle cells. *Phosphorus Metab., Symp., 2nd, 1952* Vol. 2, pp. 748-797.

Ling, G. (1960). The interpretation of selective ionic permeability and cellular potentials in terms of the fixed charge-induction hypothesis. *J. Gen. Physiol.* **43**, Suppl. 1, 149-174.

Ling, G.N. (1962). "A Physical Theory of the Living State: the Association-Induction Hypothesis." Ginn (Blaisdell), Boston, Massachusetts.

Ling, G.N. (1965a). The physical state of water in living cell and model systems. *Ann. N. Y. Acad. Sci.* **125**, 401-417.

Ling, G.N. (1965b). The membrane theory and other views for solute permeability, distribution, and transport in living cells. *Perspect. Biol. Med.* **9**, 87-106.

Ling, G.N. (1966a). All-or-none adsorption by living cells and model protein-water systems: Discussion of the problem of 'permease induction' and determination of secondary and tertiary structures of proteins. *Fed. Proc., Fed. Amer. Soc. Exp. Biol.* **25**, 958-970.

Ling, G.N. (1966b). Cell membrane and cell permeability. *Ann. N. Y. Acad. Sci.* **137**, 837-859.

Ling, G.N. (1969). A new model for the living cell: A summary of the theory and recent experimental evidence in its support. *Int. Rev. Cytol.* **26**, 1-61.

Ling, G.N., and Bohr, G. (1970). Studies on ion distribution in living cells. II. Cooperative interaction between intracellular potassium and sodium ions. *Biophys. J.* **10**, 519-538.

Ling, G.N., and Cope, F.W. (1969). Potassium ion: Is the bulk of intracellular K^+ adsorbed? *Science* **163**, 1335-1336.

Ling, G.N., and Ochsenfeld, M.M. (1966). Studies on ion accumulation in muscle cells. *J. Gen. Physiol.* **49**, 819-843.

Ling, G.N., Ochsenfeld, M.M., and Karreman, G. (1967). Is the cell membrane a universal rate-limiting barrier to the movement of water between the living cell and its surrounding medium? *J. Gen. Physiol.* **50**, 1807-1820.

Ling, G.N., Miller, C., and Ochsenfeld, M.M. (1973). The physical state of solutes and water in living cells according to the association-induction hypothesis. *Ann. N. Y. Acad. Sci.* **204**, 6-50.

Lundegårdh, H. (1935). Theorie der Ionenaufnahme in lebende Zellen. *Naturwissenschaften* **23**, 313-318.

McLaughlin, S.G.A., and Hinke, J.A.M. (1966). Sodium and water binding in single striated muscle fibers of the giant barnacle. *Can. J. Physiol. Pharmacol.* **44**, 837-848.

Martin, G.S., Venuta, S., Weber, M., and Rubin, H. (1971). Temperature-dependent alterations in sugar transport in cells infected by a temperature-sensitive mutant of Rous sarcoma virus. *Proc. Nat. Acad. Sci. U.S.* **68**, 2739-2741.

Martin, K. (1971). Some properties of an SH group essential for choline transport in human erythrocytes. *J. Physiol. (London)* **213**, 647-664.

Masiak, S.J., and LeFevre, P.G. (1972). Failure of equilibrium dialysis to show selective monosaccharide binding by erythrocyte membranes. *J. Membrane Biol.* **9**, 291-296.

Matthews, D.W. (1972). Rates of peptide uptake by small intestine. *Peptide Transp. Bacteria Mammal. Gut, Ciba Found. Symp.* pp. 71-92.

Mawe, R.C., and Hempling, H.G. (1965). The exchange of C^{14} glucose across the membrane of the human erythrocyte. *J. Cell. Comp. Physiol.* **66**, 95-104.

Meister, A. (1973). On the enzymology of amino acid transport. *Science* **180**, 33-39.

Mendelsohn, J., Skinner, A., and Kornfeld, S. (1971). The rapid induction by phytohemagglutinin of increased α-aminobutyric acid uptake by lymphocytes. *J. Clin. Invest.* **50**, 818-826.

Miller, D.M. (1965). The kinetics of selective biological transport. II. Equations for induced uphill transport of sugars in humzn erythrocytes. *Biophys. J.* **5**, 417-423.

Miller, D.M. (1968a). The kinetics of selective biological transport. III. Erythrocyte-monosaccharide transport data. *Biophys. J.* **8**, 1329-1338.

Miller, D.M. (1968b). The kinetics of selective biological transport. IV. Assessment of three carrier systems using the erythrocyte-monosaccharide transport data. *Biophys. J.* **8**, 1339-1352.

Miller, D.M. (1969). Monosaccharide transport in human erythrocytes. *In* "Red Cell Membrane Structure & Function" (G.A. Jamieson and T.J. Greenwalt, eds.), pp. 240-290. Lippincott, Philadelphia, Pennsylvania.

Miller, D.M. (1971). The kinetics of selective biological transport.V. Further data on the erythrocyte-monosaccharide transport system. *Biophys. J.* **11**, 915-923.

Miller, D.M. (1972). The effect of unstirred layers on the measurement of transport rates in individual cells. *Biochim. Biophys. Acta* **266**, 85-90.

Minkoff, L., and Damadian, R. (1973). Energy requirements of bacterial ion exchange. *Ann. N. Y. Acad. Sci.* **204**, 249-260.

Mitchell, P. (1954). Transport of phosphate through an osmotic barrier. *Symp. Soc. Exp. Biol.* **8**, 254-261.

Mitchell, P. (1962). Molecule, group and electron translocation through natural membranes. *Biochem. Soc. Symp.* **22**, 142-169.

Mitchell, P. (1967). Translocations through natural membranes. *Advan. Enzymol.* **29**, 33-87.

Mitchell, P., and Moyle, J. (1958). Group-translocation: A consequence of enzyme-catalyzed group-transfer. *Nature (London)* **182**, 372-373.

Møller, J.V. (1971). Investigations on the existence of a specific retention of D-glucose by the human erythrocyte membrane. *Biochim. Biophys. Acta* **249**, 96-100.

Moore, T.J., and Schlowsky, B. (1969a). Effects of erythrocyte lipid and of glucose and galactose concentration on transport of the sugars across a water-butanol interface. *J. Lipid Res.* **10**, 216-219.

Moore, T.J., and Schlowsky, B. (1969b). Glucose and galactose transport across a water-butanol-phospholipid interface. *Chem. Phys. Lipids* **3**, 273-279.

Mueller, P., and Rudin, D.O. (1967a). Development of K^+-Na^+ discrimination in experimental bimolecular lipid membranes by macrocyclic antibiotics. *Biochem. Biophys. Res. Commun.* **26**, 398-404.

Mueller, P., and Rudin, D.O. (1967b). Action potential phenomena in experimental bimolecular lipid membranes. *Nature (London)* **213**, 603-604.

Mueller, P., and Rudin, D.O. (1968). Action potentials induced in bimolecular lipid membranes. *Nature (London)* **217**, 713-719.

Mueller, P., Rudin, D.O., TiTien, H., and Wescott, W.C. (1962). Reconstitution of cell membrane structure *in vitro* and its transformation into an excitable system. *Nature (London)* **194**, 979-980.

Naftalin, R.J. (1970). A model for sugar transport across red cell membranes without carriers. *Biochim. Biophys. Acta* **211**, 65-78.

Naftalin, R.J. (1971). The role of unstirred layers in control of sugar movements across red cell membranes. *Biochim. Biophys. Acta* **233**, 635-643.

Naftalin, R.J. (1972). An alternative to the carrier model for sugar transport across red cell membranes. *In* "Biomembranes" (F. Kreuzer and J.F.G. Slegers, eds.), Vol. 3, pp. 117-126. Plenum, New York.

Neville, M.C. (1973). Cellular accumulation of amino acids: Adsorption revisited. *Ann. N. Y. Acad. Sci.* **204**, 538-563.

Osterhout, W.J.V. (1933). Permeability in large plant cells and in models. *Ergeb. Physiol. Exp. Pharmakol.* **35**, 967-1021.

Osterhout, W.J.V. (1935). How do electrolytes enter cells? *Proc. Nat. Acad. Sci. U.S.* **21**, 125-132.

Oxender, D.L. (1972). Membrane transport. *Annu. Rev. Biochem.* **41**, 777-814.

Oxender, D.L., and Christensen, H.N. (1959). Transcellular concentration as a consequence of intracellular accumulation. *J. Biol. Chem.* **234**, 2321-2324.

Oxender, D.L., and Christensen, H.N. (1963). Distinct mediating systems for the transport of neutral amino acids by the Ehrlich cell. *J. Biol. Chem.* **238**, 3686-3699.

Packard, M.A., and Paterson, A.R.P. (1972). Nucleoside transport in human erythrocytes: Equilibrium exchange diffusion of uridine. *Can. J. Biochem.* **50**, 704-705.

Papahadjopoulos, D. (1971). Na^+-K^+ discrimination by "pure" phospholipid membranes. *Biochim. Biophys. Acta* **241**, 254-259.

Pardee, A.B. (1968). Membrane transport proteins. *Science* **162**, 632-637.

Pardee, A.B. (1970). On deciding whether a membrane protein is involved in transport. *In* "Permeability and Function of Biological Membranes" (L. Bolis *et al.*, eds.), pp. 86-93. North-Holland Publ., Amsterdam.

Park, C.R., Post, R.L., Kalman, C.F., Wright, J.H., Jr., Johnson, L.H., and Morgan, H.E. (1956). The transport of glucose and other sugars across cell membranes and the effect of insulin. *Ciba Found. Colloq. Endocrinol.* [*Proc.*] **9**, 240-265.

Parsons, B.J. (1969). Binding of sugars to isolated brush borders. *Life Sci., Part II* **8**, 939-942.

Patlak, C.S. (1956). Contributions to the theory of active transport. *Bull. Math. Biophys.* **18**, 271-315.

Patlak, C.S. (1957). Contributions to the theory of active transport. II. The gate type non-carrier mechanism and generalizations concerning tracer flow, efficiency, and measurement of energy expenditure. *Bull. Math. Biophys.* **19**, 209-235.

Peters, J.H., and Hausen, P. (1971). Effect of phytohemagglutinin on lymphocyte membrane transport. II. Stimulation of "facilitated diffusion" of 3-O-methylglucose. *Eur. J. Biochem.* **19**, 509-513.

Peterson, R.N., Boniface, J., and Koch, A.L. (1967). Energy requirements, interactions and distinctions in the mechanisms for transport of various nucleosides in *Escherichia coli. Biochim. Biophys. Acta* **135**, 771-783.

Post, R.L., and Albright, C.D. (1961). Membrane adenosine triphosphatase system as a part of a system for active sodium and potassium transport. *In* "Membrane Transport and Metabolism" (A. Kleinzeller and A. Kotyk, eds.), pp. 219-227. Academic Press, New York.

Post, R.L., Merritt, C.R., Kinsolving, C.R., and Albright, C.D. (1960). Membrane adenosine triphosphatase as a participant in the active transport of sodium and potassium in the human erythrocyte. *J. Biol. Chem.* **235**, 1796-1802.

Pressman, B., and Haynes, D.H. (1969). Ionophorous agents as mobile ion carriers. *In* "The Molecular Basis of Membrane Function" (D.C. Tosteson, ed.), pp. 221-246. Prentice-Hall, Englewood Cliffs, New Jersey.

Pressman, B.C. (1968). Ionophorous antibiotics as models for biological transport *Fed. Proc., Fed. Amer. Soc. Exp. Biol.* **27**, 1283-1288.

Redwood, W.R., Müldner, H., and Thompson, T.E. (1969). Interaction of a bacterial adenosine triphosphatase with phospholipid bilayers. *Proc. Nat. Acad. Sci. U.S.* **64**, 989-996.

Reeves, J.P., Hong, J.-S., and Kaback, H.R. (1973). Reconstitution of D-lactate-dependent transport in membrane vesicles from a D-lactate dehydrogenase mutant of *Escherichia coli. Proc. Nat. Acad. Sci. U.S.* **70**, 1917-1921.

Regen, D.M., and Morgan, H.E. (1964). Studies of the glucose-transport system in the rabbit erythrocyte. *Biochim. Biophys. Acta* **79**, 151-166.

Regen, D.M., and Tarpley, H.L. (1974) Anomalous transport kinetics and the glucose carrier hypothesis. *Biochim. Biophys. Acta* **339**, 218-233.

Reiser, S., and Christiansen, P.A. (1968). Formation of a complex between valine and intestinal mucosal lipid; its possible role in valine absorption. *J. Lipid Res.* **9**, 606-612.

Riggs, T.R., Walker, L.M., and Christensen, H.N. (1958). Potassium migration and amino acid transport. *J. Biol. Chem.* **233**, 1479-1484.

Rodbell, M. (1967). Metabolism of isolated fat cells. VI. The effects of insulin, lipolytic hormones, and theophylline on glucose transport and metabolism in "ghosts". *J. Biol. Chem.* **242**, 5751-5756.

Roseman, S. (1972). Carbohydrate transport in bacterial cells. *In* "Metabolic Pathways" (L.E. Hokin, ed.), 3rd ed., Vol. 6, pp. 41-89. Academic Press, New York.

Rosenberg, T. (1948). On accumulation and active transport in biological systems. I. Thermodynamic considerations. *Acta Chem. Scand.* **2**, 14-33.

Rosenberg, T. (1954). The concept and definition of active transport. *Symp. Soc. Exp. Biol.* **8**, 27-41.

Rosenberg, T., and Wilbrandt, W. (1952). Enzymatic processes in cell membrane penetration. *Int. Rev. Cytol.* **1**, 65-92.

Rosenberg, T., and Wilbrandt, W. (1955). The kinetics of membrane transports involving chemical reactions. *Exp. Cell. Res.* **9**, 49-67.

Rosenberg, T., and Wilbrandt, W. (1957). Uphill transport induced by counterflow. *J. Gen. Physiol.* **41**, 289-296.

Rosenberg, T., and Wilbrandt, W. (1962). The kinetics of carrier transport inhibition. *Exp. Cell Res.* **27**, 100-117.

Rosenberg, T., and Wilbrandt, W. (1963). Carrier transport uphill. I. General. *J. Theor. Biol.* **5**, 288-305.

Rotunno, C.A., Kowalewski, V., and Cereijido, M. (1967). Nuclear spin resonance evidence for complexing of sodium in frog skin. *Biochim. Biophys. Acta* **135**, 170-173.

Rouhi, A., Blass, J., and Macheboeuf, M. (1952). Sur la formation de complexes entre la lécithine et l'urée. *C. R. Soc. Biol.* **146**, 847-849.

Rouhi, A., Blass, J., and Macheboeuf, M. (1953). Solubilisation de l'urée dans l'ether de pétrole par la lécithine et le cholestérol. *C. R. Acad. Sci.* **236**, 539-541.

Rouhi, A., Blass, J., and Polonovski, J. (1954). Solubilisation des amino-acides dans les extraits lipidiques. *Bull. Soc. Chim. Biol.* **36**, 1417-1423.

Schafer, J.A., and Heinz, E. (1971). The effect of reversal of Na^+ and K^+ electrochemical potential gradients on the active transport of amino acids in Ehrlich ascites tumor cells. *Biochim. Biophys. Acta* **249**, 15-23.

Schatzmann, H.J. (1962). Lipoprotein nature of red cell adenosine triphosphatase. *Nature (London)* **196**, 677.

Scheer, B.T. (1958). Active transport: Definitions and criteria. *Bull. Math. Biophys.* **20**, 231-244.

Schneider, P.B., and Wolff, J. (1965). Thyroidal iodide transport. VI. On a possible role for iodide-binding phospholipids. *Biochim. Biophys. Acta* **94**, 114-123.

Schultz, J.S. (1971). Passive asymmetric transport through biological membranes. *Biophys. J.* **11**, 924-943.

Schultz, J.S. (1972). Asymmetry in biological permeability barriers with carrier transport. *In* "Biomembranes" (F. Kreuzer and J.F.G. Slegers, eds.), Vol. 3, pp. 57-77. Plenum, New York.

Schultz, S.G. (1969). The interaction between sodium and amino acid transport across the brush border of rabbit ileum: A plausible molecular model. *In* "The Molecular Basis of Membrane Function" (D.C. Tosteson, ed.), pp. 401-420. Prentice-Hall, Englewood Cliffs, New Jersey.

Schultz, S.G., and Curran, P.F. (1969). The role of sodium in non-electrolyte transport across animal cell membranes. *Physiologist* **12**, 437-452.

Schultz, S.G., and Curran, P.F. (1970). Coupled transport of sodium and organic solutes. *Physiol. Rev.* **50**, 637-718.

Schultz, S.G., and Curran, P.F. (1974). Sodium and chloride transport across isolated rabbit ileum. *Curr. Top. Membranes Transp.* **5**, 225-281.

Schultz, S.G., Fuisz, R.E., and Curran, P.F. (1966). Amino acid and sugar transport in rabbit ileum. *J. Gen. Physiol.* **49**, 849-866.

Sen, A.K., and Widdas, W.F. (1960). A new method for determining the half-saturation of the facilitated transfer of glucose across the human erythrocyte membrane and for studying the effect of inhibitors, *J. Physiol. (London)* **152**, 32P-33P.

Sen, A.K., and Widdas, W.F. (1962). Determination of the temperature and pH dependence of glucose transfer across the human erythrocyte membrane measured by glucose exit. *J. Physiol. (London)* **160**, 392-403.

Sha'afi, R.I., Rich, G.T., Sidel, V.W., Bossert, W., and Solomon, A.K. (1967). The effect of the unstirred layer on human red cell water permeability. *J. Gen. Physibl.* **50**, 1377-1400.

Shamoo, A.E. (1974). Isolation of a sodium-specific ionophore from $(Na^+ + K^+)$-ATPase preparations. *Ann. N. Y. Acad. Sci.* **242**, 389–405.

Shamoo, A.E., and Albers, R.W. (1973). Na^+-selective ionophoric material derived from electric organ and kidney membranes. *Proc. Nat. Acad. Sci. U.S.* **70**, 1191-1194.

Shemyakin, M.M., Antonov, V.K., Bergelson, L.D., Ivanov, V.T., Malenkov, G.G., Ovchinnikov, Y.A., and Shkrob, A.M. (1969). Chemistry of membrane-affecting peptides, depsipeptides and depsides (structure-function relations). In "The Molecular Basis of Membrane Function" (D.C. Tosteson, ed.), pp. 173-210. Prentice-Hall, Englewood Cliffs, New Jersey.

Silberman, R., and Gaby, W.L. (1961). The uptake of amino acids by lipids of Pseudomonas aeruginosa. J. Lipid Res. 2, 172-176.

Simon, S.E. (1959). Ionic partition and fine structure in muscle. Nature (London) 184, 1978-1982.

Simon, S.E., Johnstone, B.M., Shankly, K.H., and Shaw, F.H. (1959). Muscle: A three phase system. The partition of monovalent ions across the cell surfaces. J. Gen. Physiol. 43, 55-79.

Sistrom, W.R. (1958). On the physical state of the intracellularly accumulated substrates of β-galactoside-permease in Escherichia coli. Biochim. Biophys. Acta 29, 579-587.

Sjodin, R.A., and Henderson, E.G. (1964). Tracer and non-tracer potassium fluxes in frog sartorius muscle and the kinetics of net potassium movement. J. Gen. Physiol. 47, 605-638.

Skou, J.C. (1961). The relationship of a (Mg²⁺ + Na⁺)-activated, K⁺-stimulated enzyme or enzyme to the active, linked transport of Na⁺ and K⁺ across the cell membrane. In "Membrane Transport and Metabolism" (A. Kleinzeller and A. Kotyk, eds.), pp. 228-236. Academic Press, New York.

Skou, J.C. (1965). Enzymatic basis for active transport of Na⁺ and K⁺ across cell membrane. Physiol. Rev. 45, 596-617.

Skou, J.C. (1969). The role of membrane ATPase in the active transport of ions. In "The Molecular Basis of Membrane Function" (D.C. Tosteson, ed.), pp. 455-482. Prentice-Hall, Englewood Cliffs, New Jersey.

Skou, J.C. (1973). The relationship of the (Na⁺ + K⁺)-activated enzyme system to transport of sodium and potassium across the cell membrane. J. Bioenerg. 4, 1-30.

Smyth, D.H. (1972). Peptide transport by mammalian gut. Peptide Transp. Bacteria Mammal. Gut, Ciba Found. Symp. pp. 59-70.

Soderman, D.D., Germershausen, J., and Katzen, H.H. (1973). Affinity binding of intact fat cells and their ghosts to immobilized insulin. Proc. Nat. Acad. Sci. U.S. 70, 792-796.

Solomon, A.K. (1952). The permeability of the human erythrocyte to sodium and potassium. J. Gen. Physiol. 36, 57-110.

Solomon, A.K., Lionetti, F., and Curran, P.F. (1956). Possible cation-carrier substances in blood. Nature (London) 178, 582-583.

Sols, A., and de la Fuente, G. (1961). Transport and hydrolysis in the utilization of oligosaccharides by yeasts. In "Membrane Transport and Metabolism" (A. Kleinzeller and A. Kotyk, eds.), pp. 361-377. Academic Press, New York.

Stein, W.D. (1958). N-terminal histidine at the active center of a permeability mechanism. Nature (London) 181, 1662-1663.

Stein, W.D. (1964). A procedure which labels the active centre of the glucose transport system of the human erythrocyte. In "Structure and Activity of Enzymes" (T.W. Goodwin, J.I. Harris, and B.S. Hartley, eds.), pp. 133-137. Academic Press, New York.

Stein, W.D. (1967a). "The Movement of Molecules across Cell Membranes." Academic Press, New York.

Stein, W.D. (1967b). Some properties of carrier substances isolated from bacterial and erythrocyte membranes. Biochem. J. 105, 3P-4P.

Stein, W.D. (1968). The transport of sugars. *Brit. Med. Bull.* **24,** 146-149.

Stein, W.D. (1969). Intra-protein interactions across a fluid membrane as a model for biological transport. *In* "Membrane Proteins. Proceedings of a Symposium Sponsored by the New York Heart Association," pp. 81-90. Little, Brown, Boston, Massachusetts.

Stein, W.D., and Danielli, J.F. (1956). Structure and function in red cell permeability. *Discuss. Faraday Soc.* **21,** 238-251.

Stein, W.D., Lieb, W.R., Karlish, S.J.D., and Eilam, Y. (1973). A model for active transport of sodium and potassium ions as mediated by a tetrameric enzyme. *Proc. Nat. Acad. Sci. U.S.* **70,** 275-278.

Storelli, C., Vögeli, H., and Semenza, G. (1972). Reconstitution of a sucrase-mediated sugar transport system in lipid membranes. *FEBS (Fed. Eur. Biochem. Soc.) Lett.* **24,** 287-292.

Strasberg, P., and Elliott, K.A.C. (1967). Further studies on the binding of γ-aminobutyric acid by brain. *Can. J. Biochem.* **45,** 1795-1807.

Stuart, W.D., and DeBusk, A.G. (1971). Molecular transport. I. *In vitro* studies of isolated glycoprotein subunits of the amino acid transport system of *Neurospora crassa* conidia. *Arch. Biochem. Biophys.* **144,** 512-518.

Sweet, C., and Zull, J.E. (1969). Activation of glucose diffusion from egg lecithin liquid crystals by serum albumin. *Biochim. Biophys. Acta* **173,** 94-103.

Sweet, C., and Zull, J.E. (1970). Interaction of the erythrocyte-membrane protein, spectrin, with model membrane systems. *Biochem. Biophys. Res. Commun.* **41,** 135-141.

Swift, T.J., and Barr, E.M. (1973). An oxygen magnetic resonance study of water in frog skeletal muscle. *Ann. N. Y. Acad. Sci.* **204,** 191-196.

Taverna, R.D., and Langdon, R.G. (1973). D-Glucosyl isothiocyanate, an affinity label for the glucose transport proteins of the human erythrocyte membrane. *Biochem. Biophys. Res. Commun.* **54,** 593-599.

Thomas, L. (1972). Demonstration of glucose inhibitable phlorizin binding to renal brushborder proteins separated by gel electrophoresis. *FEBS (Fed. Eur. Biochem. Soc.) Lett.* **25,** 245-248.

Thomas, L. (1973). Isolation of *N*-ethylmaleimide-labelled phlorizin-sensitive D-glucose binding protein of brush-border membrane from rat kidney cortex. *Biochim. Biophys. Acta* **291,** 454-464.

Thomas, L., and Kinne, R. (1972). Studies on the arrangement of a glucose sensitive phlorizin binding site in the microvilli of isolated rat kidney brushborder. *FEBS (Fed. Eur. Biochem. Soc.) Lett.* **24,** 242-244.

Thomas, L., Kinne, R., and Frohnert, P.P. (1972). *N*-Ethylmaleimide labeling of a phlorizin-sensitive D-glucose binding site of brush border membrane from the rat kidney. *Biochim. Biophys. Acta* **290,** 125-133.

Torres-Pinedo, R., Garcia-Castiñeiras, S., and Alvarado, F. (1972). False D-glucose binding to fractions of disrupted intestinal brush borders. *Fed. Proc., Fed. Amer. Soc. Exp. Biol.* **31,** 239 Abs.

Tosteson, D.C. (1968). Effect of macrocyclic compounds on the ionic permeability of artificial and natural membranes. *Fed. Proc., Fed. Amer. Soc. Exp. Biol.* **27,** 1269-1277.

Troshin, A.S. (1961). Sorption properties of protoplasm and their role in cell permeability. *In* "Membrane Transport and Metabolism" (A. Kleinzeller and A. Kotyk, eds.), pp. 45-53. Academic Press, New York.

Troshin, A.S. (1966). "Problems of Cell Permeability" (M.G. Hell, transl.; W.F. Widdas, ed.), rev. and suppl. ed. Pergamon, Oxford.

Tucker, A.M., and Kimmich, G.A. (1973). Characteristics of amino acid accumulation by isolated intestinal epithelial cells. *J. Membrane Biol.* **12**, 1-22.

Ussing, H.H. (1949). Transport of ions across cellular membranes. *Physiol. Rev.* **29**, 127-155.

Ussing, H.H. (1952). Some aspects of the application of tracers in permeability studies. *Advan. Enzymol.* **13**, 21-65.

Ussing, H.H. (1954). Active transport of inorganic ions. *Symp. Soc. Exp. Biol.* **8**, 407-422.

Vanatta, J.C. (1963). Reversibility of sodium binding by phospholipids extracted from kidney. *Fed. Proc., Fed. Amer. Soc. Exp. Biol.* **22**, 333.

Vanatta, J.C. (1966). Toad bladder extract which binds sodium: Role in sodium transport. *Proc. Soc. Exp. Biol. Med.* **123**, 945-949.

Van Breemen, D., and Van Breemen, C. (1969). Calcium exchange diffusion in a porous phospholipid ion-exchange membrane. *Nature (London)* **223**, 898-900.

Varon, S., Weinstein, H., and Roberts, E. (1964). Exogenous and endogenous γ-aminobutyric acid of mouse brain particulates in a binding system *in vitro*. *Biochem. Pharmacol.* **13**, 269-279.

Venuta, S., and Rubin, H. (1973). Sugar transport in normal and Rous sarcoma virus-transformed chick-embryo fibroblasts. *Proc. Nat. Acad. Sci. U.S.* **70**, 653-657.

Verhoff, F.H., and Sundaresan, K.R. (1972). A theory of coupled transport in cells. *Biochim. Biophys. Acta* **255**, 425-441.

Vidaver, G.A. (1964a). Glycine transport by hemolyzed and restored pigeon red cells. *Biochemistry* **3**, 795-799.

Vidaver, G.A. (1964b). Some tests of the hypothesis that the sodium-ion gradient furnishes the energy for glycine-active transport by pigeon red cells. *Biochemistry* **3**, 803-808.

Vidaver, G.A. (1966). Inhibition of parallel flux and augmentation of counter flux shown by transport models not involving a mobile carrier. *J. Theor. Biol.* **10**, 301-306.

Vidaver, G.A., and Shepherd, S.L. (1968). Transport of glycine by hemolyzed and restored pigeon red blood cells. Symmetry of properties, *trans* effects of sodium ion and glycine, and their description by a single rate equation. *J. Biol. Chem.* **243**, 6140-6150.

Vilkki, P. (1962). An iodide-complexing phospholipid. *Arch. Biochem. Biophys.* **97**, 425-427.

Vögeli, H., Storelli, C., and Semenza, G. (1972). Ein Disaccharidase-assoziiertes Zuckertransport-System. *Hoppe-Seyler's Z. Physiol. Chem.* **353**, 687-688.

Vogt, W. (1957). Pharmacologically active lipid-soluble acids of natural occurrence. *Nature (London)* **179**, 300-304.

Voříšek, J. (1972). The cooperative character of phenylalanine binding by a protein fraction isolated from baker's yeast membranes. *Biochim. Biophys. Acta* **290**, 256-266.

Voříšek, J. (1973). Study of the specificity of amino acid binding protein from *Saccharomyces cerevisiae* membranes. *Folia Microbiol. (Prague)* **18**, 17-21.

Ward, H.A., and Fantl, P. (1963). Transfer of hydrophilic cations from an aqueous to a lipophilic phase by phosphatidic acids. *Arch. Biochem. Biophys.* **100**, 338-339.

Wasserman, R.H., Corradino, R.A., and Taylor, A.N. (1969). Binding proteins from animals with possible transport function. *J. Gen. Physiol.* **54**, 114s-134s.

Weber, M.J. (1973). Hexose transport in normal and in Rous sarcoma-transformed cells. *J. Biol. Chem.* **248**, 2978-2983.

Weinstein, H., Varon, S., Muhleman, D.R., and Roberts, E. (1965). A carrier-mediated transfer model for the accumulation of ^{14}C γ-aminobutyric acid by subcellular brain particles. *Biochem. Pharmacol.* **14**, 273-288.

Widdas, W.F. (1952). Inability of diffusion to account for placental glucose transfer in the sheep and consideration of the kinetics of a possible carrier transfer. *J. Physiol. (London)* **118**, 23-39.

Widdas, W.F. (1954). Facilitated transfer of hexoses across the human erythrocyte membrane. *J. Physiol. (London)* **125**, 163-180.

Wilbrandt, W. (1956). The relation between rate and affinity in carrier transport. *J. Cell. Comp. Physiol.* **47**, 137-146.

Wilbrandt, W. (1959). Permeability and transport systems in living cells. *J. Pharm. Pharmacol.* **11**, 65-79.

Wilbrandt, W. (1972a). Coupling between simultaneous movements of carrier substrates. *J. Membrane Biol.* **10**, 357-366.

Wilbrandt, W. (1972b). Carrier diffusion. *In* "Biomembranes" (F. Kreuzer and J.F.G. Slegers, eds.), Vol. 3, pp. 79-99. Plenum, New York.

Wilbrandt, W., and Rosenberg, T. (1961). The concept of carrier transport and its corollaries in pharmacology. *Pharmacol. Rev.* **13**, 109-183.

Wilbrandt, W., Frei, S., and Rosenberg, T. (1956). The kinetics of glucose transport through the human red cell membrane. *Exp. Cell. Res.* **11**, 59-66.

Wiley, W.R. (1970). Tryptophan transport in *Neurospora crassa*: A tryptophan-binding protein released by cold osmotic shock. *J. Bacteriol.* **103**, 656-662.

Winkler, H.H., and Wilson, T.H. (1966). The role of energy coupling in the transport of β-galactosides by *Escherichia coli. J. Biol. Chem.* **241**, 2200-2211.

Winter, C.G., and Christensen, H.N. (1964). Migration of amino acids across the membrane of the human erythrocyte. *J. Biol. Chem.* **239**, 872-878.

Wolff, D.J., and Siegel, F.L. (1972). Purification of a calcium-binding protein from pig brain. *Arch. Biochem. Biophys.* **150**, 578-584.

Wong, J.T.-F. (1965). The possible role of polyvalent carriers in cellular transports. *Biochim. Biophys. Acta* **94**, 102-113.

Wong, P.T.S., and MacLennan, D.H. (1973). Restoration by fatty acids of active transport in a lactose mutant of *Escherichia coli. Can. J. Biochem.* **51**, 538-549.

Woolley, D.W. (1958). Serotonin receptors. I. Extraction and assay of a substance which renders serotonin fat-soluble. *Proc. Nat. Acad. Sci. U.S.* **44**, 1202-1210.

Woolley, D.W., and Campbell, N.K. (1960). Serotonin receptors. II. Calcium transport by crude and purified receptor. *Biochim. Biophys. Acta* **40**, 543-544.

Woolley, D.W., and Campbell, N.K. (1962). Tissue lipids as ion exchangers for cations and the relationship to physiological processes. *Biochim. Biophys. Acta* **57**, 384-385.

Wyssbrod, H.R., Scott, W.N., Brodsky, W.A., and Schwartz, I.L. (1971). Carrier-mediated transport processes. *In* "Handbook of Neurochemistry" (A. Lajtha, ed.), Vol. 5, Part B, pp. 683-819. Plenum, New York.

Yu, B.P., DeMartinis, F.D., and Masoro, E.J. (1968). Relation of lipid structure of sarcotubular vesicles to Ca^{++} transport activity. *J. Lipid Res.* **9**, 498-500.

Zierler, K.L. (1961). A model of a poorly-permeable membrane as an alternative to the carrier hypothesis of cell membrane penetration. *Bull. Johns Hopkins Hosp.* **109**, 35-48.

Ion Transport and Short-Circuit Technique

WARREN S. REHM

Department of Physiology and Biophysics
University of Alabama in Birmingham, The Medical Center
University Station, Birmingham, Alabama

I. INTRODUCTION

The primary purpose of this chapter is to review the use of the well known short-circuit technique of Ussing and Zerahn (1951) for studying active and passive ion transport. This technique has been used primarily for deciding between passive and active ion transport and for the determination of the effect of agents and procedures on the rate of active ion transport of the type in which the net active transport can be equated to the net transport of charge. The author's primary purpose is to present the basic concepts, and in no way is this review to be considered encyclopedic in nature. There are a large number of papers in which the short-circuit technique is used, and, in light of our purpose only a few will be mentioned, so that the absence of citation of a given paper is in no manner a reflection on its merits.

In general, techniques that are well understood by the great majority of workers, such as the measurement of isotopic fluxes, will not be discussed (Ussing, 1948; Levi and Ussing, 1949). We will start with a simple conceptual model, which is an abstraction from the real world, and present the basic concepts with reference to this simple model. We will then use models of increasing complexity, which more adequately represent actual tissues.

It is the aim of the author to make the presentation simple enough so that readers without prior knowledge in this area can gain a substantial insight into the problems of passive and active ion transport. The presentation ought to be of interest to investigators who use the short-circuit current technique since it is the author's contention that there are a number of erroneous conceptions in the minds of many investigators who use this technique, and it is hoped that this review will succeed in bringing these problems into focus. For example, many workers seem to assume that, during short-circuiting, the membrane possessing the active transport mechanism is short-circuited. Most of the workers in this field use the simple Ussing–Zerahn equivalent circuit and imply that it is adequate to represent complex tissues. Furthermore, it is generally accepted that if the transport of an ion is essentially zero under short-circuit conditions, then the conclusion is warranted that the ion in question is passive. We hope to convince the reader that the above conclusion is not justified and that other information is necessary before the conclusion that a given ion is passive is warranted. There is also considerable misunderstanding among workers using the simple Ussing–Zerahn circuit concerning the relationship between the conductance of this circuit and that of the actual tissue. These and other problems will be analyzed in this review.

The main thrust of this review is to demonstrate that (1) if all the ion transports are exclusively via conductive pathways the criteria for passive ion transport of the Ussing school are valid for both single- and double-membrane models, but (2) if there are nonconductive pathways for ions that cannot be represented by an equivalent-circuit conductive pathway (see discussion of Figs. 7 and 8), then the accepted criteria for passive transport are not valid—other information is needed. In the first portion of the review the assumption will be made that there are conductive pathways for all the ions involved, and in the latter portion the fallacy of using the criteria for systems in which there are nonconductive pathways will be presented. In the author's opinion, the results of this analysis demand a substantial reorientation of the concepts in this area. We have made liberal use of Appendices for the presentation of background material.

II. ANALYSIS OF IONIC TRANSPORT ON THE BASIS OF A SIMPLE CONCEPTUAL MODEL

A. Analysis on the Basis of Open-Circuit Data

Figure 1 represents an idealized conceptual model. It consists of a single membrane, comparable to a limiting membrane of a cell, with pathways for Na^+ and Cl^- but for no other ions. We will assume that the membrane is impermeable to water and that the Na^+ and Cl^- pathways are conductive (see Appendix A). In our review we will assume (except for the considerations in Appendix A) that water drag does not play an appreciable role in the active transport of ions. Our initial model is such that, with

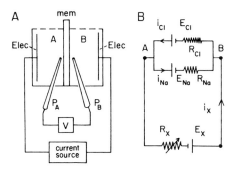

FIG. 1. A. Diagram representing a conceptual single-membrane (mem) model. The potential difference is measured by voltmeter V via probe electrode P_A and P_B making contact with the bathing solutions in compartments A and B. Current can be sent from the current source via the electrodes (elec). B. Equivalent circuit of the membrane in diagram A (the membrane has separate pathways for Cl^- and Na^+) consisting of two limbs. The i's, E's, and R's refer, respectively, to currents, emfs, and resistances. The membrane is connected to an external circuit E_X and R_X. The current flow in the external circuit is designated by i_X.

10 mM NaCl on side A and 100 mM on side B, NaCl is transported from side A to side B and that under these conditions the potential difference (PD) across the membrane measured via the probe electrodes (Fig. 1A) is 100 mV with side A positive to side B (i.e., $V_{AB} = 100$ mV). Our convention is such that when V_{AB} is positive it means that the side represented by the first subscript is positive to that represented by the second subscript and vice versa (e.g., if $V_{AB} = 100$ mV then $V_{BA} = -100$ mV). We assume that there is a substrate in the bathing fluid which serves as a source of energy for transport, but the details of how this substrate is utilized is not germane to our analysis (see Appendix C). We define a passive ion as one whose movement is determined entirely by its electrochemical potential gradient (Rosenberg, 1948; Ussing, 1949). Therefore the movement of a passive ion under open-circuit conditions would be given by the Nernst–Planck electrodiffusion equation (see Appendix B):

$$J_{Cl} = -D_{Cl}(d[Cl]/dx) + u_{Cl}(d\Psi/dx)[Cl] \tag{1}$$

$$J_{Na} = -D_{Na}(d[Na]/dx) - u_{Na}(d\Psi/dx)[Na] \tag{2}$$

where J is the rate of movement of a given ion per unit area; D, Fick's diffusion coefficient; u, the electric mobility [(cm/sec)/(volt/cm)]; and $d\Psi/dx$, the gradient of the electrical potential in the X direction; the meaning of the subscripts is self evident. Examination of the equation for Cl$^-$ reveals that both the diffusion term (dc/dx is positive) and the electric field term ($d\Psi/dx$ is negative, i.e., $V_{BA} = -100$ mV) would result in the movement of Cl$^-$ from side B to side A, but since it moves from A to B there must be other forces acting on the Cl$^-$. In other words, the transport of Cl$^-$ is against its electrochemical potential gradient and this has been classically defined as active transport (Rosenberg, 1948; Ussing, 1949). The only other force (since we have assumed water drag is negligible) that one can conceive of would be a force that could be classified as a force between molecules, i.e., a force between some substance in the membrane and Cl$^-$. In order to accomplish the uphill transport of Cl$^-$, there would not only have to be a force of attraction between some substance in the membrane and the Cl$^-$, but also there would have to be a movement of the complex from the one side of the membrane to the other and furthermore [in the absence of appropriate gradients of other ions (or molecules), see Appendix A] there would have to be a cyclic change in the affinity between the substance and the Cl$^-$. There must be a source of energy for the production of the cyclic change in affinity (see Appendix C). Substances that have these characteristics are referred to as carriers. Some investigators use the word carrier to imply a mobile type of substance that remains in the membrane but can diffuse across it (e.g., Patlak, 1956).

The term carrier can be used in a broader sense to represent the general concept of active transport as given above regardless of whether there are mobile carriers or whether molecular rearrangements result in a translocation of the ion from one side to the other.

In the model Cl^- is actively transported, and this is represented by the R_{Cl} limb of the equivalent circuit of Fig. 1B. If Cl^- were a passive ion, then its equilibrium emf would be about 60 mV oriented in the opposite direction to E_{Cl} of Fig. 1B. With our simple model and with both bathing fluids well stirred and with separate pathways for Cl^- and Na^+ it is necessary that the active Cl^- mechanism be electrogenic (see Appendix C).

Now examination of the electrodiffusion equation for Na^+ reveals that the force of diffusion would tend to transport it from side B to side A while the force of the electric field would tend to transport it in the opposite direction. With an activity ratio of about 10-fold, one would need a potential difference greater than 60 mV to move the Na^+ from side A to side B (see Appendix B). Since for our initial model $V_{AB} = 100$ mV, the movement of Na^+ could be entirely due to the force of the electric field driving the Na^+ up a 10-fold gradient.

On the basis of the facts given so far, there is no need to postulate that there are any other forces than the force of diffusion and the force of the electric field acting on the Na^+. However, the possibility has not been eliminated that there may be an active transport mechanism tending to drive the Na^+ in the direction of A to B (or even in the direction B to A, which would not be potent enough to overcome its electrochemical potential gradient under open-circuit conditions). Now if we wish to determine whether an active transport mechanism is present for Na^+, there are essentially two approaches. One could solve the electrodiffusion equation and see whether the rate of movement of Na^+ could be explained quantitatively on the basis of the concentration gradient and the PD. Ussing and his colleagues have used this as one approach (see Appendix B). The other approach, which is the primary subject of this review, is the use of the Ussing–Zerahn (1951) short-circuit technique to determine whether an ion is passively transported, and this will be taken up in the following section.

B. Analysis on the Basis of Short-Circuit Data

In Fig. 1B an equivalent circuit is shown for the model with an active Cl^- mechanism in parallel with a Na^+ pathway; an external circuit (R_xE_x) connects A to B.

From the laws of networks it follows that

$$i_x = i_{Cl} - i_{Na} \tag{3}$$

$$V_{AB} = R_x i_x - E_x \tag{4}$$

$$V_{AB} = E_{Cl} - R_{Cl} i_{Cl} \tag{5}$$

and

$$V_{AB} = E_{Na} + R_{Na} i_{Na} \tag{6}$$

The conventions used in this paper can be easily determined by comparison of the above equations with the circuit in Fig. 1B.

From equations (3) through (6) it follows that

$$i_{Cl} = (E_{Cl} - V_{AB})/R_{Cl} \tag{7}$$

$$i_{Na} = -(E_{Na} - V_{AB})/R_{Na} \tag{8}$$

and

$$i_x = [(E_{Cl} - V_{AB})/R_{Cl}] + [(E_{Na} - V_{AB})/R_{Na}] \tag{9}$$

In the analysis it will be assumed, unless otherwise stated, that the E's and R's are not changed by current flow (i.e., they are not functions of the current density). This assumption is made simply for the purpose of clarity; it is not in general essential. If the R's and E's are functions of current flow then the values of the E's and R's will be the steady-state values. For the moment we assume that the resistance of the bathing solutions between the probe electrodes and the membrane itself is negligible. Now what Ussing and Zerahn have done was to abolish the concentration and the electrical gradients across the membranes, i.e., place the same NaCl concentration on both sides (e.g., 100 mM) and short-circuit the membrane.

For the moment let us assume that they were clever enough actually to short-circuit the membrane; i.e., in Fig. 1B (with $E_x = 0$) the resistance in the external circuit (R_x) is reduced to essentially zero so that $V_{AB} = 0$ [from Eq. (4) with $E_x = 0$ and i_x finite, $V_{AB} = 0$ when $R_x = 0$]. Then with $V_{AB} = 0$ it follows from Eqs. (7), (8), and (9) that

$$i_{Cl} = E_{Cl}/R_{Cl} \tag{10}$$

$$i_{Na} = -E_{Na}/R_{Na} \tag{11}$$

and

$$i_x = (E_{Cl}/R_{Cl}) + (E_{Na}/R_{Na}) \tag{12}$$

With the potential difference across the membrane equal to zero, and with ionic concentration gradients abolished, there would be no net movement of Na^+ if Na^+ were a passive ion. If it turns out that there is a net movement of Na^+ under these conditions, then we conclude that Na^+ is actively transported. If there were net movement of Na^+ through the membrane from B to A, then E_{Na} would be positive (it would have the orientation shown in Fig. 1B). If there were net movement in the opposite direction, then E_{Na} would be negative (its orientation would be opposite to that shown in Fig. 1B). If there were no net movement of Na^+, then E_{Na} would be zero and the equivalent circuit for the short-circuit condition would consist of the chloride limb with an emf and a resistor, and a Na^+ limb consisting of only a resistor. It should be apparent to the reader that the original open-circuit data were not sufficient to warrant the conclusion that Na^+ was a passive ion.

It should be emphasized that the open circuit data for Cl^- clearly establish that Cl^- is actively transported. Many workers seem to believe that the open circuit data are not sufficient to warrant a decision as to whether or not a given ion is actively transported and that only under short-circuit conditions can the decision be made that an ion is actively transported. It should be clear to the reader that this latter orientation is erroneous. The apparent advantage (see below) of the short-circuit technique is that it seemingly enables one to establish whether or not an ion is passively transported when, on the basis of open-circuit data alone, passive transport could not be proved.

Now returning to the problem of short-circuiting, it is technically practically impossible to reduce the resistance of the external circuit to zero, so that Ussing and Zerahn (1951) placed a battery in the external circuit (E_x of Fig. 1B) to overcome the external resistance so that the PD across the membrane could easily be reduced to zero. In other words, the external battery is simply used to drive the short-circuit current through an obligatory external resistance.

The simple circuit in Fig. 1B with an active transport limb in parallel with a passive limb for the companion ion is the Ussing–Zerahn equivalent circuit. Within the framework of their assumptions they found, for the *in vitro* frog skin, active Na^+ transport and passive Cl^- transport, just the opposite of that of the model we have used for illustration. In our model we had $V_{AB} = 100$ mV, and now if we change our model so that $V_{AB} = -100$ mV ($V_{BA} = 100$ mV) and with NaCl still transported uphill from A to B (10 mM to 100 mM), the reader should be able to convince himself that Na^+ must be actively transported and now, if under the short-circuit conditions (same solution on both sides of the membrane) there is no net movement of Cl^-, it should be clear to the reader that Cl^- is a passive ion (for the simple model in Fig. 1).

C. Short-Circuit Current and Net Transport of Ions

Another important aspect of the short-circuit technique is illustrated by comparing the magnitude of the Cl^- current with the short-circuit current. Let us assume that our model consists of an active Cl^- limb and a passive Na^+ limb. Under short-circuit conditions there will be no current flow in the Na^+ channel so that all the current will be conducted via the Cl^- channel and therefore the chloride current must equal the short-circuit current. Now if the short-circuit current is not equal to the Cl^- current, there must be some other ion that is actively transported. In our original model we specifically stated that there were pathways for only Cl^- and Na^+ and that all other ions were excluded from the membrane. We will expand our model now and assume that there may be other ions that may cross the membrane.

The Cl^- current could be larger or smaller than the short-circuit current depending upon whether there is active anion and/or active cation transport. This is illustrated in Fig. 2, where we assume active transport of a cation from A to B (the R_b limb) and active transport of an anion in the same direction (the R_a limb). It will be recalled that the direction of the current for an anion is opposite to the direction of its transport. In the circuit of Fig. 2, i_{sc} is given by

$$i_{sc} = i_{Cl} + i_a - i_b \tag{13}$$

Theoretically if the rate of anion transport i_a were equal to the rate of the cation transport i_b (i.e., $i_a - i_b = 0$), then the short-circuit current would precisely equal the Cl^- current and we might erroneously conclude that Cl^- was the only ion being actively transported. This possibility

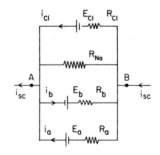

Fig. 2. Equivalent circuit for single membrane with separate pathways for Na^+, Cl^-, a second cation (R_b limb) and a second anion (R_a limb). The i's, E's, and R's refer, respectively, to currents, emfs, and resistances.

seems highly unlikely, but this must be kept in mind and open-circuit and short-circuit data should be obtained on the other ions.

Another possibility can be illustrated by reference to Fig. 2. If only a cation (in the R_b limb) and Cl^- were actively transported ($i_a = 0$) then from Eq. (13) $i_{sc} = i_{Cl} - i_b$ and obviously i_{Cl} and i_b could both be much larger than i_{sc}. Under open-circuit conditions i_{Cl} may be approximately equal to i_b (assuming that the transport of passive ions is much less than the rate of transport of the actively transported ions), and hence i_{sc} for this situation may be much smaller than i_{Cl} and i_b in the open-circuit condition. Stated another way, the magnitude of the short-circuit current would not be a good measure of the magnitude of the "electrical activity" of ion transport under these conditions (contrary to the contention of Durbin, 1967).

It should also be noted that there may be active transport across the tissue of an important ion at a rate much less than that of the ion under consideration. For example, if the Cl^- ion is transported at a rate of, let us say, 100 $\mu A/cm^2$ and another ion is actively transported at a rate of, say, 5 $\mu A/cm^2$, then owing to the errors involved in the measurements of Cl^- transport one could easily overlook the transport of the other ion; i.e., the short-circuit current would be essentially equal to the net Cl^- current within the limits of the errors of the method. Again it should be emphasized that appropriate data should be obtained for all the ions before one concludes that the only ions actively transported are those under consideration.

D. Problem of the Resistance of the Bathing Fluids

Up to now we have implied that our probe electrodes are very close to the two surfaces of the membrane; i.e., we assumed that the resistance between the probe electrodes and the membrane was essentially zero. We will analyze the problem of the resistance of the solutions between the probe electrodes and the tissue with our simple model, and the results of this analysis can be applied later to the more complex models that represent living tissues.

Figure 3 illustrates a technique (Rehm, 1962) for placing virtual probe electrodes near the surface of the stomach without actually having probe electrodes in the chambers. With many tissues, areas of only about 1–2 cm² are used, and with probe electrodes of practical size projecting into the interior of the chambers the current flow from the current sending electrodes is distorted, and this makes assessment of the true resistance between the probe electrodes and the tissues difficult.

The diagram in Fig. 3B illustrates this latter point. It should be clear to the reader that if probe electrodes like those shown in Fig. 3B, were

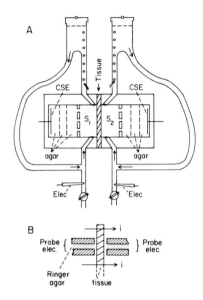

Fig. 3. A. Diagram of apparatus for mounting a tissue with no probe electrode projecting into chambers. Elec = electrodes for measuring PD; CSE = current-sending electrodes, agar layers used to keep products of electrolysis from entering compartments S_1 and S_2. Circulation is provided by gaslifts. Note that inflow and outflow openings are near tissue and opposite each other and with reasonably good flow rates provide excellent mixing. B. Probe electrodes near tissue. The arrows indicate current i flowing through tissue.

used, and were placed very close to the tissue, then the change in PD due to current flow would be smaller than that obtained with ideal probes of infinite thinness and hence a false picture of the resistance between the probe electrode and the tissue would result. It should also be obvious to the reader that, if the probe electrodes were very close to the tissue or, in the limiting case, made contact with the tissue, then with increasing applied current the rest of the tissue would be short-circuited long before the portion of the tissue between the probes was short-circuited—the error in the determination of the magnitude of the short-circuit current could be very substantial. The magnitude of error would obviously be a function of the geometry and size of the probe electrodes. It is difficult for this author to understand why apparently sophisticated investigators still make use of probe electrodes projecting into chambers.

Figure 4 shows a circuit in which the resistance between a virtual probe electrode and the membrane is $R_p/2$ for each side (a total of R_p for both sides with the same solution on both sides). Now if the potential difference between points A and B is brought to zero then AB would be short-circuited, but the membrane will not be short-circuited. The question arises as to the magnitude of the error when V_{AB} is brought to zero rather than the PD across the membrane. The PD between A and B (Fig. 4) would be given by the following equation

$$V_{AB} = V_{AX} + V_{XY} + V_{YB} \qquad (14)$$

where V_{AX} and V_{YB} are the PDs between the probe electrodes and the tissue and V_{XY} is the actual PD across the tissue. The values for V_{AX} and V_{YB} would be given by

$$V_{AX} = V_{YB} = -(R_p/2)i_x \qquad (15)$$

and substitution of Eq. (15) into Eq. (14) yields

$$V_{AB} = V_{XY} - R_p i_x \qquad (16)$$

When $V_{AB} = 0$, we have

$$V_{XY} = R_p i'_{sc} \qquad (17)$$

where i'_{sc} represents the current needed to bring V_{AB} to zero. When the membrane is actually short-circuited (i.e., $V_{XY} = 0$) then from Eq. (16) we have

$$V_{AB} = -R_p i_{sc} \qquad (18)$$

where the unprimed i_{sc} is the true short-circuit current. The equivalent circuit of Fig. 4A can be represented by the equivalent circuit in Fig. 4B where E_{eq} represents the open-circuit PD between X and Y, and R_t is the

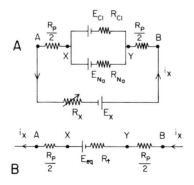

FIG. 4. A. Equivalent circuit for single-membrane model with separate pathways for Cl⁻ and Na⁺; $R_P/2$ represents resistances of bathing fluid between tissues and probe electrodes at A and B; E_X, R_X are in external circuit connecting to the bathing solutions via electrodes like those shown in Fig. 3A. B. Equivalent circuit for Fig. 4A.

actual resistance of the tissue. Then using the circuit of Fig. 4B we have

$$V_{AB} = -R_p i_x + E_{eq} - R_t i_x \qquad (19)$$

and when $V_{AB} = 0$ we have (i_x now is represented by i'_{sc})

$$i'_{sc} = E_{eq}/(R_p + R_t) \qquad (20)$$

When $V_{XY} = 0$ then

$$i_{sc} = E_{eq}/R_t \qquad (21)$$

and the ratio of the two short-circuit currents, from Eqs. (20) and (21), is:

$$i_{sc}/i'_{sc} = (R_p + R_t)/R_t \qquad (22)$$

The question arises as to the magnitude of the error if we use $V_{AB} = 0$ rather than $V_{XY} = 0$. From Eq. (22) it is clear that if $R_t \gg R_p$ then the error would be small in using i'_{sc} rather than i_{sc}. For example, if R_t equals 2000 ohm cm² and R_p is 40 ohm cm² then i_{sc}/i'_{sc} equals 1.02 (i.e., there would be only a 2% error). With $R_p = 40$ ohm cm² (for 1 cm² tissue), the resistance on each side would be 20 ohm cm² and the distance (effective) between the position of the virtual probe electrodes and the tissues would be obtained by using the well known equation:

$$R = \rho(\Delta X/A) \qquad (23)$$

Where R is the resistance, ρ the specific resistivity, ΔX the thickness of the system, and A the area. For each side, $R = 20$ ohms, $\rho \simeq 100$ ohms-cm (for amphibian Ringers) and $A = 1$ cm² we find $\Delta X \simeq 0.2$ cm or 2 mm between the tissue and the position of each virtual probe electrode. Attempts to reduce the magnitude of R_p by decreasing the distance between actual probe electrodes and the tissue as indicated above (see Fig. 3B) may lead to substantial error. However, with the use of very thin probe electrodes placed at appropriate distances from the tissue, it still would be true that if R_t were 2000 ohms cm² the use of i'_{sc} in place of i_{sc} would still result in only a small error. On the other hand, if the resistance of a tissue is of the order of 100 ohms cm² [as it is for tissues like the frog gastric mucosa, renal tubules, and intestine] and the resistance between the tissue and probes were 40 ohms cm², then the error would be appreciable ($\simeq 40\%$). In fact, in studying changes in short-circuit current very misleading results might be obtained by using i'_{sc} instead i_{sc}.

As will be shown below, it is possible for agents that produce a change in i_{sc} to produce little or no change in i'_{sc}. Let us assume that some agent or procedure changes E_{eq} to $E_{eq} + \beta E_{eq}$ and R_t to $R_t + \alpha R_t$ of Fig. 4B (where β and α may be either positive or negative), then from Eq. (21)

i_{sc} for the new conditions will be

$$(i_{sc})_a = (E + \beta E)/(R_t + \alpha R_t) \tag{24}$$

and for i'_{sc} from Eq. (20) for the new conditions we have

$$(i'_{sc})_a = (E + \beta E)/(R_t + \alpha R_t + R_p) \tag{25}$$

where the subscript a indicates the new state. We define r' and r as follows and by use of Eqs. (20) and (25), and letting $R_p = 0.4R_t$ we have

$$r' = (i'_{sc})_a/(i'_{sc})_o = [1.4(1 + \beta)]/(1.4 + \alpha) \tag{26}$$

and by use of Eqs. (21) and (24) we obtain

$$r = (i_{sc})_a/(i_{sc})_o = (1 + \beta)/(1 + \alpha) \tag{27}$$

where the subscript o indicates the original state. For purposes of illustration, let us assume that E_{eq} and R_t change, but that i'_{sc} does not change; i.e., $r' = 1$. Let us further assume that $\beta = -0.5$ (the open-circuit PD decreases to half its value), then from Eq. (26) for r' to $= 1$, $\alpha = 1.4\beta$ and hence $\alpha = -0.7$. Substituting these values for α and β in Eq. (27) yields $r = 1.67$, i.e., a 67% increase in the true short-circuit current when i'_{sc} does not change. The reader may object to the foregoing example on the grounds that the values used for α and β are not realistic. However, in the frog gastric mucosa inhibited by thiocyanate, removal of the thiocyanate may result in changes in PD and resistance comparable to the above values (Rehm et al., 1963). By the use of appropriate values of α and β, the reader can easily convince himself that i'_{sc} could change in one direction while the value of i_{sc} changes in the opposite direction. On the basis of the above, it is clear that for studies on tissues with low resistances, it is important to maintain V_{XY} at 0; i.e., V_{AB} should be maintained at $-R_p i_{sc}$. A number of papers describe automatic devices for short-circuiting and in some there is an automatic correction for the resistance of the solutions (e.g., LaForce, 1967; Wright, 1967; Forte, 1968; Issacson et al., 1971; DiBona and Civan, 1973; Flemström et al., 1973).

Up to now we have described the essential principles of the short-circuiting technique with a very simple model. This model would be adequate for tissues like the perfused squid axon or barnacle muscle in which there is radial symmetry and which can be adequately represented by a single membrane. However, with epithelial tissues, where there are two opposing membranes, the use of a more complex circuit would seem to be indicated, and in the following sections we will reassess the results of the analysis based on the simple circuit.

III. IMPLICATIONS OF THE SHORT-CIRCUITING TECHNIQUE
WITH A MORE REALISTIC CIRCUIT

Figure 5A represents a model with two membranes. Again we assume that there is active Cl^- and passive Na^+ transport. We will assume the active Cl^- limb is the R'_{Cl} limb in M_2. Passive conductive channels for Na^+ are present in M_1 and M_2 and in M_1 for Cl^-. For the moment we will assume that the membrane is not permeable to other ions. Each membrane of Fig. 5A can be represented by a resistance and an emf, and Fig. 5B is an equivalent circuit for Fig. 5A. The relationships between Fig. 5A and Fig. 5B are given by the following equations:

$$E_{eq} = [R_{Cl}/(R_{Na} + R_{Cl})]E_{Na} - [R_{Na}/(R_{Na} + R_{Cl})]E_{Cl} \quad (28)$$

$$E'_{eq} = [R'_{Na}/(R'_{Na} + R'_{Cl})]E'_{Cl} + [R'_{Cl}/(R'_{Na} + R'_{Cl})]E'_{Na} \quad (29)$$

$$R_1 = (R_{Na}R_{Cl})/(R_{Na} + R_{Cl}) \quad (30)$$

and

$$R_2 = (R'_{Na}R'_{Cl})/(R'_{Na} + R'_{Cl}) \quad (31)$$

Now with two limiting membranes the question arises whether we should short-circuit the membrane containing the active transport mechanism or should short-circuit across the tissue (i.e., between A and B). It is assumed that the reader now knows how to obtain the true short-circuit

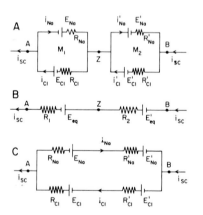

Fig. 5. A. Double membrane model; Z represents cytoplasm, A and B bathing solutions. Passive conductance limbs for Na^+ and Cl^- in membrane M_1 and active Cl^- pump (R'_{Cl} limb) and passive Na^+ limb in membrane M_2. The i_{sc} denotes short-circuit current. The i's, E's, and R's refer, respectively, to currents, emfs, and resistances. B. Equivalent circuit for Fig. 5A. C. Circuit of Fig. 5A disconnected at Z, and Na^+ and Cl^- limbs connected so that there is no connection in the middle.

current in contrast to i'_{sc} (if he is dealing with a low-resistance tissue), so we have omitted the solution resistances between the tissue and the probes in Fig. 5. The answer to the above question is that the tissue should be short-circuited, not the active membrane. If the following analysis is clear to the reader, he should be able to demonstrate to his own satisfaction that short-circuiting across the active membrane would result in general in net transport of passive ions across the tissue.

If we short-circuited across the cell layer, then in most cases we would probably not short-circuit the active membrane (membrane containing the active transport mechanism). With the active membrane not short-circuited, there would be, in general, initially a net movement of passive ions across the active membrane. We will show in this section that, while this may be true initially, under steady-state short-circuit conditions the ion gradients will be rearranged so that there will be no net transport of passive ions across the active membrane even though the PD across the active membrane is not reduced to zero. Now we may assume that the E's and the R's of membranes 1 and 2 are not changed by current flow, or we may assume that they are changed. If they are changed then the values for the E's and R's are for the steady state conditions. With the tissue short-circuited, we have from Fig. 5B

$$V_{AB} = 0 = E'_{eq} - E_{eq} - (R_1 + R_2)i_{sc} \qquad (32)$$

so that

$$i_{sc} = (E'_{eq} - E_{eq})/(R_1 + R_2) \qquad (33)$$

With $V_{AB} = 0$ we have

$$V_{ZB} = V_{ZA} \qquad (34)$$

and

$$V_{ZB} = E'_{eq} - R_2 i_{sc} \qquad (35)$$

and, combining Eqs. (33) and (35), we have

$$V_{ZB} = (R_2 E_{eq} + R_1 E'_{eq})/(R_1 + R_2) \qquad (36)$$

so unless the numerator of Eq. (36) is zero, V_{ZB} will not be zero during short-circuiting (the active membrane will not be short-circuited). For example, Curran and Cereijido (1965) found for the frog skin that during short-circuiting the PD between a microelectrode in the epithelial cells and an external electrode is not zero.

Now under steady-state short-circuit conditions the net transport of Na^+ from A to Z must equal that from Z to B so that $i_{Na} = i'_{Na}$ of Fig. 5A. With $V_{AB} = 0$ it follows that

$$V_{ZA} = E_{Na} - R_{Na}i_{Na} \qquad (37)$$

and

$$V_{ZB} = E'_{Na} + R'_{Na}i'_{Na} \tag{38}$$

and since $V_{ZA} = V_{ZB}$ and with $E_{Na} = E'_{Na}$ we find from Eqs. (37) and (38) that

$$i_{Na} = -(R'_{Na}/R_{Na})i'_{Na} \tag{39}$$

and since i_{Na} and i'_{Na} must have the same sign in the steady state and R_{Na} and R'_{Na} are both positive, the only condition that satisfies Eq. (39) is for i_{Na} and i'_{Na} to be identically zero. With i'_{Na} and i_{Na} equal to zero

$$V_{ZB} = E'_{Na} = V_{ZA} = E_{Na} \tag{40}$$

So that even when V_{ZB} is not zero, there will be no net transport of passive ions and the PD across the active membrane (and also across the opposing membrane) will equal the emf for the passive ions.

The analysis can be easily extended to any number of passive ions so that during the steady-state short-circuit condition all the emfs of the passive ions will be identical and will be equal to the PD across the membrane (for more detailed treatment, see Curran and Cereijido, 1965; Rehm, 1968).

IV. EFFECT OF A DIFFUSION BARRIER WITH A RESISTANCE COMPARABLE TO THAT OF THE CELL LAYER

So far we have ignored the presence of the diffusion barriers in living tissues. We have implicitly assumed that the cell layer was short-circuited when the tissue was short circuited. Most tissues have diffusion barriers (e.g., connective tissues or smooth muscle layers) on at least one side of the cell layer, and the question arises whether to short-circuit the tissue including the diffusion barrier or to attempt to short-circuit the cell layer itself. Figure 6 represents a cell layer, here represented by a two-limb circuit, in series with a connective tissue barrier (between B and S). Now for tissues with cell layers possessing resistances much larger than those of the diffusion barrier, one can ignore the presence of the diffusion barrier. The argument is essentially similar to that which justifies the use of i'_{sc} rather than the true short-circuit current (i.e., ignoring the resistance between the probes and the tissues). However, for tissues like the gastric mucosa the situation is different. For the *in vivo* gastric mucosa of the dog, the resistance of the submucosa and external muscle layers is about the same as the resistance of what we usually refer to as the gastric mucosa (i.e., about 5 ohms for a flap of 18 cm^2 area) which includes the

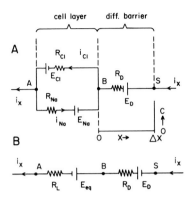

FIG. 6. A. Cell layer represented by a two-limb (Cl^- and Na^+) equivalent circuit in series with the diffusion barrier (e.g., connective tissue and/or smooth muscle). R_D and E_D represent resistance and diffusion potential in barrier. Concentration of NaCl [C] in barrier, C versus distance X from cell layer to end of barrier. The i's, E's, and R's refer, respectively, to currents, emfs, and resistances. B. Equivalent circuit for Fig. 6A.

cell layer, the lamina propria, and the muscularis mucosa (Rehm *et al.*, 1955). It should be clear to the reader that, in a situation like this, whether or not the gastric mucosa or whole stomach is short-circuited would make a substantial difference in the magnitude of the short-circuit current.

We will show that for *in vitro* tissues V_{AS}, not V_{AB}, should be maintained at zero, and for *in vivo* tissues (with intact blood supply) V_{AB}, not V_{AS}, should be maintained at zero. We imply for our present purposes that there are no contraindications for the use of the short-circuit technique for *in vivo* (or *in vitro*) tissues, such as too complex a geometry or too high a secretory rate (see Appendix D). For *in vitro* tissues we will assume that there is no bulk flow or the effect of bulk flow is negligible in the diffusion barrier; i.e., only the force of diffusion and the force of the electric field act on the ions. For simplicity we start by assuming that initially the fluid in the diffusion barriers is of uniform composition and is the same as that at A and S (the bathing fluids at A and S are well stirred and of large volume so that their composition does not appreciably change during an experiment). We send current i_x and maintain V_{AS} at zero. With $V_{AS} = 0$, V_{SB} will always equal V_{AB}. When current is applied V_{SB} will obviously be positive and hence V_{AB} will be positive, so that the cell layer (A to B) will not be short-circuited; and, since initially the Na^+ concentration is the same at A and B, E_{Na} will initially be zero and i_{Na} will be given by V_{AB}/R_{Na}. Initially Na^+ will be transported from A to B, and similarly the current flow through the diffusion barrier will transport Na^+ from S to B. For simplicity, we assume that only Na^+ and Cl^- are present in the solutions. It follows that both Na^+ and Cl^- will increase at B (obviously by the same amount). The increase in NaCl concentration at B will continue until $V_{AB} = E_{Na}$ so that i_{Na} will decrease to zero and J_{Na} [see Eq. (41)] in the diffusion barrier will decrease to zero. There are several ways to demonstrate the validity of the above statement. One could replace the diffusion barrier with a barrier containing separate channels for Na^+ and Cl^- and then the proof would be essentially the same as that given for the circuit in Fig. 5A. Another approach is to point out that under the steady-state condition the conclusion that there can be no net transport of passive ions follows from first principles, since for a passive ion with its electrochemical gradient between A and S abolished and with no coupling between other solutes (other than via the electric field) and solvent, its net transport must vanish. This must be true regardless of the complexities of the diffusion barriers; otherwise it can be shown that the laws of thermodynamics would be violated.

On the basis of the above, let us assume that in the steady state the net transport of Na^+ (J_{Na}) in the diffusion barrier is zero and we then obtain an equation relating V_{SB} (and V_{AB}) to the concentration of NaCl at B. Using the Nernst–Einstein equation ($D = RTu/F$), we have

$$J_{Na} = -[(RTu_{Na}/F) \cdot (dC_{Na}/dx)] - u_{Na}(d\Psi/dx)C_{Na} \tag{41}$$

where the symbols have been previously defined. With $J_{Na} = 0$ it is easily shown that

$$V_{SB} = RT/F \ln(C_B/C_o) = E_{Na} \tag{42}$$

where C_B is the concentration of Na^+ (or Cl^-) at B and C_o is the concentration at S and also at A. Now since E_{Na} equals V_{SB}, it follows that i_{Na} must be zero ($E_{Na} = V_{AB}$). In other words, when the net movement of Na^+ in the barrier is zero, the net movement via the R_{Na} limb is also zero. Now if we inject NaCl at B (or if we are clever enough to remove some NaCl at B), it can easily be shown that there will be a transient state in which Na^+ will move from B to A and from B to S (or vice versa for a decrease in the concentration at B) until its rates of movement in the diffusion barrier and via R_{Na} are zero.

In contrast to the *in vitro* situation, in the *in vivo* condition blood flow through the diffusion barrier would have essentially the same effect as though there were good stirring under the *in vitro* condition in the diffusion barrier, so that the composition of the barrier would remain essentially unchanged. Now if we maintain V_{AS} at zero, then there would be net transport of Na^+ from A to B, and if we were not aware of these considerations we might erroneously conclude that Na^+ was actively transported (i.e., from the left-hand side to the blood). Again it should be emphasized that the above considerations are trivial if the resistance of the cell layer is much larger than that of the barrier. Only when we deal with tissues in which the diffusion barrier resistance is an appreciable fraction of the total resistance are the above considerations relevant.

By way of summary, it should be clear to the reader that if sampling takes place at A and S and there is no bulk flow in the diffusion barrier, then V_{AS} should be maintained at zero, but not V_{AB}. On the other hand, when sampling takes place at A and B (bulk flow in the barrier as with blood flow) then V_{AB}, not V_{AS}, should be maintained at zero.

By the way of further illustration, if we placed a layer of agar (made from the bathing solution) of appropriate thickness in series with the frog skin and the layer of agar had a resistance comparable to that of the frog skin, then we could use the Ussing–Zerahn technique by maintaining the PD at zero across the whole system, since we would sample from one side of the tissue and on the opposite side from the junction between the agar and bathing fluid; we should not maintain the PD across the skin at zero.

A corollary of the foregoing (hinted at above) is that if the active membrane is short-circuited, i.e., if V_{BZ} in Fig. 5A is maintained at zero, then the reader should be able to demonstrate to his own satisfaction that the concentration of NaCl at Z (in the cytoplasm) will in general change and there will be a net transport of Na^+ from B to A. Another way of looking at this is that with $V_{BZ} = 0$ then in general V_{AB} will not equal zero and in order to maintain V_{BZ} at zero, V_{BA} will have to be maintained at a positive value and this will drive Na^+ from B to A. Therefore the tissue should be short-circuited, not the active membrane.

V. A COMPARISON OF THE SIMPLE USSING–ZERAHN CIRCUIT WITH A MORE REALISTIC CIRCUIT

Both the simple circuit and the double-membrane circuit have been used for analysis of the results with the short-circuit technique. The question arises as to the relationship between R_{Na} and R_{Cl} of the simple Ussing–Zerahn model to the R's of the double membrane model. For the simple model (Fig. 1B) the resistance R_{AB} is given by

$$R''_{AB} = (R''_{Na}R''_{Cl})/(R''_{Na} + R''_{Cl}) \tag{43}$$

where R''_{Na} and R''_{Cl} with the double primes refer to the resistances of the simple model. For the double-membrane model (Fig. 5A) the resistance R_{AB} would be given by

$$R_{AB} = [(R_{Na}R_{Cl})/(R_{Na} + R_{Cl})] + [(R'_{Na}R'_{Cl})/(R'_{Na} + R'_{Cl})] \tag{44}$$

In general $R_{AB} \neq R''_{AB}$, and it will be shown in this section that R''_{AB} is greater than or equals R_{AB} (provided certain assumptions are valid) and that the following equation gives the relationship between the R's of the two circuits

$$R''_{AB} \equiv \frac{(R_{Na} + R'_{Na})(R_{Cl} + R'_{Cl})}{R_{Na} + R'_{Na} + R_{Cl} + R'_{Cl}} \tag{45}$$

where $R''_{Na} = R_{Na} + R'_{Na}$ and $R''_{Cl} = R_{Cl} + R'_{Cl}$.

Before we proceed, let us examine a detailed proposal made by the Ussing school for active Na^+ transport (Koefoed–Johnsen and Ussing, 1958). This is illustrated in Fig. 7A. According to this scheme, the active process consists of a neutral forced exchange between Na^+ and K^+. The pond side of the skin is essentially impermeable to K^+, so in the steady state the amount of K^+ moving from cell to interstitial fluid via a passive K^+ conductance pathway equals the amount of K^+ moving via the neutral active mechanism from the interstitial fluid to the cell. For each Na^+ transported from the cell to the interstitial fluid, a unit charge (K^+) moves across the membrane. This scheme requires two membranes so it would not be applicable to our original model of Fig. 1. The model in Fig. 1 would require an electrogenic pump (see Appendix C). We define an electrogenic pump as one in which there is a net transport of charge via the pump mechanism. Figure 13 shows a model of a Cl^- electrogenic pump. The mechanism shown in Fig. 7 is not that of an electrogenic pump since the active process is neutral (no net transport of charge via the active

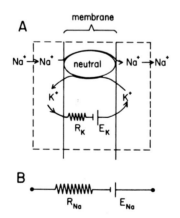

FIG. 7. A. Neutral active Na^+–K^+ exchange pump in membrane in parallel with passive K^+ conductive pathway. Dashed lines enclose system, and system is treated like a black box with Na^+ entering from left and exiting to the right. R refers to resistance; E, to emf. B. Equivalent circuit for the system in Fig. 7A.

process). We have enclosed the membrane and adjacent regions with a dotted line in Fig. 7A and the system may be treated as a *black box* so that the equivalent circuit in Fig. 7B would be adequate for certain purposes regardless of the detailed actual molecular mechanism of the active Na^+ transport as long as the overall result is a transport of a unit charge across the membrane for each transported Na^+. In other words, all the ion transport can be formally considered as conductive.

For the sake of simplicity, we will analyze the model in Fig. 5A. This analysis is based on that of Kidder and Rehm (1970); for more details that reference should be consulted. The reader should have no trouble in applying the analysis to a model with active Na^+ and passive Cl^- transport. The actual resistance of the model in Fig. 5A (assuming that the cytoplasmic resistance is negligible) is given by Eq. (44). Now let us apply a step current across the circuit of Fig. 5A (for our present purposes within the framework of our assumptions, the magnitude or direction of the current is unimportant). Let us for convenience send the current from A to B (opposite to the direction of i_{sc}). Now we assume that the membrane is impermeable to all ions except Na^+ and Cl^- and that these ions move exclusively via conductive pathways (it should be recalled that electrogenic pumps are conductive pathways). Now if $R_{Na}/R_{Cl} \neq R'_{Na}/R'_{Cl}$, then the initial change in i_{Na} will not be the same as the change in i'_{Na} (and similarly for i_{Cl} and i'_{Cl}), since from the laws of networks the initial change in the currents would be given by:

$$\Delta i_{Na} = [R_{Cl}/(R_{Cl} + R_{Na})]i_x = [1/(1 + R_{Na}/R_{Cl})]i_x \quad (46)$$

and

$$\Delta i'_{Na} = [R'_{Cl}/(R'_{Cl} + R'_{Na})]i_x = [1/(1 + R'_{Na}/R'_{Cl})]i_x \quad (47)$$

With $i_{Na} \neq i'_{Na}$ initially, then the Na^+ and Cl^- concentration in the cell with obviously change with time which will result in a change in E's until in the new steady state $i_{Na} = i'_{Na}$ and $i_{Cl} = i'_{Cl}$. During this transition period (with i_x fixed), V_{AB} will change owing to the change in the E's and shifts in the i's. We stated previously that our analyses were made on the assumption that the R's and E's do not change unless otherwise stated. In the present situation, we assume that the R's do not change, but obviously the E's do. In the steady state

$$V_{AB} = -E_{Na} + E'_{Na} + (R_{Na} + R'_{Na})i_{Na} \quad (48)$$

and

$$V_{AB} = E'_{Cl} + E_{Cl} - (R_{Cl} + R'_{Cl})i_{Cl} \quad (49)$$

Since obviously $E_{Na} = E'_{Na}$ and

$$i_{Na} = i_{Cl} + i_x \quad (50)$$

we have from Eqs. (48), (49), and (50)

$$V_{AB} = (R_{Na} + R'_{Na})i_{Cl} + (R_{Na} + R'_{Na})i_x \tag{51}$$

and solving for i_{Cl} yields

$$i_{Cl} = [V_{AB} - (R_{Na} + R'_{Na})i_x]/(R_{Na} + R'_{Na}) \tag{52}$$

and eliminating i_{Cl} from Eqs. (49) and (52) gives

$$V_{AB} = \frac{(E'_{Cl} + E_{Cl})(R_{Na} + R'_{Na})}{\Sigma R} + \frac{(R'_{Cl} + R_{Cl})(R'_{Na} + R_{Na})}{\Sigma R} i_x \tag{53}$$

where $\Sigma R \equiv R_{Cl} + R'_{Cl} + R_{Na} + R'_{Na}$

Now it is easily shown that under the steady-state open-circuit condition that the first term of Eq. (53) = V_{AB} (i.e., when $i_x = 0$), so that from Eqs. (45) and (53) we have

$$V_{AB} - (V_{AB})_{oc} = \Delta V_{AB} = R''_{AB}i_x \tag{54}$$

where the subscript oc indicates open-circuit. It is apparent that if $R_{Na}/R_{Cl} = R'_{Na}/R'_{Cl}$ that $\Delta i_{Na} = \Delta i'_{Na}$ (and $\Delta i_{Cl} = \Delta i'_{Cl}$), so that there will be no change in cell concentrations then $R''_{AB} = R_{AB}$ [Eqs. (45) and (44)]. Now for purposes of illustration, assume $R_{Na} = 200$ ohms, $R_{Cl} = 100$ ohms, $R'_{Na} = 400$ ohms, and $R'_{Cl} = 200$ ohms, then the true resistance would be 200 ohms $(66.7 + 133.3)$ and $R_{AB} = R''_{AB}$. The response of V_{AB} to i_x will be a simple step response (disregarding the time taken for the charging of the membrane's dielectric capacitors, and assuming the R's are not functions of current). Now suppose $R_{Na} = 50$ ohms, $R_{Cl} = 450$ ohms, $R'_{Na} = 450$ ohms, and $R'_{Cl} = 50$ ohms then the true resistance $R_{AB} = 90$ ohms, but $R''_{AB} = 250$ ohms. For R''_{AB} to be the true resistance, we would have to modify the circuit of Fig. 5A, by breaking it at Z and remodeling it so that there are no connections at Z between the Na^+ and Cl^- limbs, i.e., the circuit of Fig. 5A would have to be changed to that of Fig. 5C. So Fig. 5C represents the simple Ussing–Zerahn equivalent circuit. For it to represent the actual circuit on the basis of the above analysis there would have to be the specific relationships between the R's as described above. We must emphasize for even this latter to be true within the framework of the appropriate relationships the assumption has to be made that all ion transports are via conductive channels. We will see in Section VI that these assumptions may not be applicable.

Note on problem of determining the resistance of a tissue during short-circuiting by breaking the current and measuring the open-circuit PD: We had originally planned to present an analysis of the problem of determining the resistance during short-circuiting by briefly opening the circuit. However, it soon became apparent that including an analysis of the resistance problem would make this chapter far too long. The problems of measuring the resistance of living tissue should be dealt with in a separate chapter [see Rehm et al. (1973a) for an introduction to the complex problem of measuring tissue resistance].

VI. THE HEINZ–DURBIN USE OF THE SHORT-CIRCUIT TECHNIQUE FOR ELUCIDATION OF THE MOLECULAR MECHANISM OF HCL FORMATION BY THE GASTRIC MUCOSA

I hope it is clear by this time that it is generally accepted that the use of the short-circuit current enables one to determine whether or not an ion is passively transported and also to determine whether or not there is active transport of ions other than those under investigation and to determine the effect of agents or procedures on the active transport rate. As we have emphasized in this review, implicit in the analysis of data based on the short-circuit technique is the assumption that all transports are via conductive channels. Practically all the investigations published, and there are a large number, have used the short-circuit technique with the former objectives in mind. However, there have been imaginative extensions of the use of the technique in attempts to elucidate the molecular mechanisms of ion transport. A challenging one is that used by Heinz and Durbin (1958; also see Durbin and Heinz, 1958) in studies on *in vitro* frog gastric mucosa. We will analyze the Heinz–Durbin approach in order to demonstrate that the use of this short-circuit technique may lead to unjustified conclusions (Rehm and Sanders, 1972). After our analysis of the Heinz–Durbin approach, we will apply a similar analysis to a model in which there is an active neutral Na^+ and Cl^- mechanism. In fact, as will become evident below, one of the major thrusts of this review has resulted from an extension of the Heinz–Durbin approach.

Before proceeding with the Heinz–Durbin analysis it should be pointed out that up to now we have, for the most part, dealt with models in which only one ion was actively transported. In a number of tissues there is active transport of more than one ion. For example Zadunaisky et al. (1963) found active transport of both Na^+ and Cl^- in the skin of the South American frog from the outside to the blood side. Gonzalez et al. (1967) found for the turtle bladder active Na^+ and Cl^- transport from the lumen to the blood side. It should also be pointed out that the Ussing school reported that epinephrine applied to the blood side of the frog skin resulted in active Cl^- transport from the blood to the pond side, which they attributed to activation of the secretion of the mucous glands (Koefoed–Johnsen *et al.*, 1952b; Ussing, 1965). The frog gastric mucosa is also a tissue in which there is active transport of two major ions. Hogben (1951, 1952) showed that the net Cl^- transport equals the sum of the short-circuit current and the H^+ secretion, i.e.,

$$i_{Cl} = i_{sc} + i_H \qquad (55)$$

Returning to the Heinz–Durbin approach, in which they used the simple Ussing–Zerahn circuit shown in Fig. 8A to represent the whole gastric mucosa, we will show that their reasoning is valid within the framework of certain assumptions for both the simple circuit (Fig. 8A) and for the double-membrane circuit (Fig. 8B). We will also show that, with the present picture of the ion transport mechanisms in the nutrient membrane

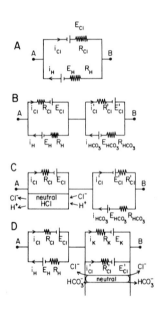

Fig. 8. A. Simple two-limb single-membrane model for gastric mucosa with active Cl^- and H^+ mechanisms in separate limbs. Subscript H refers to H^+. The i's, E's, and R's refer, respectively, to currents, emfs, and resistances. B. Double-membrane model for gastric mucosa; active Cl^- and H^+ limbs in left membrane and passive Cl^- and HCO_3^- conductive limbs in right membrane. C. Double-membrane model for gastric mucosa; neutral HCl mechanism and an electrogenic Cl^- pump are in left membrane, and passive conductive pathways for Cl^- and HCO_3^- in right membrane. D. Double-membrane model for gastric mucosa with electrogenic Cl^- and H^+ pumps in left membrane, and in right membrane a neutral $Cl^- - HCO_3^-$ exchange mechanism and passive conductive pathways for Cl^- and K^+.

of the *in vitro* frog gastric mucosa, the use of the simple circuit is not justified. It should be pointed out that they made their analysis before we showed that HCO_3^- and most of the Cl^- (Sanders *et al.*, 1972; Rehm *et al.*, 1973b) are transported across the nutrient membrane via a neutral mechanism. Heinz and Durbin were attempting to test the validity of our theory of HCl secretion according to which the Cl^- and H^+ mechanisms are both electrogenic (Rehm, 1950, 1965). They argued that the experimental data were not compatible with the theory and that therefore the HCl mechanism must be a neutral one.

They argued that if R_{Cl} and E_{Cl} of Fig. 8A were not changed by the initiation of H^+ secretion (or by an increase or decrease in the H^+ rate) then the Cl^- current under short-circuit conditions would be invariant (with $V_{AB} = 0$ and R_{Cl} and E_{Cl} fixed, i_{Cl} would be fixed and $= E_{Cl}/R_{Cl}$). Therefore in going from the resting state to the secretory state with i_{Cl} fixed, i_{sc} must decrease by an amount equal to the increase in H^+ rate, since for the secreting stomach (Hogben, 1955)

$$(i_{sc})_{sec} = i_{Cl} - i_H \tag{56}$$

and for the resting mucosa ($i_H = 0$)

$$(i_{sc})_{rest} = i_{Cl} \tag{57}$$

and with i_{Cl} fixed, then the change in i_{sc} is given by

$$\Delta i_{sc} = (i_{sc})_{sec} - (i_{sc})_{rest} = -i_H \tag{58}$$

or for changes in i_H we have

$$\Delta i_{sc} = -\Delta i_H \tag{59}$$

In other words, when secretion is initiated, i_{sc} should decrease by an amount equal to the H^+ rate, and when the H^+ rate decreases i_{sc} should increase by the decrease in i_H.

Heinz and Durbin (1958) found for a large number of mucosae during the steady state that, when the short-circuit current was plotted versus the H^+ rate, even though there was a large scatter in their data, an inverse relationship between these two quantities should be discernible if our electrogenic theory were correct. Since there seemed to be no trend in the data compatible with the relationship given in Eq. (59), they concluded that our theory was not compatible with this finding and that this finding supported the concept that the HCl mechanism is a neutral one. In order to account for the short-circuit current, they were forced to make the ad hoc postulate that there is an electrogenic Cl^- pump in parallel with the neutral HCl mechanism (see Fig. 8C).

It should be pointed out that subsequently it was found (Rehm, 1962) that histamine (a stimulant for H^+ secretion) produced a transient increase in the short-circuit current. In other words, there was an increase instead of a decrease in short-circuit current with the onset of H^+ secretion. The simplest interpretation is that histamine not only turned on the H^+ pump but also increased the potency of the Cl^- pump. The reader is referred to Rehm (1962) for the details of this, but for the purpose of illustrating the basic concepts of the Heinz–Durbin approach, we use the assumptions that Heinz and Durbin made, i.e., that E_{Cl} and R_{Cl} remain invariant in the face of changing H^+ rate.

Now the question arises if they had used a two-membrane circuit would their conclusions be valid. We will show that their analysis would be valid with a two-membrane circuit (Fig. 8B) provided that all the ions are transported via conductive mechanisms. It follows from an examination of the two-membrane circuit that under steady state open-circuit conditions

$$i_{Cl} = i_H = i'_{Cl} = i_{HCO_3} \tag{60}$$

and that under steady-state short-circuit conditions

$$i_{sc} = i_{Cl} - i_H = i'_{Cl} - i_{HCO_3} \tag{61}$$

In this analysis we have neglected for the purposes of clarity the small amounts of Na^+ and K^+ which are transported by the gastric mucosa. Also we point out for those who are not familiar with this field that it is well established that for each H^+ secreted into the lumen a HCO_3^- appears on the blood side (Hanke *et al.*, 1931; Hanke, 1937; Davies, 1951; Teorell, 1951). Implicit in the Heinz–Durbin analysis is the assumption that not only Cl^- but also HCO_3^- (see Rehm, 1967, for a detailed discussion on this point) is transported via a conductive pathway. We extend the Heinz–Durbin assumptions that E_{Cl} and R_{Cl} are invariant to the double membrane model and make the further assumption that the electrogenic Cl mechanism is well behaved (i.e., $E_{Cl} + E'_{Cl} = E^*_{Cl}$, see Appendix C). So that

$$V_{BA} = E^*_{Cl} - (R_{Cl} + R'_{Cl})i_{Cl} \tag{62}$$

and therefore under short-circuit conditions with $V_{BA} = 0$ we have

$$i_{Cl} = E^*_{Cl}/(R_{Cl} + R'_{Cl}) \tag{63}$$

So with E^*_{Cl}, R_{Cl}, and R'_{Cl} fixed, i_{Cl} would be fixed, and hence from Eq. (61) for changes in i_H we have

$$\Delta i_{sc} = -\Delta i_H \tag{64}$$

just as with the simple equivalent circuit. So the Heinz–Durbin analysis would be applicable to the two-membrane circuit with conductive pathways for all the ions. It must be obvious to the reader that implicit in

Heinz–Durbin analysis for a two-membrane circuit is the postulate that the nutrient membrane has conductive pathways for Cl^- and HCO_3^-. However, as we have indicated above we have shown that the HCO_3^- conductance of the nutrient membrane of the frog gastric mucosa is essentially zero and that there is in this membrane a neutral HCO_3^-–Cl^- exchange mechanism (or an equivalent neutral mechanism). We now examine the Heinz–Durbin analysis on the basis of a realistic picture of the nutrient membrane.

Parenthetically, it should be pointed out that the author accepted the validity of the Heinz–Durbin approach until very recently (Rehm and Sanders, 1972). We believed that if with the onset of H^+ secretion there were no change in i_{sc}, then this would constitute convincing evidence for a unitary HCl mechanism. It was not until we discovered the neutral Cl^-–HCO_3^- mechanism in the nutrient membrane of the frog gastric mucosa that we became aware of the limitations of the Heinz–Durbin approach. Once we became aware of the possible fallacies in this approach, we turned our attention to an analysis of the accepted criteria for passive transport, which will be taken up in the following section.

We will use the circuit shown in Fig. 8D, and examination of this circuit reveals the following equations to be true for the resting condition ($i_H = 0$):

$$i_{sc} = i_{Cl} = i'_{Cl} \tag{65}$$

and during H^+ secretion

$$i_{sc} = i_{Cl} - i_H = i'_{Cl} \tag{66}$$

It should be pointed out that the conductance of the nutrient membrane G_N is equal to the sum of the Cl^- (G_{Cl}) and K^+ (G_K) conductances (Spangler and Rehm, 1968; Harris and Edelman, 1964); that is

$$G_N = G_K + G_{Cl} \tag{67}$$

In other words, there is a conductance as well as a neutral mechanism for Cl^- transport across the nutrient membrane, but only a neutral mechanism for HCO_3^-. In this analysis, for simplicity we assume that the conductance of the secretory membrane is equal to the sum of the Cl^- and H^+ conductances (i.e., the K^+ conductance of this membrane is assumed to be zero), hence under steady-state short-circuit conditions the only ion that conducts current across the nutrient membrane is the Cl^- ion. Therefore

$$i_{sc} = i'_{Cl} \tag{68}$$

and V_{BA} would be given by:

$$V_{BA} = E'_{Cl} + E_{Cl} - R_{Cl}i_{Cl} - R'_{Cl}i_{sc} = 0 \tag{69}$$

and eliminating i_{Cl} from Eqs. (55) and (69) and solving for i_{sc} yields

$$i_{sc} = (E_{Cl} + E'_{Cl})/(R_{Cl} + R'_{Cl}) - [R_{Cl}/(R_{Cl} + R'_{Cl})]i_H \quad (70)$$

While the relationship between R_{Cl} and R'_{Cl} has not as yet been completely determined, one estimate (O'Callaghan et al., 1974) indicates that R'_{Cl} may be substantially greater than R_{Cl}, and for purposes of illustration we assume $R'_{Cl} = 9R_{Cl}$, and therefore we have

$$i_{sc} = [(E_{Cl} + E'_{Cl})/10R_{Cl}] - 0.1i_H \quad (71)$$

So that the changes in short-circuit current may be only one-tenth of the change in the H$^+$ rate, i.e.,

$$\Delta i_{sc} = -0.1\Delta i_H \quad (72)$$

The Heinz–Durbin data relating the short-circuit current to the H$^+$ rate for a large number of mucosa are such that the above possibility cannot be excluded. In other words, even if there were not an increase in i_{sc} with the onset of secretion (Rehm, 1962), the conclusion would not be warranted that the HCl mechanism is neutral (i.e., not electrogenic). Parenthetically, it should be pointed that both Flemström (1973) and Solberg and Forte (1971) found that under certain conditions, there is an inverse relationship between i_{sc} and the H$^+$ rate as predicted by the electrogenic theory of HCl formation.

With a neutral mechanism for the major ion transport (Cl$^-$ and HCO$_3{}^-$) across the nutrient membrane, it should be apparent to the reader that there is no simple relationship between the resistance of the simple circuit (Fig. 8A) and that of the realistic double-membrane circuit of Fig. 8D; i.e., equations similar to Eqs. (44) and (45) would not be applicable.

It should be apparent to the reader that a realistic circuit should be used whenever possible and that implicit in the use of the Ussing–Zerahn simple circuit is the assumption that the ion pathways are all conductive pathways. It could be argued that the HCl mechanism is unique in the sense that the H$^+$ which are secreted are not the H$^+$ of the bathing fluid but originate from a metabolic reaction in the luminal membrane and the HCO$_3{}^-$ originate in the cytoplasm; while the above considerations are germane to HCl secretion, they may not be applicable to other transport systems in which all of the ions involved originate from the fluids bathing the tissue. That this is not the case will be shown in the next section.

VII. FALLACY OF CONCLUDING THAT AN ION IS PASSIVE IF ITS NET TRANSPORT IS ZERO UNDER SHORT-CIRCUIT CONDITIONS

We will use the equivalent circuit shown in Fig. 9 to illustrate the point to be made in this section. In this model both Na$^+$ and Cl$^-$ are actively

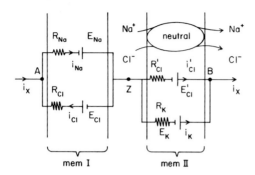

Fig. 9. Double-membrane model for hypothetical tissue. Membrane I has passive conductive pathways for Na^+ and Cl^-, membrane II has passive conductive pathways for Cl^- and K^+ and an active neutral NaCl pump. The i's, E's, and R's refer, respectively to currents, emfs, and resistances.

transported from the cytoplasm (Z) to the blood side (B) by a neutral mechanism. There are passive conductive pathways for Na^+ and Cl^- in membrane I and for Cl^- and K^+ in membrane II. Under open-circuit conditions the concentrations of Na^+ and Cl^- are maintained at low levels in the cytoplasm by the neutral NaCl pump and with isotonic Ringer's solution bathing both sides, E'_{Cl} would obviously be positive and the Cl^- gradient from B to Z would drive K^+ from B to Z until $E'_{Cl} = E_K$. Now for simplicity we assume that $E_{Na} = E_{Cl}$ and $R_{Cl} = R_{Na}$ so that V_{ZA} is zero and V_{BA} would equal E'_{Cl}. With the Cl^- concentration in B ten times that in Z, side B would be positive to side A by about 60 mV. Under short-circuit conditions V_{ZA} would no longer be zero and V_{BA} would be given by

$$V_{BA} = E'_{Cl} - E_{Cl} - R'_{Cl}i'_{Cl} + R_{Cl}i_{Cl} = 0 \qquad (73)$$

It is apparent that $E_{Cl} = E'_{Cl}$ and under steady-state conditions $i_{sc} = i'_{Cl}$ (i_K would have to be zero) and $i_{Cl} = i_{Na} - i_{sc}$ and substituting these values in Eq. (73) and rearranging yields

$$i_{Na} = [(R_{Cl} + R'_{Cl})/R_{Cl}]i_{sc} \qquad (74)$$

and

$$i_{Cl} = (R'_{Cl}/R_{Cl})i_{sc} \qquad (75)$$

For purposes of illustration, let us assume that $R_{Cl} = 50R'_{Cl}$, then from Eq. (74) we have

$$i_{Na} = 1.02i_{sc} \qquad (76)$$

which obviously is the amount of Na^+ and also of Cl^- transported under steady-state conditions via the neutral NaCl pump. From Eq. (75) we have

$$i_{Cl} = 0.02i_{sc} \tag{77}$$

which gives the amount of Cl$^-$ transported via R_{Cl} and since $i_{sc} = i'_{Cl}$ the net amount of Cl$^-$ transported via membrane II must also equal $0.02i_{sc}$ (i.e., $i_{Na} - i_{sc} = 0.02i_{sc}$).

Now considering the errors involved in measuring the net transport of ions, the experimenter working on such a system would undoubtedly conclude that $i_{Cl} = 0$ and that $i_{Na} = i_{sc}$ (i.e., that Cl$^-$ is a passive ion and that the short-circuit current equals the net transport of Na$^+$). In fact, with the usual errors in the measurement of net transport, R_{Cl} could be considerably less than $50R'_{Cl}$ before it could be clearly established that there is a net transport of Cl$^-$. In other words, with a tissue in which both Na$^+$ and Cl$^-$ are actively transported one might easily be led on the basis of the use of the short-circuit technique to the erroneous conclusion that Cl$^-$ is a passive ion.

The reader may suspect that in devising the model of Fig. 9, the author had the frog skin and other similar tissues in mind. The reader may then argue that the model would not be applicable to these tissues because of other findings, such as the findings with Cl$^-$-free (SO$_4^{2-}$) bathing media. However its applicability would depend on the assumptions one makes concerning the active transport mechanism. If the assumption be made that in the presence of both Na$^+$ and Cl$^-$ the mechanism is neutral, but in the absence of Cl$^-$ it functions as an electrogenic Na$^+$ pump (or in a manner illustrated in Fig. 7), then it would be more difficult to prove that the model is not applicable to such tissues. However, the main point the author wishes to make is that the finding that there is no "significant" net transport of an ion under short-circuit conditions does not rigorously establish that the ion is passive; other information is needed to prove that a given ion is passive. It should be emphasized that our model is not one in which there is a very large active transport of Na$^+$ compared to the active transport of Cl$^-$; according to the model with chloride present, both Na$^+$ and Cl$^-$ are transported at the same rate via the neutral active transport mechanism; i.e., under steady-state open-circuit conditions the net transport of Na$^+$ would obviously equal that of Cl$^-$. Only under short-circuit conditions is the mechanism "spinning its wheels," so to speak, with respect to Cl$^-$ transport. It should be obvious to the reader that other information, such as a knowledge of the resistance of the ion pathways in the limiting membranes, is needed in order to rigorously establish that the ion is passive. Stated another way, what seems crystal clear on the basis of a model that consists of a simple single membrane with separate pathways for Na^+ and Cl^- is by no means clear when one uses a double-membrane model, a model which is much more comparable to epithelial tissues than the single-membrane model.

At this juncture the reader might argue that the author was being unfair to the great majority of workers in this field and that they would not conclude that Cl^- was passive without the additional use of the flux-ratio test (see Appendix B). They might argue that the ratio of the two Cl^- fluxes (J_{AB}/J_{BA}) under open-circuit conditions would undoubtedly be significantly different from that for a passive ion. However, it should be noted that in an article by Andersen and Ussing (1960) these authors are clearly of the opinion that the absence of net transport of an ion under short-circuit conditions is prima facie evidence for passive transport. In the following we will show with the model in Fig. 9 that, by the right choice of the values of the parameters, the flux ratio test would indicate that Cl^- is passive.

VIII. FALLACY OF CONCLUDING THAT AN ION IS PASSIVE IF IT OBEYS THE FLUX-RATIO EQUATION

For the purpose of illustration, let us assume during the open-circuit condition that Na^+ and Cl^- are both 100 mM in the bathing solution (at A and B of Fig. 9), that Na^+ and Cl^- are both 10 mM in Z, and that $R_{Na} = R_{Cl}$. With these values $E_{Cl} = E_{Na}$ and $V_{ZA} = 0$, and assuming we could sample and add isotopes at Z the chloride flux-ratio for membrane I would be

$$J_{AZ}/J_{ZA} = C_A/C_Z \exp[(V_{ZA}F)/RT] = 10 \qquad (78)$$

It should be emphasized that the above ratio would be that obtained from a "conceptual experiment" when J_{AZ} is measured with the specific activity of Cl^- in Z maintained at essentially zero and for J_{ZA} the isotope is added to Z and specific activity in A is maintained at essentially zero. From a practical point of view, the above experiment would not be easy to perform with a reasonable degree of accuracy.

We assume that $E_K = E'_{Cl}$ and hence the ratio of K^+ in Z to that in B would be 10. The value of V_{BZ} would therefore be 60 mV, and since $V_{ZA} = 0$ it follows that $V_{BA} = 60$ mV. Now the flux-ratio for a passive anion (Cl^-) between A and B would be according to the flux-ratio equation

$$J_{AB}/J_{BA} = C_A/C_B \exp[(V_{BA}F)/RT] = 10 \qquad (79)$$

Now the question arises as to what the actual flux-ratio J_{AB}/J_{BA} would be for the model in Fig. 9, and this will be taken up below.

Now if we assume that $R'_{Cl} \ll R_{Cl}$, then when a Cl^- isotope is added to A (it is obvious that under steady-state conditions $J_{AZ} = J_{ZB} = J_{AB}$ for the isotope added to A and collected from B) and with $R'_{Cl} \ll R_{Cl}$, the gradient of the specific activity across membrane II will be very small; the main resistance to the Cl^- flux resides in membrane I. Hence sampling at B would be essentially the same as sampling at Z, and therefore J_{AZ} could be determined by sampling at B. Now when the isotope is added to B, the gradient of the specific activity in the steady state across membrane II would be essentially zero and hence J_{ZA} would be measured by adding the isotope to B and sampling at A. It is implied in the above that the magnitude of the activity of the neutral NaCl pump is not great enough to contribute appreciably to the gradient of the isotope across membrane II; i.e., the gradient across this membrane would still be essentially zero. Since for Cl^-, only a passive pathway is present in membrane I the actual flux-ratio for this membrane (with the usual assumptions, see Appendix B) must be that of a passive ion, and so we see that if $R'_{Cl} \ll R_{Cl}$, Cl^- will obey the flux-ratio equation ($J_{AB}/J_{BA} = 10$).

If we make the opposite assumption regarding R_{Cl} and R'_{Cl} and assume that $R'_{Cl} \gg R_{Cl}$, and we ignore the flux via the neutral pump for the moment, then the addition of the Cl^- isotope at A would be essentially the same as adding it to Z; similarly, after adding the isotope to B sampling from A would be the same as sampling from Z. For these conditions the flux-ratio equation will be

$$J_{ZB}/J_{BZ} = C_Z/C_B \exp[(V_{BZ}F)/RT] = 1/10 \exp(2.3) = 1 \quad (80)$$

Under these conditions ($R'_{Cl} \gg R_{Cl}$), it is obvious that J_{AB}/J_{BA} must equal J_{ZB}/J_{BZ} hence the system will not obey the flux-ratio equation when we ignore the flux via the neutral pump (the predicted flux ratio will be given by Eq. (79) regardless of the relationship between R_{Cl} and R'_{Cl}).

For the situation when $R'_{Cl} \ll R_{Cl}$, the Cl^- flux via the neutral pump would not be of significance as long as the specific activity in Z was essentially equal as that in B for isotope addition to either A or B. However, when R'_{Cl} is much greater than R_{Cl} then the Cl^- flux via the neutral pump must be taken into account. Conceptually we can isolate the neutral pump for the purpose of analysis. Suppose we have a single membrane with a neutral NaCl pump and no other pathways in the membrane. Now for a neutral NaCl pump, assuming perfect mixing of the bathing solution the rate of NaCl transport would not be a function of the PD across the membrane and the flux-ratio equation would not be applicable. However, for a neutral pump pumping NaCl from Z to B the Cl^- flux through the pump from Z to B would obviously be greater than the reverse flux, i.e., $J_{ZB} > J_{BZ}$; so unless the corresponding fluxes via R'_{Cl} were much greater than that via the pump with $R'_{Cl} \gg R_{Cl}$, the actual value of J_{ZB}/J_{BZ} would be greater than unity (Eq. 80). It is conceivable that by coincidence the flux ratio might approximate that for a passive ion.

On the basis of the above we see that with appropriate values for the parameters of the model in Fig. 9, the flux-ratio test would indicate that Cl^- is a passive ion. If the reader objects to the assumption that R'_{Cl} is very much less than R_{Cl}, he is missing the point. We are saying that for a given tissue a given ion may be actively transported and still obey the flux-ratio equation, and that other information such as the knowledge of the relationship between R_{Cl} and R'_{Cl} is needed before the conclusion is warranted that the ion is passive. We used values for R_{Na}, R_{Cl}, E_{Na}, and E_{Cl} such that V_{AZ} equals zero under open circuit conditions in order to make the illustration as simple as possible. It should be pointed out that with R'_{Cl} much less than R_{Cl} other sets of values for R_{Na}, R_{Cl}, E_{Na}, and E_{Cl} (but the calculations are more tedious) may be used, and we would still find that the flux-ratio of Cl^- is essentially that of a passive ion. However, with some sets of values of these parameters the flux ratio is not that of a passive ion; e.g., with $E_{Na} = E_{Cl} = 60$ mV and with R_{Na} much less than R_{Cl}, then the flux ratio is not that of a passive ion.

The main point to be made concerning the flux-ratio test is not that the test has all at once become suspect for determining active transport. It has been pointed out by many workers that the failure to obey the flux ratio does not justify the conclusion that an ion is actively transported [e.g., exchange diffusion (Ussing, 1965) or "single file" diffusion (Hodgkin and Keynes, 1955) may result in deviations for passive ions]. The main point is that even if an ion obeys the flux ratio for a passive ion, the conclusion that the ion is passive is not justified; other information is needed.

Parenthetically it should be noted that we devised the model in Fig. 9 to illustrate the points made in this section; our goal was not to present a viable model for tissues, like the frog skin, which under open-circuit conditions transport NaCl from the pond side to the blood side (Huf, 1935). However, with the postulate that in the absence of Cl^- the pump is an electrogenic Na^+ pump, we find that the model has certain attractive features and so far we have not been able to rigorously prove that is does not apply to tissues like the frog skin. This is probably due to the author's lack of detailed knowledge of these tissues, and it is quite probable that workers in this area can easily dispose of it (e.g., Ussing and Windhager, 1964; Zerahn, 1969).

However, several papers present data that are much more in accord with the model in Fig. 9 than with other models. For example, Watlington and Jessee (1973) reported that, in frog skins with low resistances and low PDs, the short-circuit current is about the same as that found in skins with high PDs and resistances, but with Cl^- Ringer's on both sides the net transport of Na^+ is greater than the short-circuit current, and there is a significant net flux of Cl^- from the pond side to the blood side. Using their representative values, $i_{sc} = 1.6$, $i_{Na} = 2.3$, $i_{Cl} = 0.7$ (where the units are in microequivalents cm^{-2} hr^{-1}), we can offer an explanation for their findings on the basis of the model in Fig. 9. From Eq. (75) with the above values we find that R'_{Cl}/R_{Cl} is 0.44 (i.e., R_{Cl} is about 2 times R'_{Cl}). Furthermore, if we assume the low resistance of their skins is due to an abnormally low resistance of R_{Cl} (R_{Cl} now is substantially less than R_{Na}), it is likely that V_{ZA} would be negative and hence V_{BA} would be low. C. O. Watlington (personal communication) points out that in the experiments of Zadunaisky et al. (1963) in which they found active transport of Cl^- in the skin of the South American frog, the PDs and resistances were also low; he suggested that there may be no fundamental differences in the transport mechanisms among species of frogs. Furthermore the observations of Krogh (1937) have never been satisfactorily explained on the basis of the prevailing schemes. Krogh found that intact frogs under certain conditions could absorb NaCl from the pond side when NaCl was as low as 0.01 mM. For passive absorption of Cl^- this would demand a much larger PD than was ever observed. These findings could obviously be explained on the basis of the neutral pump model of Fig. 9.

A number of workers seem to be of the opinion that there is an electrogenic Na^+ pump in frog skin and similar tissues when the tissue is bathed in Cl^- Ringers. H. H. Ussing (personal communication to the author during a meeting of the New York Academy of Sciences in 1965) pointed out that the evidence for an electrogenic H^+ pump in chloride-free media for the frog gastric mucosa is the only clear and convincing evidence for the existence of a biological electrogenic pump (Rehm and LeFevre, 1965). He felt that the evidence for an electrogenic pump in the frog skin and similar tissues with Cl^- Ringer's was based on the inability of his neutral $Na^+ - K^+$ model to explain the data. In other words, with experimental findings that were difficult to explain on the basis of the neutral K^+-Na^+ exchange mechanism, investigators assumed that the only viable alternative was an electrogenic Na^+ pump. They did not seriously consider the possibility of a neutral NaCl pump. Further analysis along these lines is beyond the scope of this review.

IX. PREDICTION OF THE SHORT-CIRCUIT CURRENT FROM THE OPEN-CIRCUIT PD AND RESISTANCE

If the E's and R's of a circuit are not functions of current then one should be able to accurately calculate the short-circuit current from the open-circuit PD and the tissue resistance; i.e., the short-circuit current should equal the open-circuit PD divided by the tissue resistance. However if the E's and/or the R's change, then this would no longer be possible. Short-circuiting in general should increase the load on at least one of the active mechanisms, so that we might anticipate a decrease in its emf and an increase in its resistance and so one would anticipate that the calculated short-circuit current would be greater than that actually found in the steady state. This is true for the *in vitro* frog gastric mucosa (Rehm, 1962). The initial i_{sc} is about that predicted from the open-circuit PD and the tissue resistance, but i_{sc} falls rapidly to a lower level; part of the decrease of the open-circuit PD is probably due to a mechanism described by Kidder and Rehm (1970). In contrast to the behavior of the frog gastric mucosa it has been found (Helman and Miller, 1971, 1973) for the frog skin that under certain conditions the steady-state short-circuit current is greater than that predicted.

By way of background information it was found by Dobson and Kidder (1968) that mounting the frog skin between the usual chambers produced "edge damage" which resulted in a decrease in PD and resistance. The effect of "edge damage" was related (as would be expected) to the total area of the skin; i.e., with very large areas the edge effect would be minimal. Helman and Miller (1971) used a tissue cement to mount the frog skin, which reduced the edge damage to a negligible amount, so that chambers with small areas could be used. Under their conditions the PD and resistance were maximal. They found in the absence of edge damage that voltage–current (V–I) plots were not linear and that the slope (dV/dI) was greater in the vicinity of the open-circuit PD than in the vicinity of the region where the PD was reduced to zero (the short-circuit condition). We confirmed them (not by using the cement technique but by using a very large chamber, i.e., an area of about 13 cm^2). A V–I plot is shown in Fig. 10 (from the unpublished work of F. M. Hoffman, S. S. Sanders, and W. S. Rehm). The currents were applied for about 1 second, and each PD is the PD 0.5 second after the start of the current (if PD values obtained at 0.15 second were used the V–I plots were shifted, but for our purpose the results were essentially the same). The predicted value of the short-circuit current calculated from dV/dI in the vicinity of the open-circuit is shown in Fig. 10 by the dotted line. Since dV/dI decreases as the PD approaches zero a knowledge of the V–I curve over the range to PD $= 0$ is needed to predict the i_{sc}. However the actual steady state short-circuit current would not only be a function of the V–I plot in Fig. 10, but would also be a function of the effect of the duration of current flow on the R's and E's of the system. As pointed out above, short-circuiting throws an increased load on the active transport mechanism, and this may

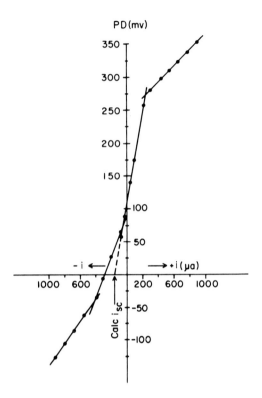

FIG. 10. Voltage–current plot for skin of bullfrog mounted in chamber (13 cm²) and bathed with high HCO_3 (25 mM) Ringer's and gassed with 95% O_2–5% CO_2. Each point represents PD 0.5 sec after start of current. Dashed line is projected from slope of dV/di in the vicinity of open-circuit PD to PD equals zero, which is the calculated short-circuit current (i_{sc}).

result in an increased resistance and decreased emf of the active mechanism. However, as it turned out, Helman and Miller found that the actual steady-state short-circuit current was greater than that predicted on the basis of dV/dI in the vicinity of the open-circuit PD, a finding that we too confirmed.

It should be pointed out that before the work of Helman and Miller (1971), other workers reported nonlinear $V–I$ relationships for frog skin (e.g., Finkelstein, 1964) and for toad bladder (e.g., Civan, 1970). Candia (1970) also reported nonlinear $V–I$ relationships for the frog skin and found in the absence of Cl^- (sulfate bathing solution) the short-circuit current was greater than that predicted for the values of dV/dI near the level of the open-circuit PD. However, with Cl^- bathing media the $V–I$ relationship in Candia's frog skin was essentially linear over the region from the open-circuit PD to zero PD; his skins may have had a significant amount of edge damage. We conclude this section by pointing out that it is not possible to accurately predict i_{sc} from the open-circuit PD and the tissue resistance determined at dV/dI in the region of the open-circuit PD.

APPENDIX A: Role of Water Drag on Ion Transport

As indicated in the main text, active transport of an ion is usually defined as transport against its electrochemical potential gradient. It is possible that water drag could produce transport of an ion against its electrochemical potential gradient and hence result in active transport. However, there is the clear implication in the concept of active ion transport that metabolic energy is directly involved in the transport. Hence, one might be tempted to redefine *active transport*, but this writer prefers not to do so. He prefers to examine sources other than metabolic energy for active transport, and water drag could be one of them.

Parenthetically it should be noted that active transport could be accomplished by means of a carrier with an affinity for two ions (or molecules) without the direct intervention of metabolic energy. The movement of one species down its concentration gradient could drive the other against its electrochemical gradient. For example, if an ion, say Cl^-, is transported via a carrier against its electrochemical potential gradient by another anion moving down its gradient via the same carrier, then we conclude that Cl^- is actively transported and the source of the energy is the gradient of the other anion. Crane's (1965) concept of the uphill movement of glucose from the intestinal lumen into the bordering cells by a carrier which couples the movement of Na^+ and glucose would be a specific example of this type of active transport; i.e., there would be no cyclic change in the affinity of the carrier for glucose and Na^+. It should be noted that, according to Crane's scheme, uphill movement of glucose is eventually dependent on metabolic energy driving Na^+ from the cytoplasm into the interstitial fluid.

Returning now to the problem of water drag, it is possible to formulate mechanisms (usually rather elaborate) on the basis of an osmotic gradient type of mechanism for active ion transport in which a carrier is not directly involved with the ion. In this type of scheme bulk flow of fluid is involved in the uphill transport and sometimes it is difficult to rigorously exclude this type of mechanism in a given situation. An example of such a scheme is given in the last part of this section. The reader is referred to Rehm *et al.* (1970) for references and more details (also see Bresler, 1967).

Now the Ussing school and others who have been interested in the problem of water drag were not looking at the type of mechanism illustrated below but instead were considering the possibility that ions and water share the same pathways (e.g., aqueous pores) across the limiting membranes of cells and that water drag on the ions in these pathways may be an important force in ionic transport. The Ussing school presented an imaginative and convincing explanation for two effects of vasopressin on the frog skin: vasopressin increased the rate of diffusion of water by only a small percentage while it produced a much greater percentage increase in the net transport of water resulting from an osmotic gradient (Ussing, 1952). They postulated that vasopressin increased by a small amount the radius of aqueous pores which would result in only a small increase in the total area available for diffusion of labeled water. On the other hand, assuming the volume flow of water through the aqueous pores is governed by Poiseuille's law, then the volume flow would be proportional to the fourth power of radius and hence a small increase in the radius would result in a relatively large percentage increase in the net transport of water due to an osmotic gradient.

The above findings and explanation argue for the presence of aqueous pores in the membrane. Furthermore, the Ussing school found that increasing the water flow increased the flux of hydrophilic molecules in the direction of the water flow and vice versa. They attempted to determine the force of water drag by a quantitative analysis of these findings on the assumption that the hydrophilic markers and water shared the same pathway in the membrane. They concluded that although water drag could be a minor force in ion transport it could not account for active ion transport in the frog skin.

Recent work has provided an alternative explanation of the vasopressin findings on the basis that water and ions move across membranes through separate pathways which would eliminate water drag as a force governing ion transport across membranes. The gist of the recent work is that the experimental findings can be explained on the basis of unstirred layers in the bathing media, in the connective tissue barriers (for *in vitro* tissues), and in the cytoplasm and in the intercellular lateral spaces (see Hays, 1972, for an excellent review of this subject). The work of Macey and Farmer (1970) and Schafer and Andreoli (1972) should also be consulted. It should be emphasized that this recent work is oriented toward the problem of water movement through the limiting membranes of cells and is not germane to the analysis by the Pappenheimer school of the movement of water and small molecules across the capillary membranes (Pappenheimer *et al.*, 1951; Pappenheimer, 1953).

Our assumption in this review is that water drag is not an important factor in ion transport.

An Example of the Osmotic Gradient Scheme for Active Na⁺ Transport

Figure 11 depicts a scheme for active Na^+ transport in which there is no interaction between Na^+ and a carrier. Figure 11A shows a region between two cells (region enclosed in dashed lines), and Fig. 11B shows an expanded diagram of the enclosed region; the membrane on the left-hand side represents the tight junction, and the region from the tight junction to the right-hand side represents the intercellular spaces. For purposes of illustration we will assume that initially the membrane (i.e., the tight junction) is permeable only to water. By means of active transport of substance G (let us assume that G is a nonelectrolyte) into the intercellular channel (near $X = 0$), a standing gradient of NaCl and G is produced in the channel (see Fig. 11B); and for the moment assume that the cellular membranes bordering the intercellular channels are not permeable to water. If the channel is long enough and the rate of transport

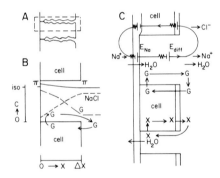

FIG. 11. Model for mechanisms for active Na^+ transport in which a Na^+ carrier is not involved. A. Dashed lines enclose tight junction between cells and the intercellular spaces which are the regions involved in active Na^+ transport. B. Expanded version of tight junction (referred to as membrane) and intercellular spaces; substance G (assumed to be a nonelectrolyte) is actively transported from cells to the intercellular space near the tight junction (i.e., near $X = 0$). Initially we assume that only the tight junction is permeable to water and the flow of water due to the osmotic gradient occurs from the left-hand medium into the intercellular spaces via the tight junctions. This flow results in gradients of NaCl and of G. The Na^+ concentration at $X = 0$ would therefore be low. C. Now in addition to H_2O permeability, we also assume the tight junction is selectively permeable to Na^+ (passive conductive pathway for Na^+) and with Ringer's on both sides this results in a Na^+ diffusion potential oriented to make right side positive. E_{diff} is diffusion potential in the intercellular spaces (see text). Circuit completed via passive Cl^- limbs in cell membranes; left cell membrane has only Cl^- conductive channels. Lower part shows mechanism for pumping H_2O from tissue to the left bathing medium without any accompanying ions (see text).

of G is high enough, then the concentration of NaCl at $X = 0$ could be quite low. We assume that G is present in the interstitial fluid at a much lower concentration than for isotonicity. We now allow the membrane to possess a conductive pathway for Na^+ (separate water and Na^+ pathways in the membrane) which would result in a Na^+ diffusion potential across the membrane oriented to make the blood side positive (Fig. 11C):

$$E_{Na} = RT/F \ln(Na^+)_l/(Na^+)_o \qquad (1A)$$

where R, T, and F have their usual meaning and $[Na^+]_l$ and $[Na^+]_o$ are the Na^+ concentrations to the left and right of the membrane, respectively. There would also be a diffusion potential in the channel given by the following equation:

$$E_{diff} = RT/F \ln[Cl^-]_{\Delta X}/[Cl^-]_o - (v\Delta X/u_{Cl}) \qquad (2A)$$

(see Rehm et al., 1970, for a development of this equation) where v is the velocity of water flow in the X direction in the channel (for simplicity it is assumed that v is not a function of the other space coordinates), u_{Cl} is the electric mobility of the Cl^- ($cm^2/volt$ sec). In general, the potential difference between $X = 0$ to the right of the membrane and $X = \Delta X$ at the end of the channel would be such that $\Psi_{\Delta X} - \Psi_o$ would be positive (i.e., E_{diff} in Fig. 11 would be positive). Thus in the absence of a completed circuit, the PD across the model would equal $E_{Na} + E_{diff}$ and the right side would be positive (see Fig. 11C). We now allow the circuit to be completed via the adjacent cells and assume that the outer cellular membrane (the left one) has a conductive pathway for Cl^- but for no other ions. The cellular inner membrane (right side) would have a conductive pathway for Cl^- and could also have conductive pathways for other ions. The net result will be a transport of NaCl from the left-hand side to the right-hand side. Since the potential difference on the right-hand side is positive, it therefore follows (with the concentration gradients across the model abolished) that Na^+ is transported against its electrochemical potential gradient. Under short-circuit conditions the short-circuit current would equal the net transport of Na^+ and there would be no net movement of Cl^-. The above-described mechanism does not involve a direct linkage between a carrier in the membrane and Na^+.

There are many objections to the above scheme, and it is probably possible to meet each objection (or most of them) with another ad hoc postulate, so that one ends up with a very elaborate scheme. For example, it might turn out that the scheme could not explain the ratio of water to NaCl transported. If too much water per unit NaCl is transported then one could postulate a second mechanism, as shown in Fig. 11C, in which a substance (labeled X) is transported from the intercellular spaces near the tight junction into the cells resulting in an osmotic gradient oriented so that water moves into the left-hand bathing media. Substance X would be recycled so that only water would move out of the tissue in this region. One could also modify the assumption concerning the water permeabilities in the region where G is transported. One could postulate that the left-hand membrane is impermeable to water and that the portion of the cellular membranes bordering the intercellular region near the membrane and also the cellular membranes on the right-hand side are permeable to water. With the latter assumptions there would be net NaCl transport without net H_2O transport; the water would be recycled via the cells and intercellular spaces.

It is difficult to rigorously exclude schemes like the above. However, since there is no real evidence for schemes like these, we have assumed in the main body of the text as pointed out above that water drag is not an important force in active transport.

APPENDIX B: The Electrodiffusion Equations and Ion Transport

The purpose of this appendix is to show how the Nernst–Planck electro-diffusion equation can be used to determine whether an ion is passively or actively transported. We will develop the Ussing flux-ratio equation on the basis of a model of very simple geometry and then outline the method used by Ussing and colleagues for models of the complexity of living tissues (Ussing, 1949, 1952; Koefoed-Johnsen and Ussing, 1953; Ussing and Andersen, 1956; Andersen and Ussing, 1957). The model shown in Fig. 12

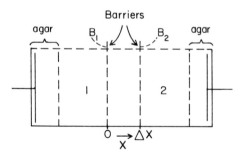

FIG. 12. Simple model consisting of chamber with 0.1 M KCl and agar made with 0.1 M KCl. Barriers B_1 and B_2 have no effect on movement of ions due to electrical or concentration gradients, but they do prevent convective flow, so chambers 1 and 2 can be well stirred. Voltage fixed between B_1 and B_2 by sending current of constant magnitude via electrodes in the agar; thickness of agar is much greater than in figure so that electrode products will not change concentrations in 1 and 2. The system to be analyzed is between B_1 and B_2 and has a thickness of ΔX.

consists of a fluid with uniform concentration throughout (e.g., 0.1 M KCl). There are two barriers (B_1 and B_2) which do not retard movements due to diffusion or to the electric field but do prevent convective flows between them so that compartment 1 and 2 can be well stirred. There are two current-sending electrodes, and for simplicity they are immersed in 0.1 M KCl agar and are at a convenient distance from compartments 1 and 2 so that the concentration changes at the electrodes will not change the concentrations in the compartments.

Now if a current of constant magnitude is sent via the electrodes, the velocity v of movement of Cl⁻ or K⁺ would be a function of the force acting on the ions and the frictional coefficients; i.e., from Coulomb's law we have

$$\text{force} = -zF(d\Psi/dx) = fv \tag{1B}$$

where z is the valence of the ion (positive for cations and negative for anions), F is the Faraday constant, $d\Psi/dx$ is the gradient of the potential in the X direction, and f is the molar frictional coefficient. Rearranging equation (1B) yields

$$\frac{v}{-zF(d\Psi/dx)} = \frac{1}{f} = u_{\mathrm{m}} \tag{2B}$$

where u_{m} is the mechanical mobility and is the inverse of the molar frictional coefficient. We can also define another mobility as follows

$$\frac{v}{-(d\Psi/dx)} = \frac{zF}{f} = u \tag{3B}$$

where u is the electrical mobility and has the units of $[(\mathrm{cm/sec})/(\mathrm{volt/cm})]$; the values for u are given in standard references. From Eqs. (1B) and (3B) we obtain

$$v = -(zF/f)(d\Psi/dx) = -u(d\Psi/dx) \tag{4B}$$

and the net transport of an ion (J) is obviously given by

$$J = vAC = -u(d\Psi/dx)AC \tag{5B}$$

where A is the area perpendicular to the X direction and C the concentration of the ion.

By similar reasoning we can obtain in the absence of an electric field the flux of an ion due to a concentration gradient (e.g., assume an isotope of K⁺ or Cl⁻ is placed in one of the compartments so that we may have a concentration gradient without an electrical gradient). The electrochemical potential ($\bar{\mu}$) for an ion (neglecting the hydrostatic pressure term) is given by the well known equation

$$\bar{\mu} = \mu^\circ + RT \ln\gamma C + zF\Psi \tag{6B}$$

where μ^0 represents the chemical potential in the standard state, C the concentration, and γ the activity coefficient. Assume for simplicity that γ is a constant (it could be equated to unity) and the gradient of the chemical potential in the X direction (i.e., with $d\Psi/dx = 0$) is

$$d\mu/dx = (RT/C)(dC/dx) \tag{7B}$$

where μ represents the chemical potential (i.e., in the absence of an electrical field). We assume the force of diffusion on an ion is equal to the gradient of its chemical potential, i.e.,

$$\text{force} = -(d\mu/dx) = fv \tag{8B}$$

where again f is the molar frictional coefficient and v the velocity. The frictional coefficient between ions and water would obviously be the same regardless of the nature of the force acting on them, hence f is still equal to zF/u. And from Eqs. (3B), (7B), and (8B) we obtain

$$v = -(RTu/CzF) \cdot (dC/dx) \tag{9B}$$

and for the net movement we have

$$J = vAC = -(RTu/zF)A(dC/dx) = -DA(dC/dx) \tag{10B}$$

where by use of the Nernst–Einstein relation $(D = RTu/zF)$ we end up with Fick's equation, where D is Fick's coefficient. Now when the above two forces act on the ions, then by the principal of superposition [linear combination of Fick's equation (10B) and equation (5B)] the equation for the transport would be given by

$$J = -(RTu/zF)A(dC/dx) - u(d\Psi/dx)AC \tag{11B}$$

for a cation, while for an anion the second term on the right of Eq. (11B) would be positive (Jacobs, 1935). We can also arrive at Eq. (11B) by differentiating Eq. (6B) with respect to X and assuming the force acting on a gram mole of ion is $-d\bar{\mu}/dx$.

We now place an isotope for either Cl^- or K^+ in one of the compartments and send current of constant magnitude. With the simple model and a constant current, the gradient of the potential is constant and is represented by $\Delta\Psi/\Delta X$ so that Eq. (11B) can be easily integrated (by separation of variables) and integration between C_1 at $X = 0$ to C_2 at $X = \Delta X$

and from $X = 0$ to $X = \Delta X$ yields

$$\ln \frac{J_{12} + u(\Delta\Psi/\Delta X)AC_2}{J_{12} + u(\Delta\Psi/\Delta X)AC_1} = \frac{zF\Delta\Psi}{RT} \qquad (12B)$$

where J_{12} means movement in the positive X direction (i.e., from 1 to 2, $J_{12} = -J_{21}$). Now when $C_2 = 0$ then we find after rearrangement that J_{12} is given by

$$J_{12} = \frac{u\Delta\Psi AC_1 \exp[-zF\Delta\Psi/RT]}{\Delta X(1 - \exp[-zF\Delta\Psi/RT])} \qquad (13B)$$

similarly with $C_1 = 0$ and solving for J_{21} ($J_{21} = -J_{12}$) yields

$$J_{21} = \frac{u\Delta\Psi AC_2}{\Delta X(1 - \exp[-zF\Delta\Psi/RT])} \qquad (14B)$$

Now J_{12} and J_{21} should be quantitatively predicted by a knowledge of u, $\Delta\Psi$, the C's and ΔX and vice versa. So if we have an unknown system between B_1 and B_2 and the above formulas enable us to quantitatively predict the values of the fluxes, then one would be tempted to conclude that our unknown system has the simple geometry of our model (or can be formally represented by it) and that the ion under investigation is passive. However, if the above equations did not predict the quantitative behavior of an ion, it would not follow that some force other than the force of diffusion and the force of the electric field were acting on the ion; i.e., the transport of the ion could still be passive. In an actual system the area and the gradient of the potential may be functions of X, and if the functions were unknown no solution of the differential equation would be possible. Ussing obtained a ratio equation before integrating (see below), and with this approach the unknown functions would cancel out and so he obtained the well known flux-ratio equation. This equation can be obtained for our simple model by dividing Eq. (13B) by Eq. (14B), and for a monovalent cation we have

$$J_{12}/J_{21} = C_1/C_2 \exp[-F\Delta\Psi/RT] \qquad (15B)$$

which is the flux-ratio equation in its simplest form. For a monovalent anion the equation is

$$J_{12}/J_{21} = C_1/C_2 \exp[F\Delta\Psi/RT]. \qquad (16B)$$

Teorell (1949) independently developed the equation (see also Behn, 1897; Hodgkin and Huxley, 1952). As pointed out above, our development would not be justified for a system with a realistic geometry, and Ussing obtained a ratio before integrating so that the flux-ratio equation would be applicable to a complex geometrical system. The reader is referred to the original papers for details. In the papers of Meares and Ussing (1958, 1959) the assumptions and development of the equation are presented in compact form, and these papers may be consulted for other references. In the development of the Ussing flux equation the assumption was originally made that there were no drag terms. In other developments by the Ussing school (again see Meares and Ussing, 1958, 1959), the drag terms were included. In a paper by Hoshiko and Lindley (1964), the formalism of irreversible thermodynamics was used. These authors showed that the equations were identical to those used by the Ussing school. In order for the Onsager reciprocal relations to hold ($L_{ij} = L_{ji}$), the Ussing school implicitly assumed that the frictional coefficient of molecules of species A acting on molecules of species B was numerically the same as the frictional coefficient of molecules B acting on molecules A—an assumption which is intuitively understandable.

It should be noted that Koefoed-Johnsen et al. (1952a) and Koefoed-Johnsen and Ussing (1953) found that predictions of the flux-ratio equation were not incompatible with experimental findings for Cl^- in the frog skin under conditions of varying potential differences. In this latter work they used the simple flux-ratio equation and ignored possible effects of water drag even though there was a large osmotic gradient across the skin (i.e., from 1/10 Ringer's to Ringer's). The scatter in the data was large so that this approach in deciding between active and passive ion transport is probably not as useful as that of the short-circuiting technique.

It should be pointed out that all the developments in this field were made for a one-dimensional model, i.e., the velocity of movement of an ion in a channel was assumed to be independent of the other space coordinates.

APPENDIX C: Electrogenic Pumps

The purpose of this section is to illustrate by means of a model the role of metabolic energy and carriers in active transport. Figure 13 is an equivalent circuit representing an electrogenic model for active Cl^- transport, and for the purposes of illustration an equation for its emf will be developed (Rehm, 1965). On the basis of this Cl^- model, the reader should be able easily to develop a comparable equation for other ions, such as Na^+. The model consists of a membrane divided into compartments I and II. We assume that only Cl^- can move from side I of the membrane to the outside and only XCl and Y^+ can cross the division between parts I and II of the membrane. Furthermore we assume that only Cl^- can enter from the side labeled "in," except that the substrates can react

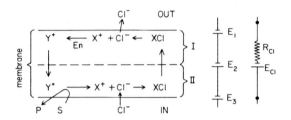

FIG. 13. Model for an electrogenic Cl⁻ pump. E_1, E_2, and E_3 are diffusion potentials and on the right an equivalent circuit for the pump is shown consisting of a resistor R_{Cl} and the electrogenic emf E_{Cl}.

with Y^+ to yield product P and X^+. It is assumed that there is a high affinity between X^+ and Cl^- and that there is essentially no affinity between Y^+ and Cl^-. The arrows are drawn for a net transport of Cl^- from the inside to the outside. Actually these arrows should be double arrows, but for simplicity we use the single arrows. Now if there would be no completed circuit, and for the moment we assume this to be the case, then the system would rapidly come to equilibrium and there would be three emfs, i.e., E_1, E_2, E_3, each being given by the Nernst equation:

$$E_1 = RT/F \ln([Cl^-]_I/[Cl^-]_o) \tag{1C}$$

$$E_2 = RT/F \ln([Y^+]_I/[Y^+]_{II}) \tag{2C}$$

$$E_3 = RT/F \ln([Cl^-]_{in}/[Cl^-]_{II}) \tag{S(3C)}$$

The sum of the three emfs will be represented by E_{Cl}, and an equivalent circuit for the electrogenic Cl^- pump is given on the extreme right of Fig. 13, consisting of a resistance and an emf. By combining Eqs. (1C), (2C), and (3C) and separating into two terms we have

$$E_{Cl} = \frac{RT}{F} \ln \frac{[Cl^-]_I[Y^+]_I}{[Cl^-]_{II}[Y^+]_{II}} + \frac{RT}{F} \ln \frac{[Cl^-]_{in}}{[Cl^-]_o} \tag{4C}$$

We will transform Eq. (4C) by the use of three equilibrium constants. For the reaction

$$Y^+ + S \rightarrow X^+ + P \tag{5C}$$

we have

$$([X^+]_{II}[P])/([Y^+]_{II}[S]) = K_1 \tag{6C}$$

for the reaction

$$X^+ \rightarrow Y^+ \tag{7C}$$

we have

$$[Y^+]_I/[X^+]_I = K_2 \tag{8C}$$

and for

$$Cl^- + X^+ \rightarrow XCl \tag{9C}$$

we get

$$([Cl^-][X^+])/[XCl] = K_3 \tag{10C}$$

By use of Eqs. (6C), (8C), and (10C), Eq. (4C) can be transformed to

$$E_{Cl} = \frac{RT}{F} \ln \frac{K_1 K_2 [S][XCl]_I}{[P][XCl]_{II}} + \frac{RT}{F} \ln \frac{[Cl^-]_{in}}{[Cl^-]_o} \tag{11C}$$

but at equilibrium $[XCl]_I = [XCl]_{II}$ and then Eq. (11C) becomes

$$E_{Cl} = E^* + \frac{RT}{F} \ln \frac{[Cl^-]_{in}}{[Cl^-]_o} \tag{12C}$$

where

$$E^* = \frac{RT}{F} \ln \frac{K_1 K_2 [S]}{[P]} \tag{13C}$$

We see that for a given ratio of S to P the magnitude of E_{Cl} is a function of the ratio the Cl^- concentration inside to that on the outside. Therefore, with Cl^- on the inside fixed, E_{Cl} should be a linear function of the log of the Cl^- on the outside with a slope of approximately 60 mV/unit log on the assumption that E^* does not change. However, changing Cl^- would probably change the ratio of substrate to product (in part due to enzymic feedback mechanisms), so that the emf of an electrogenic pump would not be expected to be well behaved or less well behaved than the emf of diffusion potentials; i.e., it would not be a simple linear function of the log of Cl^- on the outside (nor on the inside).

In the two-membrane model of Fig. 5A E'_{Cl} refers to the electrogenic pump [i.e., E_{Cl} of Eq. (12C)] and E_{Cl} of Fig. 5A to the Nernst potential on the opposite membrane. The sum of E'_{Cl} and E_{Cl} of Fig. 5A would equal E^* of Eq. (13C) since the second term of Eq. (12C) would be numerically equal to, but of opposite sign (orientation), to that of the Cl^- emf in the passive channel.

Now let us examine the situation in which there are parallel pathways in the membrane for a passive ion, i.e., Na^+ as illustrated in Fig. 14. For purposes of illustration let us assume that K_1 times $K_2 = 100$ so that when the ratio of substrate to product concentration $= 0.1$ then $K_1 K_2 [S]/[P] = 10$. Then with $Cl^-_{in} = 10$ mM and $Cl_o^- = 100$ mM, E_{Cl} would $= 0$. Now let $Na^+_{in} = Na^+_o = 100$ mM so that E_{Na} of Fig. 14 would also be zero. With both E_{Na} and $E_{Cl} = 0$ the system would be in equilibrium (i of Fig. 14 would $= 0$). Now if we increase the ratio of S to P by a small amount say 1%, then E_{Cl} would be slightly greater than

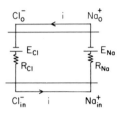

FIG. 14. Membrane with an electrogenic Cl$^-$ pump in parallel with a passive Na$^+$ conductive pathway. E, emf; R, resistance.

E_{Na} and current would flow so that Cl$^-$ would be moved from a concentration of 10 mM to 100 mM and Na$^+$ from 100 mM to 100 mM. Since S/P was increased by only 1%, the total free energy changes of the system would be approximately zero; actually it will have to decrease by a small amount, but in view of our objectives we will ignore this small decrease. For the transport of 1 mole of NaCl from "in" to "out" there will be an increase in the free energy of the NaCl system and a corresponding decrease of the free energy of the S/P system (both changes being about 1400 cal ($RT \ln 10 \simeq 1400$ cal).

Now suppose our conditions are changed and Na$^+_{in}$ is now 10 mM, so that our mechanism has to transport both Na$^+$ and Cl$^-$ from 10 mM to 100 mM, then the question arises as to what changes we have to make in the parameters of our mechanism to accomplish this. If we used a ratio of S to P of 0.1, then with $E_{Cl} = 0$ and now $E_{Na} = 60$ mV, the system would run backward (i of Fig. 14 would be negative), NaCl would be transported from "out" to "in" and P would be changed to S. Now in order to transport NaCl from 10 mM to 100 mM, we have to increase the ratio of S to P. Let us increase this ratio to unity, then $E_{Cl} = 60$ mV and the system would be in equilibrium. We now further increase the ratio S to P by say 1%, then E_{Cl} would be slightly greater than E_{Na} and i would be positive again. Ignoring the small finite decrease in free energy, we see that the decrease in free energy of the reaction S to P will now be about 2800 cal and the increase in free energy in the NaCl system will be about 2800 cal.

We can describe the change in our Cl$^-$ mechanism as follows: when Na$^+$ concentration was changed from 100 mM to 10 mM on the inside we had to increase the ratio of S to P so that the Cl$^-$ mechanism could create a potential difference sufficient to drive the Na$^+$ from 10 mM to 100 mM. In other words, we added free energy to our Cl$^-$ pump to create a electric field to transport Na$^+$ from 10 mM to 100 mM so that we performed work on Na$^+$ to move it up a concentration hill. The statement is often made that if an ion is in electrochemical equilibrium between compartments it takes essentially no work to move it from the one compartment to the other one. However, the way we described how we changed the ratio of S to P in order to move Na^+ from 10 mM to 100 mM implies that electrical work is used to transfer the Na^+ up its concentration gradient and therefore work is expended on an ion even though it be in electrochemical equilibrium.

We could also describe the work needed to transport the Cl^- as that needed to transport it against its concentration gradient and against the electric field so that the work done on the Cl^- ion would be $\simeq 2800$ cal and the work done on the $Na^+ \simeq 0$. Which description one prefers is a matter of taste. In all processes of ion transport, when one ion is transported another ion is also transported, i.e., an ion of the opposite charge in the same direction or a similarly charged ion in the opposite direction. Therefore it could be argued that it is not meaningful to talk about the work needed to transport a single ionic species when its electrochemical potential gradient is zero, but that it is meaningful to talk about the minimum free energy needed to transport NaCl from one set of concentrations (more strictly, activities) to another or from one chemical potential to another chemical potential.

It is very simple to change the scheme in Fig. 13 to an electrogenic Na^+ mechanism; Y^+ and X^+ are changed to Y^- and X^- and Na^+ is substituted for Cl^- and the orientation of E_1, E_2, E_3 are inverted, and we end up with an equation like that of Eq. (12C).

In the above illustration we increased the ratio (S/P) from its equilibrium value by 1% for purposes of illustration. With this ratio off equilibrium by only 1% the equation should give a good measure of E_{Cl}. However, the rate of ion transport with this small deviation from equilibrium would be relatively low and in order to have a rate comparable to rates needed by living tissues the system would in general have to be further from equilibrium and then obviously the equation for E_{Cl} may give us only a rough approximation. A real problem in biological tissues is that of determining the obligatory dissipation of free energy (or creation of internal entropy) and the rate of transport. For a given model a detailed knowledge of the system should enable one to calculate this relationship. It is apparent that for a given model there is going to be more obligatory dissipation of free energy the farther away from equilibrium and the higher the rate of transport. In other words, there would be two obligatory expenditures of free energy; i.e., the minimum free energy needed under equilibrium conditions and the minimum free energy to accomplish a given rate of transport for a given mechanism. In the author's opinion this is an important problem, but it could be argued that until very much more is known about actual kinetic constants of the mechanisms, it probably would not be fruitful to deal with it at present.

APPENDIX D: Problem of Using Short-Circuit Technique on Tissues with High Secretory Rates and/or Complicated Geometry

In some tissues under certain conditions it is probably not meaningful to attempt to apply the short-circuit technique. In the resting *in vivo* dog stomach Na^+ is actively transported from the lumen to the blood (Bornstein *et al.*, 1959), and it is also transported in the same direction under *in vitro* conditions for the mammalian stomach (Cummins and Vaughan, 1965; Kitahara *et al.*, 1969). In the resting *in vivo* dog stomach there is only a very small volume flow, so that it is reasonable to assume that the composition of the fluid in the lumina is essentially that of the mucosal bathing fluid. In contrast, for the *in vivo* secreting dog stomach the H^+ rate is about two orders of magnitude greater than the H^+ rate of the *in vitro* stomach, and in the *in vivo* preparation Na^+ is transported down its electrochemical potential gradient (Thull and Rehm, 1956).

In a secreting *in vivo* dog stomach, even at modest H^+ rates for this preparation, the velocity of flow of secretion in the tubular lumen is so great that the composition in the lumen is essentially uninfluenced by the composition of the fluid bathing the mucosal side (Rehm *et al.*, 1955). So if one placed a fluid on the mucosal side of the same composition as the interstitial fluid, data obtained under these conditions would not be very meaningful since we could not abolish the Na^+ gradient across the mucosal cells. In other words, there is no simple way to find out whether there is an active Na^+ mechanism present in the actively secreting *in vivo* dog stomach. Therefore, it must be obvious to the reader that the short-circuit technique cannot be meaningful applied to this tissue. We can obviously short-circuit between the mucosal and serosal surfaces, but this would have little meaning in terms of the concepts of Ussing and Zerahn. In contrast to the *in vivo* gastric mucosa it is generally assumed and with some justification that the H^+ rates in the *in vitro* mucosa are so low that the composition of fluid in the tubular lumen (except for the H^+ concentration) is not markedly different from that of the secretory bathing media (the H^+ concentration would be much less than that of an isotonic fluid and the concentration of the other ions would not be much different from that of the bathing medium).

There are other objections than the above to obtaining meaningful data by short-circuiting tissues like the *in vivo* dog stomach, but this is beyond the scope of the present paper (see Rehm, 1968, for an analysis of these other considerations).

REFERENCES

Andersen, B., and Ussing, H.H. (1957). Solvent drag on non-electrolytes during osmotic flow through isolated toad skin and its response to antidiuretic hormone. *Acta Physiol. Scand.* **39**, 228-239.

Andersen, B., and Ussing, H.H. (1960). *Comp. Biochem.* **2**, 371-402.

Behn, P. (1897). *Ann. Phys. Chem.* **62**, 54.

Bornstein, A., Dennis, W.H., and Rehm, W.S. (1959). Movement of water, sodium, chloride, and hydrogen ions across the resting stomach. *Amer. J. Physiol.* **197**, 332-336.

Bresler, E.H. (1967). On criteria for active transport. *J. Theor. Biol.* **16**, 135-146.

Candia, O.A. (1970). The hyperpolarizing region of the current-voltage curve in frog skin. *Biophys. J.* **10**, 323-344.

Civan, M.M. (1970). Effects of active sodium transport on current-voltage relationship of toad bladder. *Amer. J. Physiol.* **219**, 234-245.

Crane, R.K. (1965). Na⁺ dependent transport in the intestine and other animal tissues. *Fed. Proc., Fed. Amer. Soc. Exp. Biol.* **24**, 1000-1006.

Cummins, J.T., and Vaughan, B.E. (1965). Ionic relationships of the bioelectrogenic mechanism in isolated rat stomach. *Biochim. Biophys. Acta* **94**, 280-292.

Curran, P.F., and Cereijido, M. (1965). K fluxes in frog skin. *J. Gen. Physiol.* **48**, 1011-1033.

Davies, R.E. (1951). The mechanism of hydrochloric acid production by the stomach. *Biol. Rev. Cambridge Phil. Soc.* **26**, 87-120.

DiBona, D.R., and Civan, M.M. (1973). Pathways for movement of ions and water across toad urinary bladder. I. Anatomic site of transepithelial shunt pathways. *J. Membrane Biol.* **12**, 101-128.

Dobson, J.G., Jr., and Kidder, G.W., III. (1968). Edge damage effect in invitro frog skin preparations. *Amer. J. Physiol.* **214**, 719-724.

Durbin, R.P. (1967). "Electrical potential difference of the gastric mucosa." *In* "Handbook of Physiology" (Amer. Physiol. Soc., J. Field, ed.), Sec. 6; Vol. II, pp. 879-888. Williams & Wilkins, Baltimore, Maryland.

Durbin, R.P., and Heinz, E. (1958). Electromotive chloride transport and gastric acid secretion in the frog. *J. Gen. Physiol.* **41**, 1035-1047.

Finkelstein, A. (1964). Electrical excitability of isolated frog skin and toad bladder. *J. Gen. Physiol.* **47**, 545-565.

Flemström, G. (1973). Oscillation of H⁺ secretion and electrogenic properties in the isolated frog gastric mucosa. *Biochim. Biophys. Acta* **298**, 369-375.

Flemström, G., Öberg, P.A., and Petterson, H. (1973). A new device for automatic measurement of short-circuit current across epithelial tissues. *Upsala J. Med. Sci.* **78**, 19-21.

Forte, J.G. (1968). The effect of inhibitors of HCl secretion on the unidirectional fluxes of chloride across bullfrog gastric mucosa. *Biochim. Biophys. Acta* **150**, 136-145.

Gonzalez, C.F., Shamoo, Y.E., Wyssbrod, H.R., Solinger, R.E., and Brodsky, W.A. (1967). Electrical nature of sodium transport across the isolated turtle bladder. *Amer. J. Physiol.* **213**, 333-340.

Hanke, M.E. (1937). The acid-base and energy metabolism of the stomach and pancreas. *Science* **85**, 54-55.

Hanke, M.E., Johannesen, R.E., and Hanke, M.M. (1931). Alkalinity of gastric venous blood during gastric secretion. *Proc. Soc. Exp. Biol. Med.* **28**, 698-700.

Harris, J.B., and Edelman, I.S. (1964). Chemical concentration gradients and electrical properties of gastric mucosa. *Amer. J. Physiol.* **206**, 769-782.

Hays, R.M. (1972). The movement of water across vasopressin-sensitive epithelia *Curr. Top. Membranes Transp.* **3**, 339-366.

Heinz, E., and Durbin, R.P. (1958). Studies of the chloride transport in the gastric mucosa of the frog. *J. Gen. Physiol.* **41**, 101-117.

Helman, S.I., and Miller, D.A. (1971). In vitro techniques for avoiding edge damage in studies of frog skin. *Science* **173**, 146-149.

Helman, S.I., and Miller, D.A. (1973). Edge damage effect on electrical measurements of frog skin. *Amer. J. Physiol.* **225**, 972-977.

Hodgkin, A.L., and Huxley, A.F. (1952). Currents carried by sodium and potassium ions through the membrane of the giant axon of loligo. *J. Physiol.* (*London*) **116**, 449-472.

Hodgkin, A.L., and Keynes, R.D. (1955). The potassium permeability of a giant nerve fiber. *J. Physiol.* (*London*) **128**, 61-88.

Hogben, C.A. (1951). The chloride transport system of the gastric mucosa. *Proc. Nat. Acad. Sci. U.S.* **37**, 393-395.

Hogben, C.A. (1952). Gastric anion exchange: Its relation to the immediate mechanism of hydrochloric acid secretion. *Proc. Nat. Acad. Sci. U.S.* **38**, 13-18.

Hogben, C.A. (1955). Active transport of chloride by isolated frog gastric epithelium origin of the gastric mucosal potential. *Amer. J. Physiol.* **180**, 641-649.

Hoshiko, T., and Lindley, B.D. (1964). The relationship of Ussing's flux-ratio equation to the thermodynamic description of permeability. *Biochim. Biophys. Acta* **79**, 301-317.

Huf, E. (1935). Versuche über den Zusammenhang zwischen Stoffwechsel, Potential-bildung, und Funktion der Froschhaut. *Pfluegers Arch. Gesamte Physiol. Nenschen Tiere* **235**, 655-673.

Isaacson, L.C., Douglas, R.J., and Pepler, J. (1971). Automatic measurement of voltage and short-circuit across amphibian epithelia. *J. Appl. Physiol.* **31**, 298-299.

Jacobs, M.H. (1935). Diffusion processes. *Ergeb. Biol.* **12**, 1-160.

Kidder, G.W., III, and Rehm, W.S. (1970). A model for the long time-constant tran-sient voltage response to current in epithelial tissues. *Biophys. J.* **10**, 215-236.

Kitahara, S., Fox, K.R., and Hogben, C.A. (1969). Acid secretion, Na^+ absorption, and the origin of the potential difference across isolated mammalian stomachs. *Amer. J. Dig. Dis.* [N.S.] **14**, 221-238.

Koefoed-Johnsen, V., and Ussing, H.H. (1953). The contributions of diffusion and flow to the passage of D_2O through living membranes. *Acta Physiol. Scand.* **28**, 60-76.

Koefoed-Johnsen, V., and Ussing, H.H. (1958). The nature of the frog skin potential. *Acta Physiol. Scand.* **42**, 298-308.

Koefoed-Johnsen, V., Levi, H., and Ussing, H.H. (1952a). The mode of passage of chloride ions through the isolated frog skin. *Acta Physiol. Scand.* **25**, 150-163.

Koefoed-Johnsen, V., Ussing, H.H., and Zerahn, K. (1952b). The origin of the short circuit current in the adrenaline stimulated frog skin. *Acta Physiol. Scand.* **27**, 38-48.

Krogh, A. (1937). Osmotic regulation in the frog (R. esculenta) by active absorption of chloride ions. *Skand. Arch. Physiol.* **76**, 60-74.

LaForce, R.C. (1967). Device to measure the voltage-current relations in biological membranes. *Rev. Sci. Instrum.* **38**, 1225-1228.

Levi, H., and Ussing, H.H. (1949). Resting potential and ion movements in the frog skin. *Nature* (*London*) **164**, 928-929.

Macey, R.I., and Farmer, R.E.L. (1970). Inhibition of water and solute permeability in human red cells. *Biochim. Biophys. Acta* **211**, 104-106.

Meares, P., and Ussing, H.H. (1958). Fluxes of sodium and chloride ions across a cation-exchange resin membrane. I. Effect of a concentration gradient. *Trans. Faraday Soc.* **55**, 142.

Meares, P., and Ussing, H.H. (1959). Fluxes of sodium and chloride ions across a cation-exchange resin membrane. II. Diffusion with electric current. *Trans. Faraday Soc.* **55**, 244.

O'Callaghan, J., Sanders, S.S., Shoemaker, R.L., and Rehm, W.S. (1974). Barium and K^+ on surface and tubular cell resistance of frog stomach with microelectrodes. *Amer. J. Physiol.* **227**, 273–288.

Pappenheimer, J.R. (1953). Passage of molecules through capillary walls. *Physiol. Rev.* **33**, 387-423.

Pappenheimer, J.R., Renkin, E.M., and Borrero, L.M. (1951). Filtration, diffusion, and molecular sieving through peripheral capillary membranes. A contribution to the pore theory of capillary permeability. *Amer. J. Physiol.* **167**, 13-46.

Patlak, C.S. (1956). Contributions to the theory of active transport. *Bull. Math. Biophys.* **18**, 271-315.

Rehm, W.S. (1950). Theory of the formation of HCl by the stomach. *Gastroenterology* **14**, 401-417A.

Rehm, W.S. (1962). Acid secretion, resistance, short-circuit current and voltage-clamping in frog's stomach. *Amer. J. Physiol.* **203**, 63-72.

Rehm, W.S. (1965). Electrophysiology of the gastric mucosa in Cl^--free solutions. *Fed. Proc., Fed. Amer. Soc. Exp. Biol.* **24**, 1387-1395.

Rehm, W.S. (1967). Membrane conductivity and ion transport in gastric mucosa. *Fed. Proc., Fed. Amer. Soc. Exp. Biol.* **26**, 1303-1313.

Rehm, W.S. (1968). An analysis of the short-circuiting technique applied to in vivo tissues. *J. Theor. Biol.* **20**, 341-354.

Rehm, W.S., and LeFevre, M.E. (1965). Effect of dinitrophenol on potential, resistance, and H^+ rate of frog stomach. *Amer. J. Physiol.* **208**, 922-930.

Rehm, W.S., and Sanders, S.S. (1972). Limiting cell membranes of the gastric mucosa and cellular acid-base balance. *In* "Gastric Secretion" (G. Sachs, E. Heinz, and K.J. Ullrich, eds.), pp. 91-105. Academic Press, New York.

Rehm, W.S., Dennis, W.H., and Schlesinger, H. (1955). Electrical resistance of the mammalian stomach. *Amer. J. Physiol.* **181**, 451-470.

Rehm, W.S., Davis, T.L., Chandler, C., Gohmann, E., Jr., and Bashirelahi, A. (1963). Frog gastric mucosae bathed in Cl^--free solutions. *Amer. J. Physiol.* **204**, 233-242.

Rehm, W.S., Butler, C.F., Spangler, S.G., and Sanders, S.S. (1970). A model to explain uphill water transport in the mammalian stomach. *J. Theor. Biol.* **27**, 433-454.

Rehm, W.S., Shoemaker, R.L., Sanders, S.S., Tarvin, J.T., Wright, J.A., Jr., and Friday, E.A. (1973a). Conductance of epithelial tissues with particular reference to the frog's cornea and gastric mucosa. *Exp. Eye Res.* **15**, 533-552.

Rehm, W.S., Sanders, S.S., Shoemaker, R.L., O'Callaghan, J., Tarvin, J.T., and Friday, E.A. (1973b). Proton conductance of cell membranes. *J. Theor. Biol.* **39**, 131-153.

Rosenberg, T. (1948). On accumulation and active transport in biological systems. *Acta Chem. Scand.* **2**, 14-33.

Sanders, S.S., O'Callaghan, J., Butler, C.F., and Rehm, W.S. (1972). Conductance of submucosal-facing membrane of frog gastric mucosa. *Amer. J. Physiol.* **222**, 1348-1354.

Schafer, J.A., and Andreoli, T. (1972). Cellular constraints to diffusion. The effect of antidiuretic hormone on water flows in isolated mammalian collecting tubules. *J. Clin. Invest.* **51**, 1264-1278.

Solberg, L.A., Jr., and Forte, J.G. (1971). Differential effects of −SH reagents on transport and electrical properties of gastric mucosa. *Amer. J. Physiol.* **220,** 1404-1412.

Spangler, S.G., and Rehm, W.S. (1968). Potential responses of nutrient membrane of frog's stomach to step changes in external K⁺ and Cl⁻ concentrations. *Biophys. J.* **8,** 1211-1227.

Teorell, T. (1949). Membrane electrophoresis in relation to bio-electrical polarization effects. *Arch. Sci. Physiol.* **3,** 205-219.

Teorell, T. (1951). The acid-base balance of the secreting isolated gastric mucosa. *J. Physiol. (London)* **114,** 267-276.

Thull, N.B., and Rehm, W.S. (1956). Composition and osmolarity of gastric juice as a function of plasma osmolarity. *Amer. J. Physiol.* **185,** 317-324.

Ussing, H.H. (1948). The active ion transport through the isolated frog skin in the light of tracer studies. *Acta Physiol. Scand.* **17,** 1-37.

Ussing, H.H. (1949). The distinction by means of tracers between active transport and diffusion. *Acta Physiol. Scand.* **19,** 43-56.

Ussing, H.H. (1952). Some aspects of the application of tracers in permeability studies. *Advan. Enzymol.* **13,** 21-65.

Ussing, H.H. (1965). Transport of electrolytes and water across epithelia. *Harvey Lect.* **59,** 1-30.

Ussing, H.H., and Andersen, B. (1956). The relations between solvent drag and active transport of ions. *Proc. Int. Congr. Biochem., 3rd, 1955* pp. 434-440.

Ussing, H.H., and Windhager, E.E. (1964). Nature of shunt path and active sodium transport path through frog skin epithelium. *Acta Physiol. Scand.* **61,** 484-504.

Ussing, H.H. and Zerahn, K. (1951). Active transport of sodium as the source of electric current in the short circuit isolated frog skin. *Acta Physiol. Scand.* **23,** 110-127.

Watlington, C.O., and Jessee, F. (1973). Chloride flux across frog skins of low potential difference. *Biochim. Biophys. Acta* **330,** 102-107.

Wright, P. (1967). A simple device for electronic measurement of a short circuit current. *J. Physiol. (London)* **178,** 1P-2P.

Zadunaisky, J.A., Candia, O.A., and Chirandini, D.J. (1963). The origin of the short-circuit current in the isolated skin of the South American frog Leptodactylus ocellatus. *J. Gen. Physiol.* **47,** 393-402.

Zerahn, K. (1969). Nature and localization of the sodium pool during active transport in the isolated frog skin. *Acta Physiol. Scand.* **77,** 272-281.

Subject Index

A

Active transport
  phenomena of, problems relating to, 191–193
  at plasmalemma, 4–5
Adipose cell membranes, carrier-mediated transport in, 156
Adrenal cortex, carrier-mediated transport in, 152
Aerobacter aerogenes, carrier-mediated transport in, 128
Algae
  carrier-mediated transport in, 130–131
  chloride fluxes in, 17–23
Amphibians, carrier-mediated transport in, 133–135
Animal cells, ion transport in, comparison with plant cells, 38–40
Aspergillus, carrier-mediated transport in, 130
ATP, in energy conversion in chloroplast, 86–90
  synthesis sites in chloroplast, 90–91
Axons, of giant squid, carrier-mediated transport in, 131–132

B

Bacillus subtilis, carrier-mediated transport in, 128
Bacteria
  carrier-mediated transport in, 127–128
  nutrilite uptake by, 164–165
Bacterium lactis aerogenes, carrier-mediated transport in, 128
Bioelectric potentials, ionic activities and, 190–191

Birds, carrier-mediated transport in, 135–136
Blood/brain barrier, carrier-mediated transport in, 144–145
Blood/CSF barrier, carrier-mediated transport in, 145–146
Brain slices and cells, carrier-mediated transport in, 142–144
Brush borders, carrier-mediated transport in, 156–159

C

Candida, carrier-mediated transport in, 130
Carrier hypothesis, 109–215
  addenda to, 112–114
  adsorption and induction in, 188–193
  alternative ("noncarrier") models for, 178–184
  anomalies in, 172–178
  application of, 168–171
  biological solute distribution and translocation in, 185–193
  channel-lining "bucket-brigade" sites in, 179
  classical model for, 167–168
  competition for transfer in, 116–117
  counterflow in, 118–120
  delimitation of, 109–112
  flux-coupling phenomena and, 120–124
  high-temperature dependence of, 125–126
  internal transfer sites in, 180–182
  "introverting" hemiport sites in, 182–184
  kinetic problems in, 171

271